Politicized Physics in Seventeenth-Century Philosophy

Politicized Physics in Seventeenth-Century Philosophy

Essays on Bacon, Descartes, Hobbes, and Spinoza

Robert J. Roecklein

LEXINGTON BOOKS
Lanham • Boulder • New York • Toronto • Plymouth, UK

Published by Lexington Books
A wholly owned subsidiary of Rowman & Littlefield
4501 Forbes Boulevard, Suite 200, Lanham, Maryland 20706
www.rowman.com

10 Thornbury Road, Plymouth PL6 7PP, United Kingdom

British Library Cataloguing in Publication Information Available

Library of Congress Cataloging-in-Publication Data

Library of Congress Cataloging-in-Publication Data Available
ISBN 978-0-7391-8853-8 (cloth : alk. paper)
ISBN 978-0-7391-8854-5 (electronic)

∞™ The paper used in this publication meets the minimum requirements of American National Standard for Information Sciences Permanence of Paper for Printed Library Materials, ANSI/NISO Z39.48-1992.

Printed in the United States of America

In memory of my brother,
Joseph Edward Roecklein

Contents

Acknowledgments

I would first of all like to thank Dr. Steve Hicks, Director for the School of Humanities and Social Sciences at Penn State Erie, the Behrend College. Steve made it possible for me to obtain a timely course release which enabled me to complete work on this book. I am very grateful for his encouragement and assistance. Dr. Robert Speel, Chair of the Department of Political Science, has provided me with valuable advice on the subject of publication, and steady inspiration based on his great passion for politics. Dr. Zachary Irwin, whose expertise extends deeply into Political Theory, is a valued mentor and a great friend.

Dr. Stephen Eric Bronner of Rutgers University has graciously reviewed my work and given me significant suggestions for its improvement. Dr. Michael Thompson of William Paterson University has given encouragement at important moments in this study. Thornton Lockwood of Quinnipiac College has made it possible for me to obtain valuable feedback on my work, for which I am deeply grateful. Dr. Patrick Coby of Smith College has helped me greatly with his published comments on my book on Machiavelli. I am very thankful for the efforts of all these individuals and the contributions they have made to my current project.

Lexington Books has been a source of great inspiration. Jana Hodges-Kluck, Associate Editor of Philosophy, Classics, Sociology, and Criminology at Lexington, provided prudent, expert, and timely guidance during my work on the manuscript. Jana is a voice of calm reason and superb judgment, and she has definitely made this a much better book than it would otherwise have been.

I would like to extend a special thanks to the anonymous reader for Lexington Books who provided so detailed and meticulous an examination of my manuscript. The review supplied by the reader enabled me to make significant changes, and to avoid errors that would otherwise have gone undetected. The reviewer's examination of the text was rigorous and very fruitful. It definitely enriched my understanding of Early Modern Philosophy.

Some of my former students are currently making their way through graduate school. Jason Kinsel is one, destined to make an excellent scholar. Ryan Snyder, Brent Winslow, and Anthony Trotchenkov are undergraduates at Penn State who have made great contributions to my classroom work. The energy they bring to our discussions of the Early Modern philosophers has a direct effect on my own research and writing. They all have my thanks.

My family and close friends provide steady encouragement for my work and lend much needed support. I am grateful to Ann Marie Rice and Neil Roecklein for their love and friendship. Patrick Moloney and John Fragala are great friends of long standing. I am thankful to them for their wisdom and for the happiness they bring to my life.

This book is dedicated to my late older brother, Joseph Edward Roecklein. Joe was the family's great intellect. He lived with a sense of urgency, but also with unique style. My family and I will always mourn his loss, but we will never forget the happiness that be brought us. If this book reflects some small portion of the integrity that Joe possessed, I am satisfied.

Robert J. Roecklein
Erie, PA
September 2013

Introduction

Physics and Politics

Philosophy is an inescapable fate for human society. In this respect, surely, we are unlike any other species. By nature, human beings are equipped with perceptual faculties to know the world around them. By art, philosophy is capable of striving for a different source of knowledge. When philosophy employs its ways and means to subvert rather than to strengthen the faculties of perception given to human beings by nature, the result is unhappy.

The story of philosophy could well be told by tracing the opponents and defenders of perceptual knowledge. The sophists combated by Plato's Socrates, such as Gorgias and Protagoras, denied the veracity of the evidence made available to human beings by perception.[1] These sophists undertook to teach their students how to employ rhetoric of fierce praise and blame in order to secure the victory in any dispute whatsoever. Plato's Socrates, by contrast, insists that he "knows nothing": that indeed, he must begin in the ranks of the ordinary human beings with such information as perception affords, as the original evidence upon which reasoning must depend.[2]

Names are a very important business for political philosophy. There is no more fundamental kind of knowledge than that which enables us to assign a particular name to a particular object. The child begins to learn how to distinguish herself from the world by obtaining power in the naming process. Adults too must navigate the world based upon their ability to correctly assign name to object: What is just, what unjust; what is good, what bad; what is true, what false.

It is in the nature of human things that philosophy and authority will always be closely connected enterprises. The ordinary run of human beings must take their bearings in life from what seems to them to be so. For this huge purpose, they have mostly perception to rely upon. Philosophy, however, has from its known beginnings often claimed the power to know better than perception does. This makes philosophy a potentially dangerous thing: for it is capable of superseding the authority of perception, perhaps for ill purposes. Those dependent on ordinary perception cannot know what philosophers claim to know: and thus it is that the only healthy course of operation for human society is a lively philosophical culture, one in which all domains of knowledge are exposed to cross-examination and healthy debate.

Human beings learn to assign names to objects based upon perception on the one hand, and the ordinary usage of names on the other. Both of these processes come firmly under attack in seventeenth-century philosophy. Yet scholars today must cope with an onerous academic division of labor. The indictments leveled against perception are situated in the hard sciences. Political philosophy, long since confined to a narrowed domain of the liberal arts, appears to lack the very authority to so much as take up the issue.

The most powerful tradition of philosophy to disseminate indictments against ordinary perception is that developed by Parmenides of Elea. Parmenides insists that the senses cannot know truth, because the senses are limited to familiarity with *perishable* objects. "You must debar your thought from this way of search, nor let ordinary experience in its variety force you along this way, namely, that of allowing the eye, sightless as it is, and the ear, full of sound, and the tongue to rule; but you must judge by means of reason the much-contested proof which is expounded by me."[3]

True being, Parmenides insists, concerns that which is eternal. Among the atomist philosophies that grew up under the shadow of Parmenides's logic is that of Democritus and Leucippus. This indeed is the ancient doctrine which has made its way, despite some alterations, into the founding philosophies of modern civilization. This founding took place in the seventeenth century; and it is with these founding doctrines that this book is concerned.

The legacy of Parmenides is complicated. Parmenides was not merely a theorist of physics. He was also a political philosopher. The doctrine of indivisible reality which Parmenides founded has elements of cunning rhetoric mixed in with feats of considerable logical acumen. In Plato's *Parmenides*, Socrates's encounter with the Eleatic is a highly fraught affair. Parmenides is depicted as needling Socrates with his arguments; and Socrates must labor mightily to bring the discussion to some common ground of factual evidence.

In the seventeenth-century discussions of theories of indivisible body, the conceptual foundations of the ancient arguments are omitted. Not even Francis Bacon discusses the premises of the original atomism of Democritus and Leucippus. Those premises contain claims that are disputable. Early Modern theories of body mightily strive to sever their doctrines from the debate concerning the reality of actual change in nature, or of coming into being more specifically. Yet this is the ground towards which debate must move, if scholarship is to become open once again to the powerful critiques formulated in response to the doctrine of indivisible body. Indivisible body cannot undergo change. The theory of atoms reduces change in natural objects to the status of delusions. Against that point of view, Plato and Aristotle expended massive amounts of brainpower.

Seventeenth-century philosophers fashioned a new epistemological home for physics. A new and sharp distinction between physics and philosophical debate more generally considered is effected. Bacon and his successors are intent on cultivating a science which affects airs of non-partisanship. The trouble is that the physics which the Early Moderns harbor is extremely partisan. Political philosophy especially must assert its right to investigate the premises of the doctrines of physics, where the foundations of the latter encroach on the political domain.

At the heart of this investigation, there is the issue of ordinary perception. Ordinary perception is quite properly concerned with matters of truth; and it is the perceptual faculties that non-philosophers rely upon to negotiate matters of fact and truth. The atomistic doctrines of matter uniformly impugn the reliability of sense perception. In the Early Modern philosophies to be investigated in this work, the doctrines of perception are ruthlessly dismissive of the veracity of perception.[4] More than any other single cultural development, this heralds a political crisis. It conjures the dark vision whereby ordinary perceptions are superseded by philosophic doctrines that claim a superior insight into reality. Early Modern philosophers would like to be rather vague about their matter theories. They are not anxious to bring their theories under close inspection. Yet the premises of the physics to which they subscribe are uncertain at best. Their indictment of perception is likewise suspect. This is the nature of the discussion that I propose to conduct as regards each thinker, prior to investigating the particulars of their political doctrines.

In the contemporary academic division of labor, political philosophers do not come within a hundred miles of examining doctrines of physics. This division of labor is an obstacle to intellectual progress. Political philosophy needs to remember that it is dependent on the operative definitions of what body, what perception truly *are.* Truth, not utility, is the proper coin of philosophy. Political philosophy is obliged to concern itself with issues of truth.

There is relevant work that is being done in scholarship to begin to address these issues concerning the atomistic or Epicurean influence in the emergence of Early Modern Philosophy. Several studies have recently been completed. Important issues have been raised. These studies, however, do not penetrate into the philosophical issues of atomism, beyond the more superficial accounts made available by Early Modern philosophers themselves. Bacon pays sufficient attention to the name of Democritus, but he never mentions the nature of the deeper issues that atomism is caught up in. That atomism suggests the phenomena of change to be a mere appearance in nature, that the true or atomic bodies are not themselves capable of undergoing any sort of qualitative change (especially coming into being or passing away), is the true core of this teaching.[5] It is the mechanism of argument which directly indicts the ordinary perceptions. Accordingly, it behooves us to inquire carefully into the Early Modern doctrines to scrutinize their own matter theories, in terms of whether or not they acknowledge natural body to undergo qualitative change. Where this is denied, it is quite likely to follow that the veracity of ordinary perception will be likewise repudiated. This is the nub of the political philosophy which, even in Descartes, is operative. This aspect of politics is quite subtle. Descartes, for example, conspicuously lacks an overt teaching on political theory. Yet his method sneers at the competence of ordinary perception. Descartes unfolds quite a cultural program, and just because it is not advertised as political does not mean that it lacks political implications.

Catherine Wilson's book, *Epicureanism at the Origins of Modernity*, is focused on the grand emergence of a neo-Epicurean philosophy in the seventeenth century.[6] Wilson helps us to understand the intimate relationship between natural philosophy and science. Today, all aspects of the human experience are scientifically regarded as manifestations of the physical. Nor does science make available to us the highly conceptual debates which gave birth to the materialistic philosophy which surged to the fore during the seventeenth century. Wilson points out that several of the seventeenth-century philosophers loosely adorned their essentially Epicurean doctrines with concessions, strategic if not really substantive, to the prevailing Aristotelian religious and university authorities.[7]

Wilson has cast her study in extremely broad terms. This is not necessarily a defect of the work, but it does limit it. Wilson is most interested in the moral and political ideologies that can be seen to accompany the revival of Epicurean doctrine. However, this does not succeed in familiarizing readers with the deeper conceptual debates that underlay the history of atomism. This indeed is one of the massive intellectual feats of the Early Enlightenment: to effectively close down investigation into the depth theory of atomism, and to market the atomistic philosophy as a sort of corollary to a larger project. This helped Early Modern Philosophy to challenge the moral agenda of civilization, as Wilson has demonstrated.

In contrast to Christian Aristotelianism, Early Modern Philosophy advertises the more immediate experiences of pleasure and pain. Wilson discusses the atomistic doctrine in terms of the exemption from divine judgment that its teachings offer; she conceives of the Epicurean doctrine as distinctly humane and full of fellow feeling. One can say that Wilson is a bit dazzled by the specter of Epicurean philosophy as it organized itself in the seveneenth century. "The philosophically and morally attractive features of Epicureanism were its integration of human beings into the natural world, the postulate of human equality that it implied, and the notion that pain and pleasure, both psychological and physical, mattered, regardless of who was experiencing them and what that person's merits or status might be," Wilson writes.[8] Yet this observation neglects first things. It actually lacks insight as to what atomism *is*.

The atomistic doctrine was caused by a *problem.* That problem is the problem of genesis. Where do things come from? Atomic theory is a theory about origins. All of the early Natural Philosophers in Ionia speculated about some kind of eternal matter out of which the perishable objects could be said to emerge, and into which they could be said to return. Yet it was Parmenides, a philosophical logician, who imposed a distinctively new order upon the investigation. Parmenides argued that there must be some kind of matter that doesn't come into being at all, that there cannot be an infinite regress. Parmenides refers to this original reality as 'being.' For Parmenides, it is divisionless. Yet there is a new wrinkle in the argument that Parmenides sets forth. The earlier Natural Philosophers all believed in genesis: Thales, Anaximenes, Anaximander, even Heraclitus. Yet Parmenides, contemplating the logical necessity for some sort of ungenerated being, formulates a theory whereby the true reality is incapable of change. Parmenides's 'being' cannot come into being itself; but it is also indicated that no object can truly come into being out of Parmenides's enumerated reality.

> And it never was, nor will be, because it is now, a whole all together,
> One, continuous; for what creation of it will you look for? How, whence could
> It have sprung? . . . Justice has never released Being in its fetters and set it free
> Either to come into being or to perish, but holds it fast.[9]

Parmenides's argument sent natural philosophy into a tizzy. The arguments that had originally sought to understand the reservoir of genesis had been transformed into an argument that *denies* genesis. Philosophers such as Empedocles, Anaxagoras, Leucippus, and Democritus found themselves saddled with an entirely new problem: if the true body is something eternally changeless and ungenerated, how shall we explain the world of our experience?

It makes a great difference that Parmenides is the first link in this chain. It would be hard to construe Parmenides's attitude towards the demos as anything less than harsh. Parmenides regards the people as so many dolts and drudges. There is nothing of gentleness about a doctrine which seeks to subtract reality from the conception of genesis. In fact, the argument that change in nature is a mere illusion is a doctrine of breathtaking audacity. It suggests the severest rebuke to ordinary human faculties; and it confronts us with an image of nature that leaves little room for the humanity and tenderness that Wilson associates with the Epicurean tradition.[10]

Parmenides lays down the gauntlet: true reality, or true body, cannot be divided into parts. It is 'one.' The atomists Democritus and Leucippus are not ideologists. Like Empedocles and Anaxagoras, they are well-intentioned men, trying to establish a compromise between Parmenides's disproof of genesis, and the human experience of a world that is constantly being renewed. As Aristotle recognized, however, none of these theorists was able to account for all the relvant facts. Either nature intends to generate, or it does not.

We obtain our best portraits of Parmenides and Zeno in Plato's dialogue the *Parmenides.* Parmenides is represented in that dialogue as attempting to mock and ridicule the young Socrates's arguments. This dialogue, which has baffled so many scholars, actually contains the best analysis of the Eleatic philosophy that has ever been accomplished. Parmenides comes across as something of a bully, Zeno even more so. However, despite Parmenides's many efforts to make Socrates's doctrine of the forms the focus of debate, Socrates hoists before us the premises of Parmenides's own argument.

Wilson jumps too quickly to the reputation of the later atomistic philosophy of Epicurus as the decisive issue for the Early Modern Period. The most decisive issue for the Early Modern Period, and the key to understanding it in all of its complexity, lay with a study of Plato's investigation into the origins of this most peculiar doctrine: a doctrine which holds that reality must be a strict unity; that coming into being and passing away are mere customary delusions, an inappropriate use of language, the mistaking of appearance for reality. Parmenides and his progeny believe that they can compel the interlocutor to surrender the opinion that genesis is genuine, by power of argument. This huge pretension is a great crossroads for the history of human thought.

Wilson gives no indication of being familiar with these deeper philosophical debates. Our academic division of labor has helped see to that.[11] What Wilson manages to accomplish, nevertheless, is to indicate the manner in which the physical doctrine of atoms carries in its train an entire cultural program: one touching on religion, manners, jurisprudence, the entire way of life of a people.[12] The trouble is that we still lack the fundamental knowledge of what the doctrine of the atom *embodies*. Until we know that, we are not in a position to evaluate the several cultural concomitants of the atomist theory.

I have elsewhere argued that Epicurus is, after a fashion, very much a political philosopher.[13] His commitment to atomistic philosophy is *mercenary.* The Epicurean imperative is liberty from the customary morality of the community. Atomism, for Epicurus, made available a powerful means to this end. The Democritean physics enabled the Epicurean to produce his own regimen of facts, expressly suited to his own desires. Nevertheless, Epicurus preferred to live a life apart from the demos. Early Modern Philosophy and its atomism, as Wilson recognizes, is quite differently disposed. Early Modern Philosophy is interested in 'experimental' undertakings: it wants to build a community committed to the Epicurean moral postulates.

It is a glaring omission in Wilson's account of the emergence of neo-Epicurean philosophy in Early Modernity that it fails to recognize the pivotal role played by Machiavelli in this transformation.[14] Machiavelli it was who shifted the focus of atomistic ideology. It was Machiavelli's unique contribution to the atomist philosophy to view it as a political instrument: with the help of atomist philosophy, the ambitious theorist can undertake to view human beings as essentially formless matter. The doctrine of atomism made available to Machiavelli (and the poem of Lucretius was highly instrumental in this) the possibility of a new kind of *detachment*. If one could manage to view people through the lens of atomism, one could insulate oneself against guilt feelings that might otherwise arise from programs and policies which savor of brutality. This new attitude is pivotal to the entire movement of Early Modern Philosophy. Francis Bacon, who did so much to smooth the harsher edges of Machiavelli's teachings, fully comprehends this new attitude. Bacon's fervent attack on Aristotle's distinction between natural and violent motions in nature is a direct descendant of the Florentine's influence. This influence ripples throughout the seventeenth-century teachings: in Descartes, in Hobbes, in Spinoza and beyond.

The scope of Wilson's investigation has both strengths and weaknesses. There is a considerable discussion of the crisis of religion in Wilson's work, and this might be its most valuable contribution. Yet, on the other hand, it is the broadness of Wilson's approach that deprives it of the conceptual mastery of atomistic philosophy, the key to the whole enterprise. Wilson accepts the atomistic emergence almost at face value, as a change in worldviews and related moral emphases. The concepts necessary to evaluate the true merits of the atomistic teaching are not something that Wilson strives for. The result is a blunted critical instrument: without a fundamental reading of the essence of the atomistic teaching, Wilson is not able to get into sufficient depth about the scope and nature of the changes that are underfoot.

At the core of the atomistic doctrine, and therefore at the core of the indictment of ordinary sense perception, lurks the theory of indivisible body: Body that neither comes into being, nor passes away. When Parmenides formulates this theory, he denies that the laypersons know what the real objects are. The ordinary people form opinions about objects that are seen to come into being and to pass away; yet the true reality, it is alleged, is indivisible. It has no past or future time. Thus the ordinary opinion is worse than lost, from this vantage point.

HOLDEN'S DISCUSSION OF THE ISSUE OF INFINITE DIVISIBILITY

Thomas Holden's book *The Architecture of Matter* is an important event in the scholarship on Early Modern Philosophy.[15] No other recent book so effectively opens up the issue of the metaphysics of body as it pertains to the Early Moderns. For my own inquiries, it is very important. For it is the metaphysics of atomism which ultimately prop up and empower the critique of ordinary perception; and it is the critique of ordinary perception which enables Early Modern Philosophy to pursue the ethical strategy of governing human beings through the management of their emotions and passions. Holden does not pursue his metaphysical analyses of body into the political area. Nevertheless, he provides a thorough examination of the pervasive influence that the metaphysics of body exerts across the Early Modern teachings.

It is David Hume's work which seems to have riveted Holden's attention to the issue of the metaphysics of body. Hume, in his famous *Treatise of Human Nature,* introduces a long discussion concerning *the* issue of the metaphysics of body: divisibility.[16] Parmenides's 'being' denies divisibility; and the classical theory of atomism, from Leucippus and Democritus to Lucretius, denies divisibility to atoms. Hume is an Early Modern atomist who is, like most Early Moderns, shy about being called out as such. Hume therefore cloaks his own discussion of divisibility in a rhetoric of mathematics. Holden has seen through this ruse of Hume's. It is body that Hume is concerned about, indivisible body. Hume selects a very innovative means for delivering his teaching.[17]

One of the great early defenders of the doctrine of indivisible body was Zeno, the intimate associate of Parmenides. Parmenides, as indicated above, makes the powerful argument that reality must be a unity. Zeno, protective of his mentor, sought to defend Parmenides from attacks that were being made against his doctrine of indivisible body. Zeno, instead of simply reiterating the teaching that true being must be indivisible, sought to humiliate and subdue Parmenides's critics. Those who hold that body is divisible, Zeno argues, commit themselves to absurd implications. This is the origin, the historical origin, of the doctrine of infinite divisibility: it was born as a defense of the Eleatic doctrine of indivisible being, and Hume employs it to introduce his own theory of indivisible body or atoms.[18]

We will examine the teaching of infinite divisibility that Holden's book concentrates on in detail in a moment. Yet first it needs to be pointed out that the original atomist doctrine appears to have been prompted by Zeno's attacks. It is the consensus of the majority of scholars that Leucippus and Democritus themselves were actually inspired to develop their atomistic teaching by Zeno's attack on divisibility.[19] This brings up a very important point, one that Holden has not grasped: the attack on infinite divisibility and the doctrine of atomism are as cause and effect. Zeno's discussion of infinite divisibility was formulated to silence Parmenides's critics; and the pluralist versions of indivisible body (atomisms all) delivered to Zeno the essence of the victory that he sought.

Hume's work comes well over a century after Francis Bacon's atomism. Before Hume wrote, Descartes and Hobbes, Spinoza and Boyle, and then Locke had all incorporated the atomistic teaching into their pivotal doctrines of physics. All of these early moderns, progeny of Machiavelli in this sense, worked with the atomistic doctrine as part of a political approach to society. What Machiavelli had commenced was the transformation of philosophy into an 'experimental' medium: society was to be the object of experimentation, and the doctrine of atomism was to function as the cause of this undertaking. At the center of the modern atomistic enterprise is the effort to conceive of the people as mere metaphysical objects of formless matter. As indicated above, the disjunction of this science from the evidence available to ordinary perception is its modus operandi, and the feature with which the current study is most directly concerned.

Hume, famously, in his portrait of human kind, characterizes it as a mere 'bundle of perceptions.' We cannot say, Hume promises, that the ideas in our heads correspond with any external objects. This is indeed the teaching that Descartes had helped so much to set into motion. If the ideas or perceptions that people have cannot stand as evidence as to what the objective external bodies are, then philosophy is free to conjure up endless dizzying scenarios as to what external reality *might* be: leaving the unphilosophic multitude fatigued and baffled, a prey for their own exhausted passions, and for new forms of administration. None of this would have been possible without the atomist philosophy of body, which underwrites the indictment of perceptual evidence; which brings us back to the issue of 'infinite divisibility,' the attitude towards body which Holden has located in Hume and in many of the Early Modern teachers.

To reiterate for the reader: the arguments under examination which contemplate the infinite divisibility of body, are actually arguments designed to prove that body has *no divisions whatsoever*. The argument from infinite divisibility is an attempt to reduce any opposition to silence. Holden never understands that the doctrine of atomism was partly caused by the browbeating inflicted on those who opposed Parmenides's teaching that body or 'being' is indivisible. Holden tends to view the two positions as *opposed*; whereas in fact, the investigation into infinite divisibility was a stratagem developed by the partisans of indivisible body themselves. Plato, in the early part of the *Parmenides*, is our best authority for this matter. Socrates asks Zeno: "Is this the precise purpose of your arguments—to maintain against everything that is commonly said, that things are not a plurality? Do you regard every one of your arguments as evidence of exactly that conclusion, and so hold that, in each argument in your treatise, you are giving just one more proof that a plurality does not exist?"[20]

Zeno formulated his argument in a double-barreled format. In the first place, those who argue that body must have parts, Zeno maintains, have no choice but to concede that it must have *infinite* parts. Once we engage with the supposition that body has parts, Zeno reasons, why put a limit to the number of parts that any body can have? There is no reason to impose such a limit, Zeno continues. Hence the doctrine of 'infinite parts,' or 'infinite divisions.' From Zeno's point of view, he has already harnessed the believers in bodies as wholes-of-parts into a fatal dilemma: the implications of the conviction that body has infinite parts force the disputant into absurd and untenable consequences. Zeno insists that the believers in the reality of divisions in body must accept either one untenable result or the other.

The first result is the one (the only one) that Holden contemplates. If we stipulate that body has indivisible parts; and if we stipulate that each one of these parts retains *some* magnitude, then we are left with the inference that even a pebble must contain as much magnitude as all other objects put together, and more. For if we have an infinite number of parts, and if each part has some magnitude (no matter how small)—then when we add them all together, we come away with something infinitely *large*. This, obviously, is an unsustainable position.[21] Holden's investigation into the conviction that parts, that is, separate bodies really, already exist in the object as irreducible minimums is almost a replication of the original founding of the atomistic arguments. Division must stop somewhere, the Early Moderns reason; and that place just is the proof that Parmenides was right, in linking true reality of body to indivisibility. The actual bodies might be numerous, rather than one; they may be too small to perceive rather than too large to perceive (such as Parmenides's 'one'); but the concession to the doctrine of partless body is made.

The second 'option' that Zeno offers to his interlocutors is in fact a tactic of argument with which Epicurus sought to *terrify* his readers. If we suppose that the infinite division of body erodes the magnitude of its parts, until there is no magnitude whatsoever remaining in them—then matter or body would soon be all used up, "reducing everything to nothing."[22] In this view, we must uphold the divisionless atoms if we are to concede that some part of reality must be preserved, so that the world can go on. Holden does not know this dimension of the Zenonian arguments because he is unable to see past the boundaries of the Enlightenment teachings. None of the Early Modern Philosophers teach the ancient conceptual history of the atomistic theory. It is this loss of philosophic memory that is at the root of the crisis that the Early Modern Period rolled into operation, and from which philosophy has not yet begun to extricate itself.

Holden is not familiar with Plato's critique of the Eleatic arguments. Plato proves both that every body whatsoever *is* infinitely divisible, *and* that every body whatsoever comes into being and passes away. Plato solves Zeno's riddle in his *Parmenides*. The same arguments that defeat Parmenides's thesis defeat the atomist one. Holden fails to understand that Zeno begins with a certain definition of body. Body is homogenous, in Zeno's view. If a body has parts, from this vantage point, then it is like making slices in a cake. There are only so many slices that one can make. Soon the cake will be used up. Plato prepares his critique of Parmenides by focusing on just this postulate. There is no such thing as homogeneous body, Plato argues. A true body must be made up of *unlike* parts. A true body must be made up of *unlike natures*. The composition of body is not simply a number of divisions in a homogeneous entity. Body is heterogeneous: and its fundamental parts are 'unity' and 'being.'[23] Without unity, there is no object that we can speak of; and without the very different nature of 'being,' there is no existence for the object once we have a unity that is indicated.

Plato's critique of Parmenides stands or falls with his indictment of the latter's thesis that body is indivisible. Unity is indivisible, Plato allows; but 'unity' cannot exist on its own. 'Being' is a separate nature; and unless 'unity' acquires the part of 'being,' it cannot so much as exist. This tour de force is the deepest foundation of Plato's critique of Parmenides's doctrine. Once it is conceded, as it must be conceded, that unity cannot by itself constitute a body, Parmenides's entire argument must fall. Once it is conceded that 'being' is a separate and coequal nature along with unity, and that body must have both of them, a succession of deductions follows that cannot be resisted.

To wit: if body is a combination of the dual natures of unity and being, then any body whatsoever is a 'whole.'[24] Inside that whole, 'unity' and 'being' are 'parts.' Therefore, both 'whole' and 'part' exist *in nature, as realities. If a body is a whole-of-parts, then it must have all the parts requisite to be a whole*. It must have extremities. It must have a beginning (genesis), and an end (passing away).[25] It must accordingly have all three parts of time: past, present, and future. It must have dimensions ('extension'). It must have shape, size, color, texture, sameness with itself, difference from the other objects. All of these inferences are irresistible, if Plato's original critique of the seamless Parmenidean 'being' holds. And hold it does.

HISTORIANS OF SCIENCE ON THE SEVENTEENTH CENTURY

The purpose of this introduction is to present an overview of some of the chief issues that will be taken up for inspection. Towards this end, it is important to spend some time evaluating and remarking upon the work done over the past century in the newly arisen discipline of the history of science. The historians of science, a discipline that Thomas Kuhn did much to crystallize, remains pretty much an isolated literature to itself. It purports to represent the 'scientific revolution' of the seventeenth century, but in a context that is radically segregated apart from the discussion into the metaphysics of matter on the one hand, and any sort of political dimension to the modern scientific teachings on the other. The works of Wilson and Holden fall outside of this paradigm.[26]

The historians of science have produced a rather strong theoretical account of the seventeenth century which appears to stand in the way of the sort of analysis that I propose.[27] Thus I would like to scrutinize the bases of the historians of science, and to explain exactly why the trajectory of their inquiries is not satisfactory.

The historians of science prefer to represent seventeenth-century science as a distinctly apolitical affair. These scholars would rather that we pay more attention to the actual experimental researchers who can be seen to have made contributions to technological learning in that century. Questions such as are involved in matter theory, as it is now usually called, get a rather short shrift. The historians of science themselves may not be terribly sophisticated in terms of the knowledge of the ancient debates which played a role in the theoretical enterprises of the dominant voices of the seventeenth century. This does not however prove that such lack for importance.

When it comes time for establishing the separation of the seventeenth-century movement from the reign of Medieval Aristotelianism, the historians of science would like us to rest content with a discussion of the work of Galileo. Galileo, it is indicated, accomplished the radical break with the Aristotelian, and thus with the ancient point of view.[28] Galileo, however, was more mathematician than theorist. In proportion as the historians of science wish to give a greater share of the glory to Galileo for establishing the new foundations of seventeenth-century science, attention is diverted away from thinkers such as Hobbes and Spinoza, whose political predispositions are more easily discerned.

In my view, the argument advanced by these historians of science is untenable. Aristotle's physics had stood for centuries. Galileo only takes up a relatively modest portion of Aristotle's theory for dispute. The Galilean analysis of projectile movement is held out by these historians as effectively the moment where Aristotle's physics is defeated. First, it is not clear that Galileo's argument succeeds in refuting Aristotle's physics. Moreover, an overwhelming assault on Aristotle, pointed and political-minded, is advanced by the four philosophers with which this study is concerned. Surely it is with Bacon, Descartes, Hobbes and Spinoza, rather than with Galileo, that the new science gained the upper hand.

Francis Bacon for one is quite familiar with the ancient debates that took place between Aristotle and the atomist philosophies of Democritus and Leucippus. Bacon, furthermore, received his own decisive scientific-atomistic impulse from the newly politicized theory of atomism developed by Machiavelli.[29] The sheer conceptual intensity and intellectual impetus contained in the atomist theory of the Machiavellian lineage is vastly more powerful than the criticisms of Aristotle made from the severely limited range of argument presented by Galileo. It is a distortion to represent Galileo as the thinker who made a home for atomism and ancient matter theory in the seventeenth century.

The reader is asked to bear in mind the great power and influence that philosophy can wield, when it has imbedded itself in a theory which arrogates to philosophy the authority to determine how names shall be used. Philosophy originates in argument. If a select class is free to use names in contravention of the ordinary signification, it is hard to understand how that group could ever fail to carry the day in dispute. Hobbes envisions something very much along these lines.

"A name is taken at pleasure to serve for a mark, which may raise in our mind a thought like to some thought we had before, and which being pronounced to others, may be to them a sign of what thought the speaker had, or had not before in his mind," Hobbes writes. "And it is for brevity's sake that I suppose the original of names to be arbitrary, judging it a thing that may be assumed as unquestionable."[30] Hobbes speaks for philosophers in this passage. No other person in the human race comes upon names in such a manner, where preference or pleasure has anything to do with how things are to be called. What an apple, what a man, what just, what unjust: such names, both Plato and the non-philosopher would argue, simply indicate the 'whatness' of an object; and the proper name is given to us by ordinary usage, nothing else.

Such however is not what Hobbes has in mind. Hobbes envisions a world in which philosophers may assign names how they please, regardless of ordinary usage. "Moreover, whatsoever the common use of words be, yet philosophers, who were to teach their knowledge to others, had always the liberty, and sometimes they both had and will have a necessity, of taking to themselves such names as they please for the signifying of their meaning, if they would have it understood."[31]

THE DIVISION BETWEEN SCIENCE AND PHILOSOPHY

The separation between the categories of science and philosophy is crucial on several levels. In the first place, it obscures the fact that atomist or matter-based theories *are* philosophical theories: that they were generated by philosophical arguments, and that there is indeed a history of critiques of those arguments. In the second place are the radical political implications of the atomist-based doctrines themselves: for they relegate sense perception to the lowest rung of human information. That is a profoundly important fact to lose sight of, in the century when the rights of man were first trumpeted.

Of the four philosophers under review in this study, two of them receive a fair amount of attention from the historians of science, and two of them are all but ignored. Francis Bacon, with his passion for developing experimental techniques in science, is more congenial to the historians insofar as he can be linked to a technological program that doesn't necessarily bring a discussion of politics into the picture. Bacon is celebrated for bringing a mood of civility and 'objectivity' to scientific thought.[32]

Descartes' focus on mathematics as an allegedly pure source of new and irrefutable scientific knowledge likewise stands in good stead with these historians. Mathematics is a large part of what the new atomistic movement in the seventeenth century professes to be *about.* Descartes is comfortably treated by our historians. He is not a threatening representative for the revolutionary century in science.

Things stand somewhat differently in the cases of Hobbes and Spinoza. Hobbes and Spinoza both developed aggressive, bold, and even slashing political philosophies, for which they are now best known. Historians of science neglect Hobbes and Spinoza to the point of ignoring them: as if these two did not rest their political sciences on the very same physics that propelled Bacon and Descartes. Hobbes and Spinoza are unappealing to historians of science precisely because the political looms so large in their respective works. The fact remains that both Hobbes and Spinoza present highly developed versions of atomistic matter theory, as will be proved in the appropriate places.

THE NEW VISION OF NATURE IN EARLY MODERN SCIENCE

There is a very distinct scientific vision of reality advanced by the seventeenth-century philosophers. The world they depict is quite indifferent to human kind. It is not that nature is actually hostile to the human race, these philosophies suggest. It is simply that nature takes no notice of man. Man is on his own. For the new world is alleged to be 'matter in motion.' The senses are accused of being mere deceivers.[33] Man is said to confuse for reality certain events that take place in the lonely chamber of his brain and nervous system.

The historians of science do not succeed in presenting the matter theories of the seventeenth-century philosophers in sufficient depth. There are a couple of reasons for this. In the first place, the historians of science declaim that they themselves are not really *capable* of knowing precisely what those matter theories were. For the historian of science can allegedly only investigate subjects through his or her *own* artificial frame of reference.[34] Thus the historian of science is felt to be limited by *his or her own* presuppositions; the historian will only see what he or she has set out to see, and this construction is said to be *arbitrary*.

When the historians of science profess their inability to know any subject matter beyond the artificial frames of reference that they *stipulate*, they in fact do no more than submit to the premises of seveneenth-century philosophy *itself*.[35] For it is Bacon and Descartes, Hobbes and Spinoza who deny that individuals can ever truly have *common* objects for investigation. The very claims as to frames of reference relied upon by Kuhn and Popper, Collingwood and Shapin, are derived from philosophical theories of physics which *stand in need* of investigation. In this book I do not propose to teach my own 'frame of reference'; but to examine the philosophical claim that investigators are really *bound* to such. The decisive issue is still perception, as the historians of science indicate awareness of; and the seventeenth-century philosophies depend on theories to carry into circulation this supposed limitation of human knowledge.

The seventeenth-century philosophers did not regard their theories of matter as 'presuppositions' or 'imaginative posits' or 'paradigms.' They regarded their theories of matter in the ways characteristic of the ancient atomistic doctrine. The atomistic doctrine professes to resolve issues of truth about nature. It attempts to reconcile the permanent with the relatively transient perishable world. Aristotle in fact had great respect for Democritus as perhaps the foremost scientist of his times, although he ultimately found the atomist philosophy to be fatally flawed.[36]

There is certainly a tendency among the historians of science to present a very reductive view of that era, and to vastly understate the *philosophical* commitments of those teachings.[37] The theories of 'presuppositions' or 'paradigms' that historians of science employ to evaluate the seventeenth-century thinkers leave the deeper layers of those philosophers' arguments unexposed to view. In this book I will attempt to do better.

It was at the hands of Epicurus that the atomist arguments of Democritus and Leucippus underwent a fateful transformation. It is this Epicurean vein of philosophy which has lent to the seventeenth century its rhetorical tactics and aversion to debate. Epicurus adopted the atomist teaching with an entirely new *attitude*. For Epicurus, the purpose of philosophy is not truth.[38] Epicurean philosophy is dedicated to the psychological needs of the philosopher. It is only with the writings of Machiavelli that the Epicurean version of atomism becomes an aggressively political doctrine.[39] Yet this is the nature of the matter theory that emerged into prominence in the seventeenth-century writers.

Form and matter are concepts of physics with ancient pedigrees. In antiquity, these concepts furnished the terminology with which to discuss nature and the objects in it. In Machiavelli's writings, the term 'materia' or matter is employed to refer to the *people* as a political substratum. Form, in Machiavelli's political science, concerns the nature of the personality and mood that the philosopher-statesman seeks to impress upon the people, as if they were mere clay, and his hands, those of the potter.[40]

The historians of science barely make a peep about the 'rights of man,' which is the public face of this very same science.[41] When Spinoza makes the argument that man may no longer regard himself as living in a 'kingdom within a kingdom,' he is indicating that man may no longer regard his own morality as in fundamental ways separate from the laws that seem to attend to the business of bees, or herds of buffalo.[42] For Aristotle these distinctions were obvious and effortless. "Now that man is more of a political animal than the bees or any other gregarious animals is evident. Nature, as we often say, makes nothing in vain, and man is the only animal who has the gift of speech."[43] It is clear that in the intervening centuries, a very different foundation for knowledge has crept into place. The point is that one cannot understand either the physics or the political philosophy of the new movement separately: for in the theory of modern atomism, they are *fused*.

CULTURE AND THE ATOMISTIC SCIENCE

E.A. Burtt, in the early decades of the twentieth century, advanced the argument that the seventeenth century simply had to be implicated in a substantial new metaphysics. For the culture of western civilization had dramatically changed. Not only was man not regarded as the center of the universe, but he had come to be regarded as an orphan in nature. The new science indicated that nature had no plan for man, no place for him, in its majestic atomic indifference. The lonely, forlorn vision of man that resulted had not occurred by accident, Burtt surmises. "The really important world outside was a world hard, cold, colorless, silent, and dead," Burtt observes; "a world of quantity, a world of mathematically computable motions in mechanical regularity. The world of quality as immediately perceived by man became just a curious and quite minor effect of that infinite machine beyond."[44]

While Burtt laments the grim new role for human beings in the emergent cosmology, A.N. Whitehead seems to positively revel in the change that has come about. Ordinary people will just have to get used to the facts as they are, Whitehead indicates: the heat is not in the fire; the sound is not in the guitar; the scent is not in the rose. In accordance with the new metaphysics, these are merely illusions *in us*. "Thus nature gets credit which should in truth be reserved for ourselves," Whitehead writes, "the rose for its scent: the nightingale for his song: and the sun for his radiance. The poets are entirely mistaken. They should address their lyrics to themselves, and should turn them into odes of self-congratulation on the excellency of the human mind. Nature is a dull affair, soundless, scentless, colorless; merely the hurrying of material, endlessly, meaninglessly."[45]

Atomism brings in its train, for all four of these Early Modern thinkers, a new scientific 'skepticism' to culture. Perception is fully downgraded to the status of 'imagination,' which in ordinary speech suggests fantasy.[46] The Early Modern philosophers, for their part, profess to derive their *own* ideas from pure intellect, not from the muddied waters of perception. "But ideas which are clear and distinct can never be false; for ideas of things which are clearly and distinctly conceived are either absolutely simple or are compounded of absolutely simple ideas—that is, deduced from absolutely simple ideas."[47]

Allegedly, the heat is not in the fire. The scent is not in the rose. Yet this is just the beginning. What had commenced its preparations in the seventeenth century is a new representation of nature: as something unintelligible, mysterious, counter-intuitive, and enigmatic. This is the root of a new conception of human existence, in which man is intended to be *petrified*. Hobbes did not make *De Corpore* the first and foundational book of his trilogy for nothing. Spinoza did not devote the first two books of his *Ethics* to physics inadvertently. It is in the new theory of nature, as conceived by the Early Modern philosophers, that culture and politics are to be *shaped*.

Historians of science do not speak with much admiration for Plato and Aristotle. Nor is much attention paid to the theories of these philosophers. This is troubling. Early Modern philosophers established a very insidious trend as regards the study of Plato and Aristotle. Obloquy substitutes for argument. Machiavelli and Francis Bacon set the tone for dealing with the great philosophers of antiquity. To some degree, the historians of science partake of this attitude. "Come then, let Aristotle be summoned to the bar, that worst of sophists stupefied by his own unprofitable subtlety, the cheap dupe of words."[48] Herbert Butterfield, when he asserts with confidence that Galileo 'destroyed' Aristotelian physics, is displaying some of the same hubris.[49]

MATHEMATICS AND THE NEW PHYSICS

Mathematics, just like atomism itself, presents two faces in the early modern period. In one aspect, seventeenth-century mathematics attempts to reduce all of nature to something measurable. Galileo cannot be blamed, certainly, for wanting to convert the study of nature entirely into something that lends itself to measurement. Yet this effort of Galileo's did have indirect political consequences. For the atomistic theories of matter, which the more powerful philosophers of the seventeenth century were anxious to hide, could find a new and secure abode in the daunting and complex nomenclature of mathematical terminology. The fact of the matter is that not all aspects of body *do* lend themselves to measurement; and that Galileo was wrong to insist that this could be done.

Galileo was determined to discover mathematical laws in the fabric of nature. Galileo has been compared to Plato in this regard, though not with good reason.[50] Mathematics does not reach to the truth of what objects are, in Plato's view.[51] Thus mathematics does not possess the power to *know* objects. Of course Plato valued geometry. He regarded it as essential preparation for any would-be student in his Academy. Yet this is more for the training that the abstraction from perishable bodies furnishes to the student's mind. It is the indispensable training for learning to contemplate the forms in separation from concrete objects.[52] Galileo is not interested in what objects are in themselves. Galileo is interested in things that he can measure, period. Forms are not for measurement. They are for *knowing*.

Descartes leads the way in the Early Modern Period in defining body as 'extension.' This is all that body is, for Descartes. "The nature of body consists not in weight, hardness, color, or the like," Descartes writes, "but simply in extension."[53] Hobbes and Spinoza both take up the same definition.[54] It is worth pointing out that Plato regards extension, or dimensions and shape as they would be more commonly called, as merely several of the *many* parts characteristic of any body whatsoever. The measurable quality of a body, to Plato, cannot even be talked about before one has established the *unity* and *being* of the object. There is the nature of a 'whole' to be enumerated, and the nature of a 'part.' The reality of a whole-of-parts must first be established, in Plato's view, in order for us to even be able to talk about the reality of an object's extremities.[55]

Peter Dear makes several observations that are relevant to this inquiry into Galileo's role as the apostle for a new natural philosophy. Galileo, along with Isaac Beeckman and Descartes, belongs to a category of thinker that Dear characterizes as practicing 'physico-mathematics.' The term 'physico-mathematics' illustrates the debate: for 'physico' indicates body, whereas 'mathematicos' indicates the measurement of bodies. Could it be conceivable that the art of measurement, or mathematics, can *take over* natural philosophy? For Dear, the answer is in the affirmative. "The new category made it easier for mathematical scientists to make philosophical claims that had previously been fiercely contested. . . . A process of disciplinary materialism, whereby subject matter usually regarded as a part of physics was taken over by mathematics, operated to upgrade the status and explanatory power of the mathematical sciences."[56] Of Newton, Dear writes: "His mathematical science of optics, in other words, adopts the presumptions that had driven the increasing use of the label 'physico-mathematics' throughout the seventeenth century—it takes over topics from natural philosophy itself."[57]

The justification for this reach of mathematics into the territory previously allocated to natural philosophy is that all of natural philosophy is bound up with measurable quantity.[58] It is absolutely certain that there must be a *body* to measure before mathematics is even possible. The question, 'what is a body?' is foremost for both Plato and Aristotle. How shall mathematics reply? It makes a great deal of difference what the mathematician names the objects that he measures. What are they? If it is a body, it must have *being*. In order for being to exist at all, it must belong to some distinct *unity*. These are two irreducible elements in any body whatsoever. Yet what of geometry? Can it measure 'being'? No. Can it measure 'unity'? No. To reduce the conception of body to that which is measurable is arbitrary. It is metaphysically unsupportable. Extremities are not the metaphysical 'guts' of an object: unity and being are.

Descartes is the darling of Dear as he is for many historians of science. Yet for Descartes, 'extension' is all that body is. Descartes is quite wrong. 'Extension' or extremities can only belong to wholes-of-parts: wholes of qualitatively *different* parts.[59] We do not refer here to arbitrary divisions made up in the mind of the philosopher: we refer here to heterogeneity in the parts indigenous to any actual body whatsoever. Mathematics therefore cannot 'take over' this subject matter from natural philosophy. At least mathematics cannot take over a *true* natural philosophy. When we do recover the proper conception of body, it will be possible to return to those things, such as forms, which are essential to knowledge but irrelevant to measurement.

The Early Modern philosophers' interest in mathematics takes on a different mood in the work of Hobbes and Spinoza. Hobbes and Spinoza are very much interested in the authority to establish definitions.[60] The Early Modern philosophers seek to avail themselves of the authority of mathematics in this regard, but not due to the fact that the definitions they are interested in establishing truly have anything to do with measurement. Mathematics appeals to Hobbes and Spinoza in that it generally displays the power to enforce unanimity of judgment upon the reasoner. "For if the patterns of human action were known with the same certainty as the relations of magnitude in figures, ambition and greed, whose power rests on the false opinions of the common people about right and wrong . . . would be disarmed, and the human race would enjoy such secure peace that . . . it seems unlikely that it would ever have to fight again."[61]

Hobbes and Spinoza exploit the new mathematical pretensions of natural philosophy in a political direction. Euclid, in his geometry, presents a number of definitions which are stipulated as givens. They are to be accepted as 'givens' insofar as they are self-evident claims. For example, the argument that a whole is greater than one of its parts is one of Euclid's definitions. Hobbes and Spinoza go out of their way to wrap their arms around Euclid. Yet they do so for a rather different reason: for they seek to set forth new definitions towards a moral philosophy, and such definitions! Self-evident to the ordinary individual, they most certainly are not.

THE EARLY MODERN PHILOSOPHERS' ASSAULT ON UNIVERSALS

The political face of the proponents of the atomistic philosophy in the seventeenth century can very well be discerned by their assault upon 'universals.' The universals ('cat,' 'courage,' 'man,' 'quickness') are not just patterns that are alleged by Plato to exist in the perishable objects. In combination with the powerful attack on the authority of sense-perception, the Early Modern ridicule of the status of universals in ordinary speech is a devastating political offensive. For one thing, the Early Modern philosophers do not convey to their readers the arguments by virtue of which Plato and Aristotle defended those universals; and of course, the general population is not going to be able to defend itself in this regard.

Plato's Socrates made the discovery that ordinary speakers know something when they assign names to perishable objects so effortlessly: and what they know is something that they have previously experienced and recollected, that is, the forms.[62] A single form is present in a multitude of objects. To know the form in one object is therefore to know the form in any object.[63] Plato's Socrates only claims, however, to describe how ordinary knowledge of objects comes about: he certainly does not profess to be the *founder* of the practice. Yet in his science of definition, which Plato certainly does found, the accuracy of the ordinary assignment of name to object is offered as the original, indisputable evidence.

The attack on the Platonic forms are common to all four of the thinkers scrutinized in this book. The doctrine of forms is derided as useless, deluded, or fraudulent. It is in the work of Hobbes and Spinoza where this attack on the classical forms is best appraised. For both Hobbes and Spinoza recognize the ubiquity of such forms in ordinary discourse. The 'universal' is a direct object of attack in these two thinkers, whom the historians of science barely acknowledge.[64]

The attack on universals waged by Hobbes and Spinoza is a politically fraught one.[65] The issues involved are essentially two. The first issue concerns whether or not the non-philosophers know something when they effortlessly assign names to perishable objects without error. The second issue is related to the first. It concerns the object of knowledge itself. The conviction that universals 'exist' consists in the insight that no object is *entirely* individual or perishable. Every object perceived is of a *kind*. The argument that universals exist does more than fit hand-in-glove with the perceptual evidence. It binds perception analytically to the structure of natural languages; and it furthermore fuses the evidence from perception to the foundation of scientific definitions.

LEO STRAUSS ON THE EARLY MODERNS

Leo Strauss's work also traces the Epicurean influence on the development of modern science, in Bacon, Descartes, Hobbes and Spinoza. Strauss to some degree offers a corrective of the superficial approach of the historians of science. This is not to say that Strauss's presentation of the Early Moderns is entirely sufficient. It is important because it illuminates what is not much talked about elsewhere: namely the degree to which the Epicurean attitude saturated the Early Modern philosophers. It is only in the Epicurean heritage where it is possible to obtain a different insight as to the motive of the new natural sciences, and the forlorn condition to which it has banished man. "Man can be sovereign only because there is no cosmic support for his humanity," Strauss writes, "He can be sovereign only because he is absolutely a stranger in the universe. Since the universe is unintelligible and since control of nature does not require understanding of nature, there are no knowable limits to his conquest of nature. He has nothing to lose but his chains, and for all he knows, he may have everything to gain."[66]

Hobbes and Spinoza equally chastise the pride of man, including his conception of himself as living in a 'kingdom within a kingdom.' But in both thinkers, there is a new species of man that has been liberated: the man for whom virtue is identical with power. Spinoza boasts about creating a new model of human being. Such a one looks with ridicule upon customary ethics, and based on an alleged 'unity' in nature argues that it is by 'sovereign natural right' that big fishes eat smaller fishes, and that the more powerful human beings violate their contracts with the less powerful.[67]

Strauss, in his youthful writings, is remarkably candid and blunt about the philosophers of the Early Modern Period. His own fascination with Spinoza makes him quite conversant with the Epicurean tradition, and Machiavelli's place in it. The Epicurean heritage flows freely in early modern science.[68] Yet this is the Epicurean atomism that Machiavelli has radicalized: it is an inventive, enterprising, audacious doctrine that reduces men to masses, to 'matter,' as stuff to be remolded and impressed with artificial forms.

The purpose of modern science is not first of all to subdue nature. It is to deny our *knowledge* of it. What is obvious to a child will be called into question by the new metaphysics.[69] Perception is assailed. Ordinary judgment is indicted. All this is done by calm scientific authority. Every thing that people think they know—beginning with what a body is—is in a way suspended. Nobody has described this aspect of modern science with the intensity and precision characteristic of Strauss's early work. It is the natural science of the Baconian movement that seeks to make men feel *lost*. It is only when men feel lost that they become susceptible to a new form of government, that is, management of their *passions*.[70]

Hobbes could not be clearer than he is in the introduction to his *Leviathan*.[71] When Hobbes advises man that the best evidence for his arguments is 'self-knowledge,' he clarifies his meaning. Men are not alike in what they perceive, Hobbes indicates; they are to be considered alike in what they *feel*. When Hobbes excludes the objects of perception from that which human beings can and do know in common, he is replacing the passions commonly evoked in human beings by mutual facts for a new set of passions induced by *philosophy*.

Hobbes is not in favor of a human race which is in a position to entertain the *impulse* to justice. Hobbes undertakes to create, for the human race or certain portions of it, a new psychological profile. Here the impulse to justice is itself felt to be an insupportable *luxury,* even a fantasy; in all events it is portrayed as *an action against self-interest*. The decent and typical man of Hobbesian nature must firmly disown and disdain this impulse to justice as a species of brute self-indulgence, archaic selfishness, and indeed as contravening the security arrangement upon which society itself must depend.

It is this philosophy, not nature, which begins to puncture the confidence of ordinary opinion in its own faculties. Self-preservation, the new moral ethos ushered into existence by Hobbes and Spinoza, is a moral regimen for a *desperate* condition: the people must be made desperate, before they can be prepared to receive this artificial form.

Machiavelli's virtu consists in the ruthlessness and daring required to impose new forms on people. It is from this new status as mere matter that the accidental man of the new era of science, for whom life cannot possibly be brave or just, arises. Leo Strauss is far from sad when he makes the following commentary on Hobbes's argument. "For the man who has once come into contact with this world, joy and laughter are over. It is the fearfulness of death, rather than the sweetness of life, which makes men cling to existence."[72] Hobbes's desideratum, that man can live in lonely comfort, or perish in violent conflict, would freeze all the nobler and stronger impulses to life in man.

NOTES

1. *Theatetus* 180a–184b; *Gorgias* 472a.
2. *Laches* 186e.

3. Fragment 8. In Kathleen Freeman. *Ancilla to the Pre-Socratic Philosophers*. Cambridge: Harvard University Press, 1996, 43.

4. See Rene Descartes. *Discourse on the Method, of rightly conducting one's reason and seeking the truth in the sciences. In the Philosophical Writings of Descartes*, volume I. Translated by John Cottingham, Robert Stoothoff, and Dugald Murdoch. Cambridge University Press, 1985, p. 127 (hereafter *CSM* I): "But now since I wished to devote myself solely to the search for truth, I thought it necessary to do the very opposite and reject as if absolutely false, everything in which I could imagine the least doubt, in order to see if I was left believing any thing that was entirely indubitable. Thus because our senses sometimes deceive us, I decided to suppose that nothing was such as they led us to imagine . . ." Cf. Francis Bacon. *The New Organon*. Edited by Lisa Jardine and Michael Silverthorne. Cambridge University Press, 2000, 17: "The senses are defective in two ways: they may fail us altogether or they may deceive. First, there are many thing which escape the senses, even when they are healthy and quite unimpeded. . . . And even when the senses do grasp an object, their apprehensions of it are not always reliable. For the evidence and information given by the senses is always based on the analogy of man not of the universe; it is a very great error to assert that the senses are the measures of things." Cf. Thomas Hobbes. *Leviathan*. Edited by J.C.A. Gaskin. Oxford University Press, 1996, part I, chapter 1, article 4: "And though at some certain distance, the real and very object seems invested with the fancy it begets in us; yet still the object is one thing, the image or fancy is another. So that sense in all cases, is nothing but original fancy . . ." (p. 10). See also Baruch Spinoza. *The Ethics*. In *The Complete Works*. Translations by Samuel Shirley. Edited by Michael L. Morgan. Indianapolis: Hackett, 2002, part I, appendix, p. 243: "We see therefore that all the notions whereby the common people are wont to explain nature are merely modes of imagination and denote not the nature of anything but only of the constitution of the imagination."

5. Cf. Alexander Mourelatos. "Intrinsic and Relational Properties of Atoms in the Democritean Ontology" (42–44). In Richard Salles, editor. *Metaphysics, Soul and Ethics in Ancient Thought*. Oxford: Clarendon Press, 2005; James Warren. *The Presocratics*. Berkeley: University of California Press, 2007, 160–61; Istvan M. Bodnar, "Atomic Independence and Indivisibility." *Oxford Studies in Ancient Philosophy* 16(1998):43; R.W. Baldes, "Divisibility and Division in Democritus." *Apeiron* 12(1978):4; Friedrich Solmsen, "Abdera's Argument for the Atomic Theory." *Greek, Roman and Byzantine Studies* 29(1988):50, 61, 65. See also my *Machiavelli and Epicureanism: An Investigation into the Origins of Early Modern Political Thought*. Lanham MD: Lexington Books, 2012, chapter 1.

6. *Epicureanism at the Origins of Modernity*. Oxford: Clarendon Press, 2010.

7. Ibid., 3.

8. Ibid., 37.

9. Fragment 8. In Kathleen Freeman. *Ancilla to the Presocratic Philosophers*. Cambridge: Harvard University Press, 1996, p.43.

10. Wilson, *Epicureanism at the Origins*, 37.

11. I have examined this literature in *Plato Versus Parmenides: The Debate Over Coming-into-Being in Greek Philosophy*. Lanham MD: Lexington Books, 2010.

12. Ibid., chapter 7 passim.

13. *Machiavelli and Epicureanism*, chapter 2.

14. The literature is not insignificant on Machiavelli's relationship to Epicurean philosophy via his exposure to the poem of Lucretius. See Paul Rahe, "In the Shadow of Lucretius: The Epicurean Foundations of Machiavelli's Political Thought." *The History of Political Thought* 1(2007):30–66; Alison Brown. *The Return of Lucretius to Renaissance Florence*. Cambridge: Harvard University Press, 2010; John N. Najemy, introduction. In Najemy, editor. *The Cambridge Companion to Machiavelli*. Cambridge University Press, 2010. In the same volume, see James B. Atkinson, "Niccolo Machiavelli: A Portrait," p. 16. Michael Reeve, "Lucretius in the Early Middle Ages and Renaissance—Transmission and Scholarship." In Stuart Gillespie and Philip Hardie, editors. *The Cambridge Companion to Lucretius*. Cambridge University Press, 2007, p. 211. In the same volume see Valentina Prosperi, "Lucretius in the Renaissance," p. 214. Don Cameron Allen, "The Rehabilitation of Epicurus and the Theory of Pleasure in the Early Renaissance." *Studies in Philology* 41(1944):6; Paul Oskar Kristeller. *Renaissance*

Thought and the Arts. Princeton University Press, 1990, 36; see also Kristeller, "Marsilio Ficino and His Circle," in *Studies in Renaissance Thought and Letters*. Rome: Storia e letteratura, 1956, 37, 38, 49. See also Peter Godman. *From Poliziano to Machiavelli: Florentine Humanism in the High Renaissance*. Princeton University Press, 1998, 31, 39, 141, 144.

15. *The Architecture of Matter: From Galileo to Kant*. Oxford: Clarendon Press, 2006.

16. *A Treatise of Human Nature*. Edited by L.A. Selby-Bigge. Oxford: Clarendon Press, 1978. Part II, section iv, beginning on p. 29.

17. Ibid., section II–IV passim.

18. Ibid., p. 28.

19. W.K.C. Guthrie. *The PreSocratic Tradition From Parmenides to Democritus*. Cambridge University Press, 1965, p. 384; Daniel Graham, "Leucippus's Atomism," p. 347. In Patricia Curd and Daniel W. Graham, editors. *PreSocratic Philosophy: The Oxford Handbook*, 2008. John Burnet. *Early Greek Philosophy*. New York: Meridian, 1930, p. 332. Stephen Makin. *Indifference Arguments*. Oxford: Blackwell, 1993, p. 9. Richard Sorabji. *Time, Creation and the Continuum*. Ithaca: Cornell University Press, 1983, p. 322. See also C.C.W. Taylor. *The Atomists Leucippus and Democritus*. Toronto University Press, 1999, p. 165.

20. *The Parmenides* 127e.

21. *The Architecture of Matter*, 24.

22. "And if that which disappears were destroyed into what is not, all things would have been destroyed, since that into which they were dissolved does not exist. Further, the totality of things has always been just like it is now and always will be. For there is nothing for it to change into." *Letter to Herodotus* 39. In *The Epicurus Reader*. Edited by Brad Inwood and L.P. Gerson. Indianapolis: Hackett, 1994, 6.

23. *The Parmenides*, 142b–143.

24. *Parmenides* 145–146.

25. *Parmenides* 145b.

26. See also Helen Hattab. *Descartes on Forms and Mechanisms*. Cambridge University Press, 2009; and Kurt Smith. *Matter Matters: Metaphysics and Methodology in the Early Modern Period.* Oxford: Oxford University Press, 2010.

27. Some of the works that I have consulted include: Alexandre Koyre. *Metaphysics and Measurement.* Cambridge: Harvard University Press, 1968. Stephen Gaukroger. *Francis Bacon and the Transformation of Early-Modern Philosophy*. Cambridge University Press, 2001. Richard F. Jones. *Ancients and Moderns: A Study of the Rise of the Scientific Movement in Seventeenth Century England.* St. Louis: Washington University Press, 1961. Jurgen Renn, editor. *Galileo In Context*. Cambridge University Press, 2001. Conal Condren, Stephen Gaukroger, and Ian Hunter, editors. *The Philosopher in Early Modern Europe: The Nature of a Contested Identity.* Cambridge University Press, 2006. Timothy J. Reiss. *Knowledge, Discovery and Imagination in Early Modern Europe*. Cambridge University Press, 1997. A.R. Hall. *The Scientific Revolution 1500–1800: The Formation of the Modern Scientific Attitude.* 2nd edition. Boston: Beacon Press, 1966. Dudley Shapere, "Meaning and Scientific Change." In Ian Hacking, editor. *Scientific Revolutions*. Oxford University Press, 1981.

28. Alexandre Koyre. *Metaphysics and Measurement*. Cambridge: Harvard University Press, 1968, 2. Gary Hatfield, "Metaphysics and the New Science," 117–118. In David C. Lindberg and Robert S. Westman, editors. *Reappraisals of the Scientific Revolution*. Cambridge University Press, 1990. Herbert Butterfield. *The Origins of Modern Science*. Revised edition. New York: Free Press, 1965, 93–102. A.R. Hall. *The Scientific Revolution 1500–1800: The Formation of the Modern Scientific Attitude*. 2nd edition. Boston: Beacon Press, 1966, 100–101. Stephen Gaukroger. *The Emergence of a Scientific Culture: Science and the Shaping of Modernity 1210–1685*. Oxford University Press, 2006, 414.

29. I have examined the evidence for Machiavelli's involvement with Epicurean atomism in my book *Machiavelli and Epicureanism: An Investigation into the Origins of Early Modern Political Thought*. Lanham: Lexington Books, 2012.

30. *Works* I, p. 16.

31. *Works* I, p. 16.

32. See Stephen Gaukroger. *The Emergence of a Scientific Culture: Science and the Shaping of Modernity 1210–1685*. Oxford University Press, 2006, chapter 6 passim.

33. Hobbes can serve, for the moment, as an exemplar for all four philosophers and their attitude towards sense perception, in *Leviathan:* "And though at some certain distance, the real and very object seems invested with the fancy it begets in us, yet still the object is one thing, the image or fancy is another. So that sense in all cases, is nothing else but original fancy, caused (as I have said) by the pressure, that is, by the motion, of external things upon our eyes, ears, and other organs thereunto ordained." (Chapter 2, #4).

34. 'Facts,' for these scholars, are simply products of their theories. The facts drawn from one theory cannot be challenged by the facts drawn from a different theory. None of the theories, for that matter, are conceded to aim for truth. "Nevertheless, there is another step or kind of step, which many philosophers of science wish to take and which I refuse. They wish, that is, to compose theories as representations of nature, as statements about 'what is really out there' . . . I believe nothing of the sort can be found. On the other hand, I no longer think that anything is lost." Thomas Kuhn. "Reflections on my Critics." In Lakatos and Musgrave, editors. *Criticism and the Growth of Knowledge.* Cambridge University Press, 1999, 265. "We cannot say with assurance that the two men even see the same thing, possess the same data, but identify or interpret it differently. What they are responding to differently is stimuli, and stimuli receive much neural processing before anything is seen or any data are given to the senses" (276). Karl Popper's theory, though delivered from the quarters of a philosopher rather than a historian, is in tight league with Kuhn. "All observations are theory impregnated: there is no pure, disinterested, theory-free observation . . . I have stressed before (in section II) that theories are like sense organs. Now I wish to stress that our sense organs are like theories." In "The Rationality of Scientific Revolutions," in Ian Hacking, editor. *Scientific Revolutions.* Oxford University Press, 1981. Paul Feyerabend, the notorious rebel in the philosophy of science, is likewise docile and obedient toward the mantra that the seventeenth-century metaphysicians established. "On closer analysis, we even find that science knows no bare facts at all, but that the 'facts' that enter our knowledge are already viewed in a certain way and are therefore, essentially ideational" (3). "Facts and theories are much more intimately connected than is admitted by the autonomy principle" (20). In *Against Method.* New York: Verso, 2010.

35. R.G. Collingwood's definition of metaphysics in the early twentieth century appears to have been influential for thinkers such as Thomas Kuhn, Karl Popper, and Imre Lakatos. Collingwood's conception of metaphysics concerns assumptions *the truth of which cannot be evaluated.* Collingwood, in other words, stands on the shoulders of the seventeenth-century philosophers. They argued that nature cannot be truly known, and that thought is an artificial contrivance. Collingwood's 'presuppositions' follow suit. "Metaphysics is the science which deals with the presuppositions underlying ordinary science; where by 'ordinary science' I mean such thinking as is scientific in the sense defined in the preceding chapter . . . (11)" "The logical efficacy of a presupposition does not depend upon the truth of what is supposed, or even in its being thought true, but only on its being supposed" (28). "Absolute presuppositions are not verifiable" (32). *An Essay on Metaphysics.* Oxford: Clarendon, 1998. The lens through which Kuhn views scientific theories likewise eschews all interest in truth. When Kuhn analyzes historical periods in science, his own criteria of a 'paradigm' is what he reads back into those theories. "We do not believe that there are rules for inducing correct theories from facts, or even that theories, correct or incorrect, are induced at all. Instead we view them as imaginative posits, invented in one piece for application to nature" (12). Kuhn's conception of a theory denies interest in truth. "Not that scientists discovered the truth about nature, nor that they approach ever closer to truth" (20). "Logic of Discovery or Psychology of Research?" In Imre Lakatos and Alan Musgrave, editors. *Criticism and the Growth of Knowledge.* Cambridge University Press, 1999.

36. Aristotle's critiques of Democritus are distributed widely throughout his writings. For our purposes, the more important passages are *Physics* 188a21; 194a20; *On the Heavens* 275b30; *On Generation and Corruption* 314a1–19; *Metaphysics* 985b5, and 1009a22.

37. See, for example, Thomas Kuhn. *The Essential Tension.* Chicago: University of Chicago Press, 1977, p. 41: "If therefore one thinks of the scientific revolution as a revolution of ideas, it is the changes in these traditional, quasi-mathematical fields which one must seek to understand." "When scientific ideas are discussed without a reference to the concrete, technical problems against which they were forged, what results is a decidedly misleading notion of the

way in which scientific theories develop and impinge on their extra-scientific environment" (136). Kuhn is at the very least reticent to examine the towering political implications of the physics that was established in the seventeenth century. Nor is it appropriate to limit ourselves to an examination of the activities of laboratory researchers, when the theories from which they set sail are freighted with heavy philosophical baggage. Kuhn's point of view makes sense from the vantage point of investigations into *technology*. Yet as a broader attempt to represent the intellectual and scientific culture of early modern Europe, it is ill conceived. Cf. Steven Shapin. *The Scientific Revolution*. Chicago: University of Chicago Press, 1998. "The very idea of the Scientific Revolution, therefore, is at least partly an expression of 'our' interest in our ancestors, where 'we' are late twentieth-century scientists and those for whom what they believe counts as truth about the natural world. . . . In this sense a story about the Scientific Revolution can be an account of those changes that we think led on—never directly, or simply, to be sure—to certain features of the present in which, for certain purposes, we happen to be interested" (6–7). "I don't consider that there is anything like an 'essence' of seventeenth century science or indeed of seventeenth century reforms in science. . . . Since in my view there is no essence of the scientific revolution, a multiplicity of stories can legitimately be told, each aiming to draw attention to some real feature of that past culture" (10). Cf. Alfred North Whitehead. *Science and the Modern World*. New York: Free Press, 1967, 16: "Science has never shaken off the impress of its origin in the historical revolt of the later Renaissance. It has remained predominantly an anti-rationalistic movement, based upon naïve faith. . . . Science repudiates philosophy."

38. For a full discussion of Epicurus, see my "Epicurus, Political Philosopher." In *Machiavelli and Epicureanism: An Investigation into the Origins of Early Modern Political Thought*. Lanham: Lexington Books, 2012, chapter 2.

39. Ibid., chapter 4: "Machiavelli's *Discourses* and the Birth of Neo-Epicureanism."

40. *Discourses* I 16: "This same thing happens to a people that, being unaccustomed to living under the orders of others, not knowing how to think about offense or defense by the state, not understanding monarchs and not being understood by them, returns quickly beneath the yoke that usually is heavier than the one that a little earlier it threw off its neck. And it finds itself in these difficulties because its matter is not corrupted." Translated by Allan Gilbert in *Machiavelli: the Chief Works and Others*, volume I. Durham: Duke University Press, 1989, p. 235. Cf. *Discourses* I 18: "For Rome to maintain itself free once it was corrupted, it would therefore have had to make new laws, because different rules and ways of life have to be devised for a bad subject than for a good one; the form cannot be the same in completely opposite matter." Translated by James B. Atkinson and David Sices. *The Sweetness of Power: Machiavelli's Discourses and Guicciardini's Considerations*. Dekalb: Northern Illinois University Press, 2002, p. 75. Cf. *Discourses* III 8: "There is no question that we may conclude from the words of the historian quoted above, that if Manlius had been born in the age of Marius and Sulla, when the matter was already clearly corrupted, and he would have been able to impress the form of his ambition on it, he would have had the same results and success as Marius and Sulla." In Atkinson and Sices, *The Sweetness of Power*, 294.

41. Kuhn admits that "although other vitally important things also happened to the sciences during the sixteenth and seventeenth centuries (the scientific revolution was not merely a revolution in thought), they prove to be of a different and to some extent independent sort" (41). In *The Essential Tension*. University of Chicago Press, 1977. The point that I will be laboring to bring before the reader in this book is twofold. First, that the scientific revolution is itself political—it advances a political ideology, which is obscure enough to withhold from public scrutiny; and second, that it is the scientific theory which gives the *decisive* impulse to the overt political teaching, especially in Bacon, Hobbes, and Spinoza.

42. *Theological Political Treatise*, chapter 16, p. 528. In *Spinoza, the Complete Works*. Translated by Samuel Shirley. Edited by Michael Morgan. Indianapolis: Hackett, 2002: "This is not surprising, for Nature's bounds are not set by the laws of human reason which aim only at man's true interest and his preservation, but by infinite other laws which have regard to the eternal order of the whole of nature, of which man is but a particle."

43. *Politics* 1253a7.

44. *The Metaphysical Foundations of Modern Science*. New York: Doubleday, 1932, 239.

45. *Science and the Modern World*. New York: Free Press, 1967, 54.

46. Spinoza, *Emendation of the Intellect*, paragraph 84: "Thus we have distinguished between the true idea and other perceptions, and we have established that the fictitious, the false, and the other ideas have their origin in the imagination, that is, in certain sensations that are (so to speak) fortuitous and unconnected, arising not from the power of the mind but from external causes, in accordance as the body, dreaming or waking, receives various motions" (23). In *Spinoza, The Complete Works*.

47. Ibid., paragraph 68, p.19. Cf. Rene Descartes, in *Rules for the Direction of the Mind*: "We should attend only to those objects of which our minds seem capable of having certain and indubitable cognition." (rule 2, p. 10). "Alternatively, and this comes to the same thing, intuition is the indubitable conception of a clear and attentive mind, which proceeds solely from the light of reason" (rule 3, p. 14). "Once he has surveyed everything that follows immediately upon knowledge of the pure intellect, among what remains he will enumerate whatever instruments of knowledge we possess in addition to the intellect; and there are only two of these, namely imagination and sense perception. . . . He will see that there can be no truth or falsity in the strict sense except in the intellect alone." (rule 8, p. 30). All of these citations are from John Cottingham, Robert Stoothoff, and Dugald Murdoch, translators. *The Philosophical Writings of Descartes,* volume 1. Cambridge University Press, 1985 (hereafter *CSM* I).

48. "The Masculinity of Time." In Benjamin Farrington. *The Philosophy of Francis Bacon: An Essay on its Development from 1603–1609*. Chicago: University of Chicago Press, 1964, p. 63. "Let Plato be summoned next to the bar, that mocking wit, that swelling poet, that deluded theologian. . . . Your wisdom was a sham which you imposed by an affectation of ignorance" (64).

49. Herbert Butterfield. *The Origins of Modern Science*. Revised edition. New York: Free Press, 1965, 7.

50. Alexandre Koyre. *Metaphysics and Measurement*. Cambridge: Harvard University Press, 1968, 39.

51. *Republic* 510e.

52. *Republic* 510a–511c.

53. *Principles of Philosophy* part 2, #4. *CSM* I, p. 224.

54. Spinoza, *Ethics* II, definition 1; in *Complete Works*, p. 244. Hobbes, *Works* I, p. 102: "The definition therefore of a body may be this; a body is that, which having no dependence on our thought, is coincident or coextended with some part of space."

55. *Parmenides* 142ad.

56. Peter Dear. *Discipline and Experience: The Mathematical Way in the Scientific Revolution*. Chicago: University of Chicago Press, 1995, p. 172.

57. Dear, *Discipline and Experience*, p. 235.

58. Dear, *Discipline and Experience*, p. 223.

59. This brief discussion of the parts of 'unity' and 'being' as they pertain to the definitions of body goes to the very root of my analysis of Plato's arguments in the *Parmenides*. As indicated, these arguments have been made elsewhere in the requisite detail. In this introduction, at least, a sketch must suffice. In the chapters to follow, some basic citations from the dialogue will be displayed to make the points with greater clarity.

60. Spinoza's entire *Ethics* is presented in a geometric form.

61. Hobbes, "Epistle Dedicatory," p. 5, in *On the Citizen*. Edited by Richard Tuck and Michael Silverthorne. Cambridge University Press, 1998.

62. *Theatetus* 191d.

63. *Republic* 476a.

64. *Leviathan* chapter 4: "Of names, some are proper and singular to one thing only: as Peter, John, this man, this tree: and some are common to many things; as man, horse, tree, every one of which though but one name, is nevertheless the name of diverse particular things; in respect of all which together, it is called an universal; there being nothing in the world universal but names; for the things named are every one of them individual and singular."

65. *Ethics* II, proposition 40, scholium I.

66. Leo Strauss. *Natural Right and History*. Chicago: University of Chicago Press, 1965, 175.

67. *Ethics* IV Preface, towards the end, p. 322. In *the Complete Works*.

68. *Spinoza's Critique of Religion*. Translated by Elsa Sinclair. University of Chicago Press, 1997, 16, 45.

69. *Natural Right and History*. Chicago: University of Chicago Press, 1965, 177: "All intelligibility or all meaning has its ultimate basis in human needs," Strauss writes. This must be the result if nature is defined to be unknowable. Strauss's use of the name 'needs' is ambiguous, however. Knowledge of man as a form is forsworn by the early moderns. This knowledge swept away includes knowledge of man's ends, that is, his *happiness*. 'Needs' indicates the rump of a soul that is left after Hobbes has savaged it. 'Needs' indicate mere appetites, pleasures, desires. The ancient sophist teaching, pleasure, is to become the modern beacon.

70. Leo Strauss. *Natural Right and History*. Chicago: University of Chicago Press, 1965, 175.

71. *Leviathan*, introduction, second page. Oxford University Press, 1998.

72. *The Political Philosophy of Hobbes: Its Basis and Genesis*. Translated by Elsa Sinclair. Chicago: University of Chicago, 1984, 124–25. In the case of Hobbes, and his indictment of sense perception, we are dealing with a very different universe than the one indicated by Galileo or Kepler or even Descartes. *This* blindness, politically conceived, makes one a prey to devastating injury. It is only on these grounds that we can come to appreciate the urgency of the atomistic matter theory on which modernity rests. Yet these are still grounds about body, physics. Thither must political philosophy go, if it is to reclaim its own lost soul. Insofar as Strauss's analysis is correct, the newly forbidden impulse is justice itself. For to do justice assumes one has power to spare, *leisure* to be just, which the newly orphaned man, abandoned to his appetites, has no time for. Self-preservation is unfree time, whence one must be the slave of the self merely, in order to survive. A whole new morality is to be built around self-preservation.

Chapter One

Francis Bacon's Uncharitable Charity

The Birth of a New Rationality

Niccolo Machiavelli, in his *Discourses on Livy*, informs his reader that he has always been determined to work for the common benefit of individuals, regardless of any obstacles. Machiavelli anticipates that some, perhaps many, of his intended beneficiaries might not want what he intends to bring them. This poses no serious dilemma for Machiavelli. For he is resolved, and the impulse of consequence is a 'natural desire' in himself.

> "Although the envious nature of men has always made it no less dangerous to find new modes and orders than to seek unknown waters and lands," Machiavelli writes, "nonetheless, driven by that natural desire that has always been in me to work, without any respect, for those things I believe will bring common benefit to everyone, I have decided to take a path as yet untrodden by anyone, and if it brings me trouble and difficulty, it could also bring me reward through those who consider humanely the end of these labors of mine."[1]

Francis Bacon's invocation of 'charity' strikes a less confrontational posture, but it follows a similar logic. "This corrective spice, the mixture whereof maketh knowledge so sovereign, is Charity," Bacon writes.[2] Bacon assumes to himself the prerogative to determine what people need. It also seems clear, in a study of Bacon's work, that he is not terribly concerned about winning the people's respect for his labors. Deliberation is not the means that Bacons envisions using.[3] Since Bacon's proposed reforms extend into the moral and political domains, one might expect that deliberation would be included in the package of charitable intentions.

In the new rationality that Bacon proposes, the typical individual will occupy the status of an object of administration. In Bacon's view, it is a mistaken enterprise to engage the public in open deliberation when major issues are at stake. Bacon would rather appeal to the people's passions, instead of to their reason. This is a disposition that can be seen in a number of Early Modern philosophers.

In Bacon's defense, one must marvel at the truly beneficial things that he does want to bring to society. Control over nature is sure to bring immense new productivity. It is also sure to bring relief from disease and natural hardships. One must observe that there is a certain greatness in Bacon's ambitions. Bacon is not noted for his own concrete contributions to the science of his century; however, the new road upon which Bacon set science must certainly have contributed a great deal to the discoveries that others made in the centuries to follow.[4]

Bacon calls attention to the mechanical arts with special solicitousness. "But if my judgment be of any weight, the use of history mechanical is, of all others, the most radical and fundamental towards natural philosophy; such natural philosophy," Bacon states, "I mean as shall not vanish in the fumes of subtle or sublime speculations, but such as shall be operative to relieve the inconveniences of man's estate."[5] Society itself is an object of experiment for Bacon: and as will be discussed below, morals and politics are two of the domains in which he intends to make serious changes.[6]

This brings up the question as to what possible downside might pertain to a philosophy which is committed to pursuing extensive mechanical works. First of all, it can be pointed out that one need not have a revolution in cultural values in order to pursue an ambitious program of technological development. Even Plato regards it as inevitable that progress in society is going to involve technological improvements. It is such development in the first place that creates the need for a governing authority, in Plato's view. It seems to be the case, however, that Bacon weds his own plan for technological advance to a transformation of the nature of authority. People do not like to be maneuvered through life by indirect mechanisms of psychological and emotional pressure. Yet Bacon advertises such governmental mechanisms. This does not subtract from the value of his technological program; yet such alteration in the modes of governance cannot be excused by the technological program either.

Power, human power, is certainly a good thing. It is something that civilization strives for. It is no indictment of Bacon to observe that he is very interested in developing society's powers. My concern, as will become apparent, follows the manner in which certain aspects of this power would be deployed in cultural and political life.

> I for my part have set my face in the opposite direction. My intention is to Impart to you, not the figment of my own brain, nor the shadows thrown by Words, nor a mixture of religion and science, nor a few commonplace

> Observations or notorious experiments tricked out to make a composition
> As fanciful as a stage play. No; I am come in very truth leading you to Nature
> With all her children to bind her to your service and make her your slave.[7]

Every student of modern political science is familiar with the manner in which modern philosophers prefer to organize different factions in society, so as to set them against one another. It may be, though, that this teaching has never been evaluated with the thoroughness that it requires. Managing existing conflicts is one thing. The prospect of helping to cause conflicts in society, so that such mechanisms for government can be employed, is something quite different. I will argue, in this chapter, that Bacon does not really share a belief in the customary sense of right and wrong in his society. A great deal of the people's happiness and contentment in life depends upon their felt standards of right and wrong.[8]

Not only does Bacon propose to alter certain fundamental moral beliefs; he proposes to enact such changes through operating on the passions of the people, that is, by making the conflict between citizens a productive principle, a normal aspect of the governing process, setting their passions into conflict with one another.[9] "Moreover," Bacon writes, "in all wise human governments, those who sit at the helm can introduce and insinuate what they desire for the good of the people more successfully by pretexts and indirect ways than directly. Nay, (which perchance may seem strange), even in mere natural things you may deceive nature sooner than force her; so ineffectual and self-impeding are all things which are done directly; whereas on the other hand the indirect and insinuating way proceeds smoothly and gains its end."[10] Scholars who believe that the political nature of human beings, as Aristotle says, is a principal aspect of their personalities and is therefore directly relevant to their happiness must have strong reservations about the administrative character that Bacon recommends to governments.

As promised, it is a major feature of this study to examine the way in which political principles are anchored in doctrines usually relegated to the domain of physics. Bacon towers as an influential figure for Descartes and Hobbes. All three thinkers make direct attacks on Aristotle's theory of formal and final causes. These two causes are linked in the Aristotelian presentation. It is by intelligible forms that ordinary individuals know the objects of their world, for one thing; and it is by this familiarity that individuals recognize certain potentials, goals, and modes of fulfillment for the respective objects on the other. It is not clear why the new program of science must take these two parts of Aristotle's metaphysics for opponents. Yet to make 'final causes' an object of attack is to challenge Aristotle's theory that happiness is the true end of human beings in society.

Bacon is also a trailblazer when it comes to his thoughts on language. This is quite directly relevant to the issue of human beings as political animals. It does not appear inevitable that the new science that Bacon seeks to bring to birth should find it irritating to need to speak in the common tongue. Bacon does not merely indicate that ordinary opinions are difficult to improve or elevate. He suggests that the ordinary opinions lack competence in the areas that the new science seeks to bring its improvements to. There is no more fundamental bond between human beings than language. Bacon, however, as will be discussed below, offers a wholesale indictment of the ordinary opinions on almost every subject that one pleases. This is not part of the charitable aspect of his program. "The nature of words, being vague and ill-defined, is another source of illusion, nay almost of violence to the human understanding," Bacon writes, "Words are a kind of currency, which reflect vulgar opinions and preferences, for they combine or distinguish things according to popular notions and acceptations, which are for the most part mistaken or confused."[11]

It is part of the plan of this study to examine the physics of each philosopher, in order to examine their respective theories of matter. One of the principal reasons for doing so is because the atomistic theory of matter, or its corpuscularian cousins, tend to indict the validity of ordinary perception. Such arguments would lend support to the cultivation of an administrative language that has stopped seeking to communicate with the people in direct speech. It must be worthwhile to investigate whether such doctrines of matter and perception are at least true to begin with.

BACON ON BODY

Bacon is ambivalent about the heritage of ancient atomism. On the one hand, he celebrates the tradition founded by Democritus; and he makes it plain that Aristotle's physics is, in Bacon's view, inferior to that of the original atomists. On the other hand, Bacon does later in his career seek to separate himself from the atomistic teachings. The issue is therefore controversial amongst scholars, as to what Bacon's relationship to atomism finally is. I will argue that, despite Bacon's disclaimers, his doctrines do preserve the essential atomistic postulates.[12]

The atomist philosophy originates in the core conviction that true coming into being is not possible in nature, and that there cannot be any true passing away. The atomist postulates are carefully laid out in Bacon's very presentation of his first philosophy. "But setting all high conceits aside, my meaning is simply this," Bacon writes, "that a science be constituted, which may be a receptacle for all such axioms as are not peculiar to any of the particular sciences, but belong to several of them in common."[13] Among these 'axioms,' according to Bacon, is the claim that "all things are changed and nothing is lost"; "the quantum of nature is neither diminished nor increased";

and "things are preserved from destruction by bringing them back to their first principles is a rule in physics."[14]

The classical theory of atomism is, it is true, not *exactly* duplicated in Bacon's work. In the original atomist theory of Leucippus and Democritus, the atoms are indicated as changeless nuggets of 'being.' According to Aristotle, the atomists referred to the atoms as the 'full,' and to the void as the 'empty.'[15] The atoms of Leucippus and Democritus are not allowed to have parts: they must be indivisible, in order to be eternal. If there are parts in the atoms, these philosophers considered, then that would mean that they must have the capacity for change. What is eternal cannot be changing, in Parmenides's teaching. Leucippus and Democritus adhere to that teaching. Their atoms are like smaller variations of Parmenides's 'being.'

Bacon does not have many nice things to say about the physics of Plato and Aristotle. Nor does he spend any serious time talking about the doctrine of forms which was the great discovery in the Socratic line of philosophers. Democritus did not possess any theory of forms, but it is Bacon's belief that Democritus and Leucippus were better students of nature than either Plato or Aristotle. "The human understanding is carried away to abstractions by its own nature, and pretends that things which are in flux are unchanging," Bacon writes, "But it is better to dissect nature than to abstract; as the school of Democritus did, which penetrated more deeply into nature than the others."[16]

Bacon does not deign to take up any of the very detailed and dispassionate criticisms that Aristotle makes of the atomic theory. Yet as indicated above, Bacon does later make disparaging statements about the atomist doctrine. In *The New Organon* for instance, Bacon dismisses the atomist theory, or gives the appearance of doing so. "However, we will not end up with the atom, which presupposes a vacuum and unmoving matter (both of which are false), but to true particles as they are found to be."[17]

In fact, neither of these principles is absolutely requisite to atomism. One can have indestructible body even in a plenum. What is atomistic is the indestructible.

The core principle of the atomistic philosophy is that there must be some eternal body, *out of which* the perishable objects can be generated. If all objects were perishable, the atomists argue, then soon all the body in the universe would be depleted, and nothing would be left. Epicurus presents this aspect of the atomist teaching: "Having made this distinction, we must now consider what is not evident to our senses: first of all, that nothing is created from what does not exist. For everything would be born from everything without the need for seed. And if that which is destroyed were dissolved into what does not exist, everything would be destroyed, since that into which they were dissolved does not exist. Moreover the universe was and always will be the same as it is now. For there is nothing into which it changes."[18]

Let us return to the axioms of Bacon's 'First Philosophy.' "All things are changed and nothing is lost." This is an atomistic postulate. What changes are the objects perceived to be perishable. Yet the perishable, in the atomist philosophy, is not the true body: the true and real bodies are the atoms, and these cannot be destroyed. "Nothing is lost." "The quantum of nature is neither diminished nor increased." This goes to the root of the Epicurean philosophy of mind. The doctrine of atoms, in its Epicurean variation, soothes the philosopher's mind, by assuring him that there is no such thing as destruction really.

Bacon is speaking a bit more abstractly about body here. He is not referring to the atoms directly or by name. Yet he does indicate a reservoir of matter that is indestructible, or that cannot be diminished. Aristotle has a theory of prime matter, that is, matter in its eternal state; but eternal matter for Aristotle lacks the property of *being*. Eternal matter is merely potential being for Aristotle: it depends upon a form to *give* it being. For Plato also, coming into being and passing away are real attributes of any body whatsoever.

Plato's argument would compel Bacon to talk more specifically about any body that he likes. Any body that Bacon would like to talk about, perceptible or imperceptible, must possess the nature of *being*.[19] No body, obviously, can afford to be without the nature of being, as a *part* of it. To not have the part of being, would be not to exist. It would mean that we are talking about no body at all. Bacon could not avoid making this concession about the bodies or particles that he regards as most real. Plato's insight can then be made clear. It is not possible for any *being* to exist in the universe, Plato argues, that does not belong expressly and exclusively to some particular *unity*.[20] Any being, therefore, must be some *one* being. If it does not refer to some unity, 'being' cannot indicate *anything*.

The above postulates were developed by Plato to deal both with the theory of Parmenides and with atomist philosophies such as that of Democritus and Leucippus (which Epicurus is still upholding essentially). If the above two claims are granted, that means that any body whatsoever must be a compound of 'being' and 'unity.' These are the two original parts of any body. A body lacking one of these parts cannot exist at all, or even be spoken about properly, in Plato's view. This means that the smallest and oldest body must be a whole-of-parts. For 'unity' and 'being' are parts. The unity that has being is a body as a whole of parts. Thus the reality of a 'whole' in nature is proved. The reality of a 'part' in nature is likewise proved.[21] These are not mere names, nor figments of some philosopher's brain. They exist in reality, as has been demonstrated.

Once Plato has proved the existence of a whole of parts as the proper explication of body, he proves that the whole must have extremities. A whole cannot be a whole if it lacks for extremities.[22] The extremities of a whole include the parts of time: past, present, and future. The extremities of a body include what Descartes will later refer to as 'extension,' or shape and dimensions. Yet the extremities of a real body must also include coming into being and passing away. For without these extremities, the object would not be a *whole*. And thus it is proved: that any body whatsoever comes into being and passes away. Objects acquire being at one point, and they 'lose' it at another point.[23]

Thus the axioms of Bacon's first philosophy are false. No body can be brought back to its beginnings. Any object that has had a beginning in time, can only have an end in time. It can't have a return to its beginnings. Nor is there anything like a constant quantity of the universe. Objects come into being and perish. Generation is real, and passing out of existence is real. Needless to say, since the extremities of an object have been proven to be real (shape, size, dimensions), it likewise goes to prove that they should be perceptible; or at the very least, that those bodies which *are* perceptible are indeed real; which indicates that the ordinary people know very well what they are talking about when they indicate the objects of their common world to one another.[24]

In Bacon's earlier writings, he is less bashful about his rendezvous with atomism:

> Now an abstract principle is not an entity, and again, a mortal entity is Not a principle; so that a clearly irresistible necessity drives men's thoughts (if they want to be consistent) to the atom, which is a true entity, having matter, form, dimension, place, resistance, appetite, motion and emanation. Likewise, amid the destruction of all natural bodies, it remains constant and Eternal. For since the corruption of the greater bodies are so many and Various, it is absolutely necessary that that which remains as an unchanging Center should be either potential or extremely small, but it is not potential.[25]

Scholars may well resist the notion that Bacon's very physics trades in some of the Epicurean influences that are present in Machiavelli's pioneering doctrine. It can be said though that Machiavelli, in adapting the thesis of indestructible matter for a political philosophy, begins to characterize the very impermanent people themselves as 'matter.' Matter, for an atomist, is indestructible; therefore one cannot possibly do violence to it.

Machiavelli seems to have created a psychological corollary of atomist doctrine, whereby the philosopher becomes free to regard the secondary or compound bodies as so much indestructible matter. In other words, by applying the atomist nomenclature to the people, it becomes possible for Machiavelli to conceive of policies that would treat them, calmly, with great brutality.

Something very similar seems to attend to Bacon's exploration of matter and the experimental situation of his natural philosopher. There is no such thing as a violent movement in nature, according to Bacon: in fact, the contortions and vexations to which Bacon and his brethren would subject natural objects, according to him, count as natural movements.[26] When man tries to force an object against its natural grain, in Bacon's nomenclature, this counts as a natural movement, because the man imposing the experiment is a natural being too.[27] Now Bacon makes a defensible point, that it may be necessary to subject natural objects to extreme pressure in order to discern their hidden attributes. This has always been a part of the insight of the mechanical arts, and in itself it is unexceptionable. However, it would be a very different thing to think of human beings in such terms. Bacon at least has the habit of making statements which suggest that he thinks of human beings as part of this experimental arrangement. Human beings too, Machiavelli believed, are only revealed for what they are under extreme pressure. "For like as a man's disposition is never well known or proved till he be crossed," Bacon writes, "nor Proteus ever changed shapes till he was straitened and held fast; so nature exhibits herself more clearly under the trials and vexations of art than when left to herself."[28] It is not necessary to analyze every word that Bacon utters. It is reasonable however to bear this attitude towards nature in mind when investigating Bacon's doctrine of moral and political science.

BACON'S DOCTRINE OF PERCEPTION

Prima philosophia, or first philosophy, is the foundation of Baconian theory that serves as the principal vantage point from which an entire new academic division of labor can be surveyed. Bacon's *prima philosophia* comes to rest on his philosophy of matter.[29] The classical atomist theories, from Leucippus and Democritus to Epicurus and Lucretius, all insist that perception is unreliable as a way to know the objects external to us. For one thing, ordinary perceptions can be indicted because they gather information about objects that come into being and pass away. The true objects, from the atomist vantage point, do not have these properties. A second aspect of the atomist critique of human perception is the claim that material particles from the external objects directly collide with our sensory organs. This in turn would render perception into a solipsistic mental experience for the individual, rather than any sort of authoritative report about external objects. In Bacon's view,

> The senses are defective in two ways: they may fail us altogether or they may deceive. First, there are many things which escape the senses even when they are in health and quite unimpeded; either because of the rarity of the whole body or by the extremely small size of its parts, or by distance, or by slowness or speed, or because the object is too familiar for other reasons. And even when the senses do grasp an object, their apprehensions of it are not always

> reliable. For if the evidence and the information given by the senses is always based on the analogy of man, not of the universe, it is a very great error that the senses are the measures of things. So to meet these defects, we have sought and gathered from every side with great and faithful devotion, assistants to the senses, so as to provide substitutes in the case of total failure and correction in the case of distortion. We do this not so much with instruments as with experiments. For the subtlety of the experiments is far greater than that of the senses themselves even when assisted by carefully designed instruments; we speak of experiments which have been devised and applied specifically for the question under investigation with skill and good technique. And therefore we do not rely very much on the immediate and proper perception of the senses, but we bring the matter to the point that the senses judge only of the experiment, the experiment judges of the thing. Hence we believe that we have made the senses (from which, if we prefer not to be insane we must derive everything in natural things) sacred high priests in nature and skilled interpreters of its oracles; while others merely seem to respect and honor the senses, we do so in actual fact.[30]

Bacon's statements make it clear that the senses as available to ordinary human beings are not trustworthy. Part of the senses' weakness is said to pertain to the smallness of the bodies that would have to be experienced. These would be the particles certified as genuine by Bacon's *prima philosophia*. Even when the objects are of considerable size, Bacon insists, human senses are rarely competent to make the proper judgment. Only scientifically controlled experiment is allowed to have some perception which is competent. This 'perception' is held up by Bacon as the proper voice for everybody's perception.

The reader should bear in mind that Bacon is undertaking to present a new vantage point from which to view nature. The scientist is going to have increased control, and expanded authority to pronounce upon truth of fact. Bacon criticizes individuals for seeing the natural world through human eyes, as if they had any other eyes with which to see. "For the evidence and information given by the senses is always based on the analogy of man, not the universe; it is a very great error that the senses are the measure of things" (18). Presumably the scientific investigators also perceive their experiments with the human faculties that they possess. It is not immediately clear what Bacon means when he argues that human beings are thus prejudiced by their inclination to view the world in 'human' terms. It is quite likely, though, that Bacon's observation has to do with his objections to the notions of final cause, which are indeed quite wrapped up with the ordinary way of perceiving objects.

The core of Bacon's atomism can be seen in his determination to view nature as a *unity*. The atoms, or particles, or whatsoever name Bacon attempts to employ, represent for Bacon the whole of nature. There is no diversity in nature, of a qualitative sort. This is the reason why Bacon rejects the distinction between violent and natural movement as a postulate of physics. In accordance with the doctrines of form in Plato and Aristotle, there is endless diversity among objects in nature. The Early Modern philosophers tend to regard the distinctions between objects in nature as superficial, as mere appearances.

"Thus since nature is one and the same, its power effectual in everything, and its trueness to itself unfailing, these three things ought wholly to be set down as alike subordinate only in nature: the course of nature, the wandering of nature, and art or the application of man to natural things."[31]

BACON ON THE PROPER USE OF NAMES

Let us examine Bacon's indictment against the validity of the ordinary assignment of name to object. This operation is, in Bacon's formulation, fraught with problems. The individual, Bacon argues, is forever engaged in worshipping 'idols': either idols of the *tribe*, the *cave*, the *marketplace,* or the *theater.*[32] Bacon's complaints about the ordinary opinions are so broad that the entire category appears to be sunken into the domain of prejudice. Bacon thinks that his experimenter is needed to furnish the proper context for the operation of perception; and it seems to be a correlative doctrine that the application of names to objects needs to be similarly dependent.[33]

The competence of the individual, insofar as the assigning of names to objects is concerned, is challenged by Bacon on several fronts. The individual can err based upon the influence of his community. He can identify objects, that is, based upon the alleged prejudice of how his community views objects (idols of the tribe). Or the individual can lose objectivity due to individual eccentricities, whether behavioral, physiological or attitudinal (idols of the cave). Yet it is in the third category, the idols of the marketplace, where Bacon makes his sharpest criticisms of the ordinary opinions as to what objects are. "Men associate through talk; and words are chosen to suit the understanding of the common people. And thus a poor and unskillful code of words incredibly obstructs the understanding."[34]

Words, however, are the common bond among human beings. Scholars may have at Plato's arguments and insights, but the issue he raised remains with us: either the average person *knows* something when she assigns name to object correctly, or not. It is Bacon's attempt to supersede the common language which binds him most powerfully to Epicurean forbears. It really isn't clear why these criticisms are important to Bacon's doctrine of charitable science. It is a great inconvenience for the people, to have their basic faculties called into question. Nor is it clear why Bacon's experimenters should have an interest in such an attitude.

Language, according to Aristotle, is the distinctively human attribute. It is what makes us capable of deliberating with one another, as to better and worse causes of action for common ends. Freedom of expression certainly depends upon one's competence in the language of one's society. Bacon does not seem to have much concern for these vital aspect of language. In fact, he goes out of his way to disparage the ordinary use of speech. Bacon does so in order to underscore his attitude to deliberation insofar as it pertains to Bacon's philosophy. Deliberation is not the means of progress that Bacon envisions. It can also be pointed out that given the magnitude of Bacon's claim here, he is providing precious little evidence. People do seem to get along well enough with the language capacities that they have. Even if these capacities are imperfect, one would expect efforts to attempt to improve and build upon them. This is not Bacon's inclination.

> And lastly let us consider the false appearances that are imposed upon us by words, which are framed and applied according to the conceit and capacities of the vulgar sort; and although we think that we govern our words, and prescribe it well . . . yet certain it is that words, as a Tartar's bow, do shoot back upon the understanding of the wisest, and mightily entangle and pervert the judgment; so as it is almost necessary in all controversies and disputations to imitate the wisdom of the mathematicians, in setting down in the very beginning the definitions of our words and terms, that others may know how we accept and understand them, and whether they concur with us or no.[35]

Bacon appears to be referring to Euclid above, and the definitions that geometers set forth antecedent to making their proofs. The geometer, however, sets forth only such definitions as are uncontroversial. That a whole is greater than its parts is not a doctrine that the ordinary opinions would disagree with. Bacon, however, argues in favor of definition as a remedy for defective understanding in the majority. It can only be his class of experimental philosophers who can fill this role. The establishment of standards of truth is always important to a civilization. In the philosophies of Plato and Aristotle, the ordinary perception of the facts served as the standard of evidence for sciences of definition. Bacon is proposing to reverse that alignment. It should at least give us pause.

This is one of the reasons why the incorporation of atomistic physics looms large in this investigation. The atomistic physics appears to supply unquestionable reasons for setting the ordinary opinions aside, and for substituting expert valuations. If those atomistic postulates do not bear scrutiny, then we may not be so ready to stand idly while the ordinary opinions are indicted as incompetent standards for any knowledge of reality.

Proteus indicates a figure that can take on any shape. This is what Bacon likens nature to. Nature, or 'matter,' is said to be like Proteus.

> But if anyone gets annoyed because I call the arts the bounds of nature
> When they ought rather to be considered its liberators and champions in
> That in some cases they allow nature to achieve its ends by reducing
> Obstacles to order, then I reply that I do not much care for such fancy ideas
> And pretty words: I intend and mean only that nature, like Proteus, is
> Forced by art to do what would not have been done without it; and it
> Does not matter whether you call this forcing and enchaining, or assisting
> And perfecting.[36]

Nature, in Bacon's view, takes many different shapes. Yet in his opinion, it seems to be a unity underneath all the appearances. In order to get to the truth of nature, Bacon suggests, science must apply great pressure to it; science must 'vex' it and drive it into a corner; for it is only when nature is thus driven into extremity, that Bacon thinks it is possible to discern reality. Peter Pesic makes some interesting claims as regards Bacon's disposition towards nature. The passage in Bacon to which Pesic devotes his cares is where Bacon insists that nature must be subjected to vexation in order to yield her secrets. Pesic defends Bacon, principally by pointing out that in Bacon's view, nature is indestructible, and cannot be hurt. Matter can only be destroyed by God. It is all very well to claim that objects in nature are indestructible. The trouble with Pesic's claim is that the hypothesis in question is disputed. All bodies in nature come into being and pass away. The illusion of invulnerability which Bacon superimposes over the objects of his experimental science is not so innocent.

Pesic brings up a helpful point. Bacon allows sense perception to pronounce on truth of fact only when it is guided by experiment. In other words, science sets up the experiment: and the experiment constitutes the context within which sense perception is allowed to operate with 'logic.' For Bacon, logic contains the principles of 'first philosophy.' Therefore, logic contains the postulates of atomism which we have heretofore surveyed. This power vested in logic by Bacon enables it to overrule any principles of knowledge peculiar to any subject matter. "But our position is that true logic should enter the provinces of the individual sciences with greater authority than is in their own principles, and compel those supposed principles themselves to give an account as to what extent they are firmly established," Bacon writes.[37] Ordinary logic, or logic as ordinarily understood, is not concerned with 'invention' or 'the discovery of arts,' but Bacon's is.[38] Ordinary logic, which Bacon correctly links to Aristotle's syllogism, is reviled by Bacon as 'childish.'[39] The notion of simply changing the definitions of terms does not well conduce to the advance of knowledge. By the term 'induction,' for example, Aristotle developed the theory that we obtain our knowledge of kinds through our memory and experience of particulars. Bacon's conception of induction is radically different.

It does not appear to be the case that even Bacon's experimental philosophers will enjoy much liberty of thought. The experimental approach as Bacon develops it rather expects the experimenters to avoid investigating the premises of experiments. In other words, there is a somewhat rigid authority in Bacon's notion of the experimental method.

Scientists will be slotted to work within established paradigms of experimental research, in a minute division of labor that does not conduce to self-awareness much less to any serious reflections on the metaphysical presuppositions of Bacon's science.[40] One looks in vain for the department in Bacon's *New Organon* where debate and deliberation will be proper concerns for individuals, even for highly trained ones.

Bacon expresses disdain for meaningless argument or contentious argument. In this, few would disagree. Yet it seems that for Bacon, almost any deliberation savors of a uselessness; that cross-examination, which is quite important to philosophic knowledge, is simply being driven out of his methodology. The people cannot criticize the experimental philosophy, allegedly, because they are limited by faculties which are prejudiced in decisive ways. Highly trained experimenters do not seem to enjoy a greater degree of intellectual liberty, however. If Bacon limited his doctrine to the mechanical arts, this would not be problematic. Yet he aspires to create a new intellectual mood for the culture at large, across its every domain.

Lucretius, famously, argues in his poem that we are not entitled to think that the legs a human being possesses are meant for the purposes of locomotion. Instead, Lucretius instructs us, we must regard the legs as simply objects in nature to which we may assign the 'use' that pleases or conveniences us. "Do not suppose that the clear light of the eyes was made that we might see our way before us, or that the ends of thighs and calves were jointed and set on the foundation of the feet to help us with great strides to march along. . . . Since nothing is born in our body that we may use it, but what is born itself creates the use."[41] This is a pretty strained point of view. We would not be allowed to construe the eyes as intended to allow human beings to see the world about us, so that we could move about effectively in it, and come to know the objects in view. Bacon similarly criticizes the ordinary human conceptions as anthropomorphic 'anticipations of nature.' Yet really, it is Bacon who is injecting an eccentric human point of view into the contemplation of the natural bodies.

It was Anaxagoras who first argued that there must be a Divine Mind, which orders the objects about us.[42] Plato tells us in the *Phaedo* that this is the insight that terribly excited Socrates originally, despite the fact that in Socrates's view, Anaxagoras was unable to really bring home the argument.[43] Anaxagoras, when he observed that there is a natural order in objects, appeared to Aristotle to be speaking with sobriety, understatement, *and fidelity to the facts.*[44] Those who, like the atomists, talked about random collisions in nature, accordingly spoke like men inebriated, whose account of things did not square with the available evidence. It is one thing to mindlessly believe that everything in nature is designed for the welfare of human beings. That would indeed be a counter-productive prejudice. On the other hand, to doctrinally suppose that nature possesses no order of its own, in any of its objects, seems equally ignorant. Yet the authority of the science that Bacon envisions is not shy about making such broad evaluations. This brings us to Bacon's famous comment about eyelashes.[45] It is true. Ordinary speakers will say that the eyelashes exist to protect the eye against alien objects that might enter and disable the eye. So will farmers observe that the leaves of the plants protect their fruit. Bacon takes exception when we say this. For this is to 'anthropomorphize' nature. Is it really? The eyelids do not wait upon our concurrence, to serve as protectors of the eyes; and the leaves do shield the fruit of the trees. The inference that there is intelligence in nature can be debated, surely; but those who deny the evidence of perception as a principal part of their reasoning are that much less persuasive. It is the obvious evidence that their reasoning cannot account for.

BACON ON PLATO AND ARISTOTLE

Early Modern Philosophy owes a great deal to the work of Plato and Aristotle. If it is Bacon's aim to be charitable in general, this is not the direction of his attitude towards his philosophical forbears. Plato and Aristotle were not enamored of contention for contention's sake. They prefer to make arguments. They prefer discourse. It would be preferable, for example, that Bacon take up the arguments of Plato and Aristotle and refute them, if they are false. This is not Bacon's preference. He elects to speak harshly of Plato and Aristotle. Machiavelli had come first, in establishing this tone and attitude towards the classical Greek heritage; but Bacon helped cement this trend into place. It never seems to have truly abated since. "And herein I cannot a little marvel at the philosopher Aristotle, that did proceed in such a spirit of difference and contradiction toward all antiquity," Bacon says, "undertaking not only to frame new words of science at pleasure, but to confound and extinguish all ancient wisdom; insomuch as he never nameth or mentioneth any ancient author or opinion, but to confute and reprove; wherein for glory, and drawing followers and disciples, he took the right course."[46] Ridicule is a very powerful rhetorical tool, Bacon knows. "Let Plato be summoned next to the bar, that mocking wit, that swelling poet, that deluded theologian. Your philosophy, Plato, was but scraps of borrowed information polished and strung together. Your wisdom was a sham which you imposed by an affectation of ignorance."[47]

In Bacon's project, if we are to accept his charity, we must become most ungrateful to the philosophies of Plato and Aristotle. "I shall not shrink from telling you that I put Plato and Aristotle among the sophists," Bacon writes.[48] Bacon, in his project for a complete renewal of human knowledge, expressly rejects the principle that the ordinary opinions possess insight and knowledge which philosophy is both subject to and obligated to build upon. "The reason is that the first notions of things which the mind accepts, keeps and accumulates (and which are the source of everything else) are faulty and confused and abstracted from things without care," Bacon states. "The consequence is that the general human reason which we bring to bear on the inquiry into nature is not well founded and properly constructed."[49] Bacon indicates that his trained philosophers have access to a superior point of view. Based on Bacon's theory of *prima philosophia*, this superior knowledge consists in the proto-atomistic postulates examined above. That does not inspire a great deal of confidence.

It is not inevitable that Bacon's project of technological enhancements should aspire to supreme authority across the several dimensions of human life. The point is that Bacon does seek for such an authority for his experimental method. Even if Bacon did not present a specific political theory, which he does do, the political implications of Bacon's project would still be conspicuous. It is not so much the expansion of experimental arts, or the harnessing of alchemy and the old mechanical methods into new streamlined programs of exploration. It is Bacon's opposition to all of the customary forms of knowledge and authority which indicate the political aspirations of his project. Bacon is a terribly ambitious man. "But there is likewise another and more subtle error which has crept into the human mind; namely, that of considering art as merely an assistant to nature, having the power indeed to finish what nature has begun, to correct her when lapsing into error, or to set her free when in bondage, but by no means to change, transmute, or fundamentally alter nature. And this had bred a premature despair in human enterprises. Whereas men ought on the contrary to be surely persuaded of this; that the artificial does not differ from the natural in form or essence, but only in the efficient."[50]

BACON'S HIERARCHY OF KNOWLEDGE: LOGIC VERSUS IMAGINATION

The contours of the division that Bacon envisions between his new class of certified knowers and the commonweal should by now be visible. Those who are initiated into the new teaching of science, Bacon's teaching, are denizens of *prima philosophia*. The principles of first philosophy provide the alleged truths upon which all investigation depends. The learned class that Bacon envisions will be schooled in what Bacon denominates 'logic.' The rest of the population, however, will not participate in this education. In and of itself, this is not strange. The ordinary population is not suited to highly complicated doctrines of knowledge. No one is suggesting that Bacon ought to make *prima philosophia* a part of the standard curriculum. Yet it is something else to desire to nudge the unsophisticated part of the population towards their passions, towards 'imagination.' It is not pleasant for human beings to be caught up in cycles of repetitive passion. No matter how attractive an administrative art, it is no improvement of life for the generality of human beings. In fact, this represents a decrease in the quality of life that is available, at least to the extent that governing the people through their imaginations should become effective.

It should be clear by now what Bacon indicates with the term 'invention.' Bacon's science traffics in *nothing but* inventions. Bacon is quite clear that rhetoric is one of the two major sciences of invention that his philosophers will be busied with. "Invention is of two kinds, very different; the one of arts and sciences, and the other of speech and argument."[51] Bacon, taking careful aim at Plato's theory of forms, insists that "the spirit of man (being of an equal and uniform substance) presupposes and feigns in nature a greater equality and uniformity than really is."[52]

Aristotle's deliberative model of rhetoric is a far cry indeed from what Bacon is proposing in his science of invention. "Rhetoric is subservient to the imagination, as logic is to the understanding; and the duty and office of rhetoric, if it be deeply looked into, is no other than to apply and recommend the dictates of reason to imagination, in order to excite the appetite and will."[53] At some point in time, of course, the new public knowledge that Bacon's philosophy seeks to give birth to and the ordinary opinions must meet. Public opinion will be the product of this convergence. This is why it matters that Bacon has aimed his criticisms of the public capacity for knowledge and truth at so fundamental a level. Based on *prima philosophia*, Bacon's doctrine of the inadequacy of unguided perception, and Bacon's complaints about the public's mere use of language, exactly how is it proposed for the new science of understanding to mobilize the public passions on its side? It cannot be by education. For the new science of understanding in Bacon does not profess to share any common ground with the public. Bacon's philosophy does not admit a common use of names, a common foundation for perception, or any common root of authority, save for one. Bacon promises that his philosophy is dedicated to charitable ends. That then must be the point of approach, for the formation of the new public opinion. It is not necessary that the public understand; it is only necessary that they feel that the new science is going to work in their interest.

It is supposed to be the spirit of modernity that ordinary people have thrown off superstition and complacent trust in things that they do not understand. Bacon's conception of the formation of public opinion suggests something very different. Bacon's science makes no efforts to render its projects intelligible to the people directly; in fact it takes measures to ensure that the public cannot understand. Public opinion, it seems clear, is to settle for consequences: it is promised material advantages, and if material advantages are delivered, this is going to suffice for the legitimacy of the new science.

Bacon defends the rhetoric that he seeks to employ in public management. He insists that the new rhetoric has the purpose of enlisting the public on the side of reason. Yet that is exactly what we have been attempting to ascertain in this study. Is the atomistic doctrine, upon which Bacon lays the foundations of his science, true? Is it true that nature is a unity, more or less? That final causes are a delusion, that the ordinary understanding of kinds of object (or forms, in the language of Plato and Aristotle) are primitive confusions? Certainly these are not things that the scholar is entitled to take on faith. It is a problem with Early Modern Philosophy that it does not seem anxious to share its assumptions with the rest of us, nor to meet the objections which arise from earlier forms of learning. Bacon asks us to take his experimental science to our hearts based upon trust or faith. Yet the objections to his philosophy of matter, and subsequently all the subsequent doctrines erected upon that edifice, were not exactly obscure during the time when he wrote. Aristotle's critique of the atomistic philosophy, strictly on the level of matter theory, was quite available to Bacon. Ignoring those arguments does not inspire faith. "For the end of logic is to teach a form of argument to secure reason, and not to entrap it; the end likewise of moral philosophy is to procure the affections to fight on the side of reason, and not to invade it; the end of rhetoric is to fill the imagination with observations and images, to second reason, and not to oppress it."[54]

> Again, if the affections themselves were brought to order, and pliant and
> Obedient to reason, it is true there would be no great use of persuasions
> And insinuations to give access to the mind, but naked and simple propositions
> And proofs would be enough. But the affections do on the contrary make such
> Secessions, and raise such mutinies and seditions...that reason would become
> Captive and servile if eloquence of persuasions did not win the imagination
> From the affection's part, and contract a confederacy between the reason and
> The imagination against them.[55]

It is instructive that Bacon has singled out the affections, rather than say the appetites as the unruly energy in public life. When we talk about affections we talk about the sphere in which individuals feel love and duty, honor and loyalty. The sphere of the affections is precisely the area of life where individuals experience meaning most directly. In the conceptual diagram that Bacon furnishes above, it is the affections which are to be isolated and impoverished. It also makes it easier to understand why Bacon does not think discourse would be of much avail for the changes in society that he proposes. People would never be willing to subordinate their attachments to other claims on their emotions. By nature, they would not be willing to do so. Yet Bacon is determined that the affections pose an obstacle, and that they must be overcome through other means.

In his *De Augmentis*, Bacon spends a considerable amount of time explicating and even demonstrating the modes of communication that the new scientific regime will employ. Between scientists, the preferred form of communication will be aphorisms. Aphorisms are abridged forms of reasoning, brief statements that tend to suppress the operative premises and conclusions. Bacon composed most of the second book of his *New Organon* in the aphoristic style.

ETHICS: OR MAN 'SEGREGATE'

There are three parts of 'civil knowledge,' according to Bacon: "the knowledge of conversation, the knowledge of negotiation, and the knowledge of empire or government."[56] The 'knowledge of conversation' we have already treated of. It concerns a rhetorical address to the imagination of the public. The knowledge of empire or government will be treated in the next section. Here, we address an entirely new sector of civil knowledge, which Bacon calls 'negotiation.' To briefly characterize it, we may say that it involves the art of getting ahead in the world. In fact, this enterprise of inculcating arts of 'negotiation' is Bacon's new moral philosophy: and in this moral philosophy, man is contemplated as 'segregate,' or isolate.[57] This constitutes a pretty major change in the outlook of western civilization. The Aristotelian heritage insists on the domain of the political as the proper realm for human fulfillment. At the very least, it anchors the domain of ethics in the practice of ordinary praise and blame. Bacon is making a radical rupture in the arena of morals. He is advocating an ethics of selfishness.

There doesn't seem to be any more fundamental public issue than the distinction between right and wrong. Customary societies have always cared deeply about such issues. The English civilization that Bacon had been born into cared very deeply about these issues: in its Statutory and Common Law, in its religion, in its art. The heritage of Machiavelli again seems entirely relevant. In *Prince* XV, Machiavelli announces a morality whereby the individual is no longer at liberty to defer to common values. Machiavelli is a political theorist. He advances the view that the world is too unreliable for the individual to trust in; that the individual must tend to his own needs, even if this sets him in defiance of the ordinary discernment of virtue and vice, right and wrong. It seems that there is nothing more central to a people's chances for happiness than the moral culture to which it dedicates itself. At the very least, this should surely be a domain for public deliberation. If we really wish to suspend the authority of ordinary praise and blame, then this above all other things should be an object for talk. Bacon does not envision this.

'Civil knowledge,' Bacon argues, does not depend upon the internal feelings, dispositions, and attitudes of human beings. It depends only on 'external goodness.'[58] Bacon is appealing to common platitudes, at times, but the signification of these platitudes lose their common meaning when man is contemplated as 'segregate,' which is Bacon's, but not the popular, intent. Thus the proverb "every man is the architect of his own fortune," while it makes sense in many a customary community, nevertheless insinuates a different teaching under Baconian tutelage, because here the individual is urged to act for himself against the others.[59]

Bacon envisions a new type of moral personality. It is one that is always attempting to suit disposition to circumstance, so as to be able to take advantage. One could say that this is the moral code of the opportunist, and a number of scholars have commented on this aspect of Bacon's moral teaching.

> First, to consider how their natural and moral constitution sort with the general
> State of the times; which if they find agreeable and consonant, then in all things
> To give themselves more scope and liberty, and indulge their dispositions; but
> If there be anything differing and discordant, then in the whole course of their life
> To be more close, retired, and reserved.[60]

Bacon's *Essays* furnish a model for the sort of education that is going to be taking place within the new regime. Quentin Skinner has expressed disbelief at the suggestion that Machiavelli could have played much of a role in the emergence of British ethical philosophy. British ethical philosophy, in Skinner's analysis, is bound up with the fortunes and aspirations of individuals; whereas Machiavelli's model of political science seems infinitely more wrapped up in the prerogatives of the community of common interest.[61] The proper parallel to draw, however, is between Machiavelli's Prince and the British citizen for whom Bacon writes. The moral constitution of these two classes of being, despite their differences in power and rank, are virtually identical. Bacon has democratized Machiavelli's princely ethos. Bacon is sculpting the moral philosophy which will suit the order of his imagination. As one scholar has observed, Bacon's appeal is virtually for a science of selfishness.[62]

Bacon's new portrait of man 'segregate,' as a moral isolate, betokens a new mode of 'intelligence.' It is an innovative teaching to indicate that individuals should seek to conceal their defects of character, rather than to reform them. This study has come a long way from the original contemplation of the doctrine of atomism in Bacon's philosophy. Now it is evaluating the moral teaching in Bacon's doctrine; and Bacon does indeed seem to be retailing the sort of gospel of selfishness that was enshrined in the philosophy of Epicurus, which Cicero so bitterly criticized. Are we certain that this is the charitable path? Is it really in the interest of human happiness that selfishness be set as the preeminent goal, as well as the disposition to be rewarded, cultivated, and praised? Is it really clear that ordinary opinion is incompetent to share in this particular debate based upon the forms of knowledge which are available to it?

In Bacon's view, those will be praised, and advanced, who are "able to set forth to advantage before others, with grace and skill, his virtues, fortunes and merits (which may be done without arrogance or breeding disgust); and again, to cover artificially his weaknesses, defects, and misfortunes, and disgraces; dwelling upon the former, and turning them to the light, sliding from the latter or explaining them away by apt interpretations, and the like."[63]

Bacon's moral exemplar must be quick to leap at the chance, and to chastise within himself any humility or restraint as he may feel involuntarily. "With regard to turning and shaping the mind, we must strive with all possible endeavor to render the mind obedient to occasions and opportunities, and to be noways obstinate and refractory towards them," Bacon writes.[64] This is Bacon's moral bequest, his conception of what the human being should finally regard as 'advancement in life.'[65]

Rhetoric appeals to the emotions of individuals. At least, we may say that this is true of the rhetoric that Bacon envisages. The administrative authority that Bacon has developed through his more fundamental doctrines of *prima philosophia* and logic is quite operative now in the domain of moral teaching. The people are not to be instructed directly in a new set of morals. Rather, 'indirection' is the proposed method. The task of the experiment, in Bacon's proposed new version of man as a moral isolate (i.e., as putting himself and his self-interest ahead of every other consideration) is to cause human beings to merge with this initiative. From Bacon's point of view, it is better that the generality of human beings should find themselves stumbling into this moral ethos without thinking about it too much: "that the mind is brought to anything with more sweetness and happiness, if that whereunto you pretend be not first in the intention, but be obtained as it were by the way while you are attending to something else; because of the natural hatred of the mind against necessity and constraint."[66]

How the individuals are to be brought to this convergence with the new set of moral guidelines cannot be a direct one: for that would lay outside the parameters of the new scientific method. Discourse, open discourse, would require shared facts to talk about; it would shatter Bacon's 'first philosophy.' Instead, the migration towards a new moral philosophy must be incorporated into the educational mission of the society, as well as in a campaign of public relations that seeks to present the new moral ethos as a necessary means to the charitable goals of the new scientific civilization.

POLITICS, OR MAN 'CONGREGATE'

Bacon's division of the human being into man 'segregate' and man 'congregate' is a double-barreled attempt to arrest the role that customary virtues play in society. The human being, conceived as a political animal in Aristotle's philosophy, is conceived of as an individualistic animal in Bacon's philosophy. Certain it is that the happiness of civilization to a large extent depends upon the moral culture that it pursues. Bacon has made the judgment that the internal character of the human being needs to be consecrated to personal advantage. Politics, or 'man congregate,' in Bacon's analysis, really does not speak any longer to the nature of the human being per se. The political animal is theorized out of existence. Politics, for Bacon, is rather an affair of state. It does not involve deliberation. Yet it does involve some observations bearing upon the cultivation of the proper moral temper in civilization.

When the reader finally reaches the traditional topography of political science in Bacon, it is easily seen that the earlier language of charity and humanity has effectively disintegrated. The government in Bacon's view obtains its authority from its sovereignty. Preservation and happiness of life, Bacon tells us, have been treated of already in the area of man 'segregate.' Formal political science, then, will concern 'the extension of empire.'[67]

Bacon's state now appears as something of an independent organism. As it it lives and breathes, the state must grow. According to Bacon, the citizenry needs the tonic of war in order to avoid decaying into effeminacy and complacency. "But above all," Bacon writes, "for empire and greatness, it importeth that a nation do profess arms as their principal honor, study, and occupation."[68] "No body can be healthful without exercise," Bacon writes, "neither natural body nor body politic; and certainly to a kingdom or estate a just and honorable war is the true exercise. A civil war indeed is like the heat of a fever; but a foreign war is like the heat of exercise, and serves most of all to keep the body in health. For in a slothful peace both courage will effeminate and manners corrupt."[69]

Bacon does not profess to feel any disproportion of attachment to either noble or democrat in his political science. Yet this study has made the attempt to identify new corridors of authority that are being envisioned in Bacon's doctrines. The most potent aspect of Bacon's philosophy is its ability to claim for itself a non-partisan point of view. The Baconian point of view however traces back to the *prima philosophia*, the atomistic philosophy.

If the atomistic doctrine upon which Bacon's teaching relies is seen to be disproven, then his critiques of the spheres of human perception and the ordinary use of names will also be found problematic. If his critiques of human perception and the ordinary use of speech is problematic, then his intellectual division of labor between logic and imagination will be seen to be problematic. All of these issues would gather together in a contemplation of Bacon's analysis of morality and politics.

It is Bacon who chooses to invoke the name of Machiavelli in his writings. It was Machiavelli who authored the truly revolutionary principle that the true mark of a society's freedom consists in the degree and intensity of the conflict between society's parts or elements. The major elements of the political society, in the language of Bacon's time, were still democrats and aristocrats.

Aristotle's political science, in his *Politics*, is full of examinations of the range of relationships that can exist between these two orders of society, the great and the common. In Aristotle's view, the best forms of government, even if they are not excellent, still must partake to some degree of 'mixed' authority. This is to say that the major elements of society, the noble and the democratic, must to some degree undertake to respect one another, and to provide one another with some role in the government. An aristocratic government must accord some places of power to democratic authority; and a democratic regime must allocate some positions of authority to the aristocratic part of society. Thus Bacon is not entered upon any new domain when he contemplates the relationship between the nobility and the commons.

The question which arises is, how shall the new experimental science be figured into the organization of social authority? Bacon's ethical postulates do not square with the traditional conceptions of justice of either the aristocracy or the democracy. Both the aristocracy and the democracy assert *claims* to justice. This involves deliberation and speech: it involves recognition and authority for deliberation and speech. Bacon's experimental science seems to subtract recognition and respect from ordinary public deliberations and speech, especially in the case of the democracy. Thus when Bacon observes that one must watch the tide of discontent in the public arena, among the aristocrats and the democrats, he is not giving a satisfactory account of his political philosophy. For he does not mention the authority that his new experimental science and its non-partisan reputation are designed to play in the emerging society.

> For removing discontentments, or at least the danger of them, there is in
> Every state (as we know) two portions of subjects; the nobles and the
> Commonality. When one of these is discontent, the danger is not great; for
> Common people are of slow motion, if they be not excited by the greater
> Sort; and the greater sort are of small strength, except the multitude be
> Apt and ready to move of themselves. Then is the danger, when the greater

> Sort do but wait for the troubling of the waters amongst the meaner, that they May declare themselves.[70]

It is not a new theory that it could be useful to pit the aristocracy against the democrats, in certain areas of politics, in order to stabilize discontents. Yet Bacon does not seem to allow for any authority to countervail against the new experimental philosophy. Just as the atomistic doctrine is obscure, so the authority which is founded upon its doctrines is obscure. The evidence for the authority of the new atomistic science in moral and political affairs seems to have been established in Bacon's discussion of 'man segregate.' To elevate selfishness into a new moral creed is quite the feat.

Bacon uses the common name of 'faction' to recommend policies of government. Yet 'faction' is a name. What does it mean? It indicates a part of society that is attempting to steer the whole of society in its own interest, that is, against the common cause. When Bacon uses the name, are we sure that he means what we think it means?

It may be that I exaggerate the degree to which Bacon has been influenced by Machiavelli. Yet Machiavelli set forth a new blueprint for political science: one which thrives on internecine conflict. In Machiavelli's model of political society, intense and steady conflict between the diverse parts of society is ideal for freedom. This is not the place to go into detail about what Machiavelli means by 'freedom.' Yet it can be observed that Machiavelli's principal political insight rests upon his understanding that society must have order; that it cannot function absent some type of order; and that if society is forced to choose between brute authority and no authority whatsoever, it will by necessity opt for the former. Life must go on; and it cannot go on without order.

Bacon's celebration of selfishness as the ethic of man 'isolate,' his very argument that man is an isolated moral nature, suggests (at least indirectly) an endorsement of the omnipresence of conflict in society. Order will not emerge out of competing self-interests. Scholars may at least evaluate Bacon's political science to see whether or not it does seek to render government dependent upon the sort of desperation fostered by chronic selfishness. Bacon has also made a number of arguments which suggest an aversion to deliberation. Deliberation is the only alternative to coercive government, to government by urgent necessity rather than choice.

Ultimately, it seems that Bacon's political science aspires to do more than pit one element in society against the other elements. It is more ambitious than that. It seeks, rather, to divide the individual against himself. Certainly it is too much power for any mode of government to possess, to exert that kind of force upon its citizenry. It really does bring us back to the issues that seem less related to politics, like the authority of ordinary perception and the ordinary assignment of name to object.

Knowledge is the great antidote to passion. Passion is very easily a crippling state of existence for human beings. There are many passions that we cherish, but we do not cherish being governed by our passions. For this leads us into bondage. Appetites are mingled into the passions; and the passions change. Passions make it hard for us to discern the actual good from the actual bad in situations. And our ability to know, to know facts is one of our most fundamental guardians against the rule of the passions.

In later Early Modern philosophies, such as those of Locke and Hume, great pains will be taken by those philosophers to insist that human knowledge simply can't confidently extend even to so meager a thing as knowledge of the common objects right in front of us. Now, this certainly isn't going to interfere with individuals navigating their world by their perceptual faculties; nor does any philosopher intend for this to happen. Yet the calling into question of what appear to the people to be obvious facts will certainly raise the temperature of the passions. It will certainly imbue the polity with tides of anxiety that are new. In that sense, scientific authorities which enjoy a unilateral position of authority to pronounce upon any truth of fact whatsoever, can greatly affect the warp and woof of political and cultural life.

Bacon's promise began with the gospel of charity, the promise of the conquest of nature. Why must this involve an equivalent shifting of the bases of public authority, towards the passions of the people? It is not charitable to convert people's emotions into divisions within themselves. Yet Bacon seems to contemplate this.

> But to speak the real truth, the poets and writers of history are the best doctors
> Of this knowledge, where we may find painted forth with great life and dissected,
> How affections are kindled and excited, and how pacified and restrained, and how
> Again contained from act and further degree; how they disclose themselves,
> Though repressed and concealed; how they work; how they vary; how they are
> Enwrapped one within another; how they fight and encounter one with another;
> And many other particularities of this kind; amongst which this last is of special
> Use in moral and civil matters; how, I say, to set affection against affection, and
> To use the aid of one to master another.[71]

NOTES

1. *Discourses on Livy*. Translated by Harvey C. Mansfield and Nathan Tarcov. Chicago: University of Chicago Press, 1996. I Pr.

2. *The Advancement of Learning*, book 1. In Francis Bacon, *The Major Works*. Edited by Brian Vickers. Oxford: Oxford University Press, 2002, p. 124. "To conclude, therefore, let no man, upon a weak conceit of sobriety, or an ill-applied moderation, think or maintain that a man can search too far or be too well studied in the book of God's word or in the book of God's Works, divinity or philosophy; but rather let men endeavor an endless progress of proficiencies in both, only let men beware that they apply both to charity, not to swelling; to use, and not to ostentation" (126).

3. The distinct zeal of Bacon's prose has misled some commentators as to the nature of his project. Stephen Gaukroger thinks that Bacon's fierce dismissal of classical Greek philosophy

is due to the 'esotericism' of the Greeks. By contrast, Gaukroger thinks, Bacon intends to supply us with a new, inclusive, public philosophy, with the natural philosopher as a sort of apostle. "He is motivated by two kinds of issue. The first is the fact that natural philosophy is to be of use to the state; if it is to add to the productive capacity of the nation, then it must be reformed from an esoteric discipline into a public one" (57). "The answer is that the natural philosopher is not simply someone with expertise for Bacon, but someone with a particular kind of standing, a quasi-moral standing, which results from the replacement of the idea of the sage as a moral philosopher by the idea of the sage as a natural philosopher" (112). In *Francis Bacon and the Transformation of Early Modern Philosophy*. Cambridge: Cambridge University Press, 2001. To Gaukroger's points two objections need to be raised. In the first place, and most conspicuously, Bacon dismisses out of hand the ordinary use of names—whether as ideas of the tribe, the cave, the marketplace, or the theater. This is the direct result of his decision to reject the notion of forms in perishable objects, which are the things indicated by common names. This hardly qualifies as a rejection of the 'esoteric' tradition. Bacon correctly recognizes Plato and Aristotle as philosophers who *shared* principles of truth with 'the vulgar.' Secondly, the zeal of Bacon's speech is a direct parallel to the incendiary style of Gorgias or Polus. Bacon, denying that the perceptions make the common objects knowable to us, opts for an indeed esoteric doctrine of the 'interpretation' of nature, which model of knowing the 'public' cannot share in at all. Cf. Perez Zagorin, *Francis Bacon*. Princeton: Princeton University Press, 1999, 14.

4. See Thomas Kuhn. *The Essential Tension*. Chicago: University of Chicago Press, 1977, 46: "If Baconianism contributed little to the development of the classical sciences, it did give rise to a large number of new scientific fields, often with their roots in prior crafts," Kuhn writes. "If the possession of a body of consistent theory capable of producing refined predictions is the mark of a developed scientific field, the Baconian sciences remained underdeveloped throughout the seventeenth and much of the eighteenth centuries" (47). Kuhn suffers from the narrowness of his perspective. For Bacon's scientific theory fully entails a new political and cultural order for society. In these aspects Bacon's influence may have been felt far more directly, and much sooner, than Kuhn allows.

5. *The Dignity and Advancement of Learning* (*De Augmentis*), book II, p. 415. In *the Works of Francis Bacon*. Edited by James Spedding, Robert Leslie Ellis, and Douglas Denton Heath. Volume 8. Boston: Taggard and Thompson, 1843. All of my references to *De Augmentis* will be from the Spedding edition.

6. There can be little doubt that the trend in the secondary literature over the last several decades is to regard Bacon's most progressive accomplishment as an attempt to reconstruct the idea of rationality into one of 'making' or craftsmanship. Paolo Rossi was one of the earlier exponents of this point of view. "Knowing is for Bacon a kind of making," Rossi writes, "I entirely agree with Perez-Ramos's well grounded and admirable portrait of Bacon as an exponent of the 'Maker's knowledge tradition': a tradition which postulates an intimate relationship between objects of cognition and objects of construction, and regards all knowing as a kind of making or capacity to make" (38). In "Bacon's Idea of Science." Markuu Pelonen, editor. *The Cambridge Companion to Francis Bacon*. Cambridge: Cambridge University Press, 1996. Perez-Ramos himself thinks that the Baconian 'makers' tradition involves abstracting away the human subject from inquiry. "What sort of man-independent reality is this? This context suggests that true reality is only unveiled when the human perceiver, whose cognitive equipment Bacon has already declared faulty . . . is put into brackets, or, so to speak, abstracted from the very process of cognition" (102–103). In "Bacon's Forms and the Maker's Knowledge Tradition," from the same volume. The analyses of Rossi and Perez-Ramos are strained, however. Bacon's experimental method of causation cannot be reduced to the proposition that cognition is 'making.' For Bacon's method excludes and rejects many hypotheses; it dismisses as false many of the experiments it conducts—all of which is to say that Bacon is guided by a doctrine of true and false which is itself not constructed. In my view, Bacon's attempt to redefine knowledge into a kind of making is a political dimension in his theory: for he is seeking not so much how to make, as how to subordinate to the experimenter's will the objects in nature. In order to pursue his project, Bacon denies to these objects their customary names. This with-

holding of recognition from natural objects enables the experimenter to treat said objects, in the Machiavellian sense, as formless matter, which they really are not.

7. "The Masculine Birth of Time or Three Books on the Interpretation of Nature." Translated by Benjamin Farrington in *The Philosophy of Francis Bacon.* Chicago: University of Chicago Press, 1966. 62.

8. The sheer scope of Bacon's ambitions, his proclamation that charity and alleviating man's estate are the proper goals of science, are liable to lead many readers to pardon Bacon for seeking to create a new class of privileged knowers. Benjamin Farrington is only too ready to climb aboard Bacon's train. "And true it is that Bacon exerted himself to banish the age-old pessimism of the philosophic tradition and rouse the human race from passive acceptance of ignorance, poverty and ill-health. He labored to create both the knowledge and the will to attempt a transformation of the world." *The Philosophy of Francis Bacon.* Chicago: Phoenix Books, 1966. Farrington's evaluation is hasty. As the text of this chapter indicates, Bacon does not believe that what he regards as ignorance in the human race can be abolished. Bacon is resolved to govern the majority through a rhetoric that has no place for common objects of deliberation. Bacon's opposition to Plato and Aristotle is not principally motivated by their comparative disdain for mechanical arts; rather, Bacon opposes their conception of truth, and he reviles the virtues they prize. Cf. Sidney Warhaft, "Science Against Man in Bacon." *Bucknell Review* 7(1958): 158.

9. Those scholars who espouse enthusiasm for the characterization of Bacon as the embodiment of the 'Maker's tradition' of knowledge tend to spend less time talking about the transformation in ethics that Bacon proposes. This phenomenon reflects a confusion in the secondary literature. Bacon, they say, derives his natural science partly from the Alchemist tradition and partly from the Pre-Socratics, but his ethics and politics, are said not to derive from these sciences. Nor does Bacon's science, as Rossi indicates, derive from a new *ethic. Francis Bacon: From Magic to Science.* Translated by Sacha Rabinovich. London: Routledge and Kegan Paul, 1968, 37: "Bacon's whole outlook was based on the belief that a new era in the history of mankind was at hand. If man, confronted by his new destiny, were to recall science from the gloom of antiquity, rather than seek it in the bright light of nature, it would indeed be sinful . . ." It would be truer to say that the revival of atomism, of Lucretian atomism, first in Machiavelli and then in Bacon, involves the assertion of an ethic which is anything but a 'bright light.' The atomist disposition *withdraws respect* from the reputation of perishable objects, since the atoms allegedly indicate immunity to destruction in true bodies. That Bacon links 'Cupid,' or Eros, to the atom, is to make boundless desire the first truth in the universe, one to which formal ethics, in Bacon, will ultimately be made to bow. In "On Principles and Origins According to the Fables of Cupid and Coelum," 199: "This chaos, then, which was coeval with Cupid, signified the uncreated mass or congregation of matter. But matter itself, its power and nature, and in fine the principles of things, had been shadowed forth in Cupid himself. He is presented without a parent, that is, without a cause." Eros, to Bacon, is first in nature: and Eros is not in the neighborhood of 'charity.' *The Oxford Francis Bacon VI. Philosophical Studies C.1611–C.1619.* Edited, with introduction by Graham Rees. Oxford: Clarendon Press, 1996.

10. *De Augmentis*, book 2, p. 452.

11. "Thoughts and Conclusions, or the interpretation of Nature or a Science of Productive Works," 80–81. Translated by Farrington, in *The Philosophy of Francis Bacon.*

12. Graham Rees caused something of a stir when he argued that Francis Bacon had never been an atomist at any point in his career. "The main concerns of this article will be to show that there is no evidence that Bacon committed himself to the doctrine of atoms or void separately or together at any time, and above all, that his 'atomism' is best understood in its wholly subordinate relationship to speculative philosophy" (552). "'Atomism' and 'Subtlety' in Francis Bacon's Philosophy." *Annals of Science* 37(1980):549–571. Rees goes on to claim that Bacon's interest in atomism is not part of his 'positive science,' but of his 'speculative philosophy.' Rees has some evidence on his side. The classical atomic theory in Democritus and Leucippus involves eternally indivisible bodies, incapable of undergoing change of any kind. Bacon's atoms have 'appetites' and even powers of something that Bacon characterizes as 'perception.' 'Matter' is the terminology that Bacon prefers. Yet the point of Bacon's positive

science is to 'vex' nature, to 'harass' it. This means imposing changes upon matter. Thus the classical atomist view of indivisible bodies accidentally 'generating' compounds through collision is not quite in keeping with Bacon's theory. However, the alterations to the atomist theory that Bacon makes (it is not a subject he chooses to discuss in direct detail) are superficial ones. Bacon's matter *is indestructible. That* is the Eleatic paradigm to which Democritus and Leucippus responded. There must be some indestructible reality in order that anything could exist. Furthermore, Bacon has followed Machiavelli in transferring the attitude coeval with atomism—that of indifference towards the perishability of existing objects—into a new cultural and political ethos that is quite aggressive towards the re-imagined perishable world. In any event, Rees's argument has not affected the scholarly view that Bacon is some sort of atomist, for the most part. That view long preceded Rees's article, and has subsisted subsequent to it. See Charles T. Harrison, "The Ancient Atomists and English Literature of the Seventeenth Century." *Harvard Studies in Classical Philology* 45(1934):4; Stephen Gaukroger, *Francis Bacon and the Transformation of Early Modern Philosophy*. Cambridge: Cambridge University Press, 2001, 143; Paolo Rossi, *Francis Bacon: From Magic to Science*. Translated by Sacha Ravinovich. London: Routledge and Kegan Paul, 1968, 14; Robert Hugh Kargon, *Atomism in England From Hariot to Newton*. Oxford: Clarendon Press, 1966, 43–44; F.H. Anderson, *The Philosophy of Francis Bacon*. Chicago: Phoenix Books, 1966, 1948, 70; Reid Barbour, "Bacon, Atomism and Imposture: The True and Useful in History, Myth and Theory," 41–42. In Julie Robinson Solomon and Catherine Gimelli Martin, editors. *Francis Bacon and the Reconfiguring of Early Modern Thought: Essays to Commemorate the Advancement of Learning*. Aldershot: Ashgate, 2005; Stephen H. Daniel, "Myth and the Grammar of Discovery in Francis Bacon," *Philosophy and Rhetoric* 15(1982):228. Richard Kennington denies that Bacon is an atomist, but his evidence is limited to Bacon's statement in *The New Organon* to that effect, which is even a weaker claim than Rees's for the reasons adduced above. "Bacon's Critique of Ancient Philosophy in *New Organon* I," in Daniel O. Dahlstrom, editor. *Nature and Scientific Method*. Washington D.C.: Catholic University Press, 1991, 246.

13. *De Augmentis*, book 3, 472.
14. Ibid., p. 473.
15. *Metaphysics* 985b4.
16. *The New Organon*, 45 (book 1, #51). See also *The Advancement of Learning*, book 2, 193: "If then it be true that Democritus said that 'the truth of nature lieth hid in certain deep mines and caves,' and if it be true likewise that the Alchemists do so much inculcate, that Vulcan is a second nature, and imitate that dexterously and compendiously which nature worketh by ambages and length of time; it were good to divide natural philosophy into the mind and the furnace, and to make two professions or occupations of natural philosophers, some to be pioneers and some smiths; some to dig, and some to refine and hammer . . ."
17. *New Organon*, 108 (book 2, #8).
18. Epicurus. "Letter to Herodotus," paragraph 39. In *The Essential Epicurus*. Translated by Eugene O'Connor. New York: Prometheus, 1993.
19. *Parmenides* 142ab.
20. *Parmenides* 142cd.
21. *Parmenides* 142e.
22. *Parmenides* 145ac.
23. *Parmenides* 153ac.
24. *Parmenides* 155de.
25. "On Principles and Origins According to the Fables of Cupid and Coelum, or the Philosophy of Parmenides and Telesio and Especially that of Democritus as it is Treated of in the Fable of Cupid," 253. In *The Oxford Francis Bacon, VI. Philosophical Studies C.1611–C.1619*. Edited with an introduction by Graham Rees. Oxford: Clarendon Press, 1996.
26. *De Augmentis*, 410 11.
27. Bacon really demonstrates the attitude of his science in this way. What Bacon seeks to obliterate is the conception of violence; this is done by denying the concept that objects *have* a nature. "Whereas men ought on the contrary to be surely persuaded of this; that the artificial does not differ from the natural in form or essence, but only in the efficient; in that man has no power over nature except that of motion; he can put natural bodies together, and he can separate

them" (410–411, *De Augmentis*, book 2). This leads us to Bacon's 'experimentation' with language. For nature gives us the natural bodies of lion, oak, and gold; but Bacon *refuses* to accept these as the true bodies; he regards these as 'compounds' of the true 'simple natures'; through which linguistic apparatus he can splinter the forms of the true objects, and divide them and order them in opposition to themselves. Peter Pesic attempts to defend Bacon's claim, but he ignores the false postulate in Bacon's theory: that true natural objects are immune to destruction. "Wrestling with Proteus: Francis Bacon and the 'Torture' of Nature." *Isis* 90(1999):82, 86. Robert K. Faulkner has a clearer understanding of Bacon. "The starting point of Baconian method," Faulkner writes, "is less in human perception than in human will" (9). *Francis Bacon and the Project of Progress*. Lanham: Rowman & Littlefield, 1993.

28. *De Augmentis*, 415.

29. *De Augmentis*, book 3, chapter I, 472–474.

30. Francis Bacon. "The Great Renewal," p. 18. In Lisa Jardine and Michael Siverthorne, editors. *The New Organon*. Cambridge University Press, 2000.

31. "A Description of the Intellectual Globe," 105. In Graham Rees, editor. *The Oxford Francis Bacon VI: Philosophical Studies C.1611–C.1619*. Oxford: Clarendon Press, 1996.

32. *The New Organon*, book I, xxxix, xli, xlii, xliii, xliv.

33. Julie Robin Solomon (*Objectivity in the Making: Francis Bacon and the Politics of Inquiry* Baltimore: Johns Hopkins, 1998) thinks that Bacon is to some degree a product of historical developments. Solomon indicates that commercial and social changes occurring in Jacobean England contributed to the formation of Bacon's scientific goals, and that Bacon's scientific conceptions in turn enables the historical trends. "Bacon offered up the persona of the merchant, and the apparatus of mercantilism, as well as the cognitive and productive practices of the man of commerce and manufacture, as models for natural philosophy" (15). "This supported and fractured domestic economy, generated by the state's mercantilist intervention into families of its most productive citizens, forms the affective substructure of Bacon's discourse on knowledge: one that alienates private desire from knowing" (101). Bacon, however, is motivated principally by *doctrine*, that is, by philosophy: the atomist estrangement from the ordinary use of names that he seeks betokens a *boundless* new liberty for blending personal desire into inquiry. Just consider Bacon's definition of 'sapience': "Again, for the pleasure and delight of knowledge and learning, it far surpasseth all other in nature: for shall the pleasures of the affections so exceed the pleasures of the senses, as much as the obtaining of desire or victory exceedeth a song or dinner; and must not of consequence the pleasure of the intellect or understanding exceed the pleasures of the affections? We see in all other pleasures, there is satiety, and after they be used, their verdure departeth; which sheweth well they be but deceits of pleasure, and not pleasures. . . . But of knowledge there is no satiety, but satisfaction and appetite are perpetually interchangeable." *Advancement of Learning* book I, 167.

34. *The New Organon*, book I, xliii.

35. Francis Bacon. *The Advancement of Learning*. In *Major Works*. Edited by Brian Vickers. Oxford University Press, 2002, book 2, 228.

36. "A Description of the Intellectual Globe," 101. In the *Oxford Francis Bacon*, VI.

37. "Plan of the Great Renewal," 17. In *The New Organon*.

38. Ibid., 16.

39. Ibid., 17.

40. The much lauded 'collaborative' nature of the scientific enterprise for Bacon fails to recognize the limits of collaboration. Bacon's *prima philosophia*, the postulates of Eleatic atomism, are non-negotiable. As such, they seek to impose orthodoxy across the disciplines. Even emotions such as anger and shame will be subjected by Bacon to the reduction into 'simpler' natures. Bacon is quite clear about the fact that his method requires careful screening of disciples. It is a submissive nature that Bacon seeks for in his new experimental method. "Our method of discovery in the sciences is designed not to leave much to the sharpness and strength of individual talent," Bacon writes, "it more or less equalizes talents and intellects" (*New Organon*, book I, LXI, p. 50). This is to say, it equalizes or standardizes intellectual labor in the new rationality; but the gap between the new science and ordinary opinion could not possibly be wider. Not even F.H. Anderson is inclined to be critical of Bacon's indictments of the ordinary use of speech. On the other hand, Anderson does not appear to be especially

sensitive to the issues implicit in it. *The Philosophy of Francis Bacon*. Chicago: University of Chicago Press, 1948, 102–103.

41. *On the Nature of the Universe*, 4.825.

42. Fragment 12. In Freeman, *Ancilla to the Pre-Socratic Philosophers*, 84.

43. *Phaedo* 97c.

44. *Physics* 198b17.

45. *Advancement of Learning*, book 2, 198–99: "For to say that the hairs of the eye lids are for a quickset to defend them from the extremities of cold and heat; or that bones are for the columns or beams, whereupon the frames of the bodies of living creatures are built; or that the leaves of trees are for the protecting of fruit; or that the clouds are for the watering of the earth; or that the solidness of the earth is for the station and mansion of living creatures, and the like, is well enquired and collected in metaphysic; but in physic they are impertinent. . . . And therefore the natural philosophy of Democritus and some others, who did not suppose a mind of reason in the frame of things, but attribute the form thereof able to maintain itself to infinite essays or proofs of nature, which they term 'fortune,' seemeth to me (as far as I can judge by the recital and fragments which remain unto us) in particularities of physical causes more real and better inquired than that of Aristotle and Plato; whereof both intermingled final causes, the one as a part of theology, and the other as a part of logic, which were the favorite studies respectively of both those persons."

46. *The Advancement of Learning*, book 2, in *Major Works*, 193–94.

47. "The Masculine Birth of Time," 64, in Farrington, *Philosophy of Francis Bacon*.

48. "The Refutation of Philosophy," 111, in Farrington, *Philosophy of Francis Bacon*.

49. "The Great Renewal," 2. In Bacon, *The New Organon*, edited by Lisa Jardine and Michael Silverthorne. Cambridge University Press: 2000.

50. *De Augmentis*, book 2, 410.

51. *De Augmentis*, book 5, chapter ii, p. 64 (volume 9).

52. *De Augmentis*, book 5, chapter iv, p. 99.

53. *De Augmentis*, book 6, chapter iii, p.131.

54. Ibid., p.132.

55. Ibid,. p.133.

56. *De Augmentis*, book 8, 233.

57. Bacon's conceptualization of the ethical human being as 'man segregate' is an attempt to directly reverse Aristotle's teaching. First, Bacon defines soul as something bodily. This reverses Aristotle's ethical teaching. "A living creature consists in the first place of soul and body, and of these two, the one is by nature a ruler and the other the subject" (*Politics* 1254a34). "A social instinct is implanted in all men by nature," Aristotle writes, "but he who is unable to live in society, or has no need, must be either a beast or a god: he is no part of a state" (1253a30). Bacon's experimental definition of the human being as morally segregate, makes human nature seem antagonistic to the just and the right. For Bacon's self-serving man is virtually insensible of a genuine social instinct: he is truly of the Lucretian persuasion. Ian Box recognizes the Machiavellian undertones in Bacon's theory, but he does not comprehend Bacon's ethics on the conceptual level. Thus Box urges us to think that Bacon, in ethical analyses, "reflects a Christian love for humanity more than a strictly civil concern for one's fellow citizens" (277, "Bacon's Moral Philosophy," in *the Cambridge Companion to Francis Bacon*). In Box's view, Bacon's science is in service to humanity as a whole, while his politics remain dedicated to the moral realism of nation states ('Politics and Philosophy: Bacon on the Values of War and Peace.' *Seventeenth Century* 7(1992):115). Yet, the conception of man as 'segregate' is a scientific axiom, that is, part of the experimental method itself. Man 'segregate' is not the Christian vision.

58. *De Augmentis*, book 8, 233.

59. *De Augmentis*, book 8, 269.

60. *De Augmentis*, book 8, 278.

61. Quentin Skinner. *Liberty Before Liberalism*. Cambridge: Cambridge University Press, 2006, 59–61. According to Skinner, Machiavelli was a devotee of something called 'neo-Roman' liberty, or liberty based on a republic that is self-governing. Hobbes rebels against this

tradition in Skinner's argument, by positing that individual liberty is compatible with an unfree, non-republican state.

62. Perez Zagorin discusses Bacon's relationship to morality in *Francis Bacon*. Princeton University Press, 1999, 14–15.

63. *De Augmentis*, book 8, 281.

64. Ibid., 287.

65. Ibid., 297.

66. Ibid., 223.

67. *De Augmentis*, book 8, 298–99.

68. "Of True Greatness of Kingdoms and Estates." In *Essays*. *Francis Bacon, the Major Works*, 401.

69. *De Augmentis*, book 8, 308.

70. 'Of Seditions and Troubles,' in *Essays*, 369. In *Francis Bacon, the Major Works.*

71. *De Augmentis*, book 7, 220–221.

Chapter Two

Descartes and the Science of Authority

Plato believed that there is a serious difference between ordinary opinion and philosophical knowledge. The ordinary opinion is guided mostly by images, and memories of images. However, Plato's philosophers, as the divided line of book VI of the *Republic* makes clear, also begin with images. Plato's philosophers begin with the same images and memories that ordinary opinion does. When Plato ascends to higher knowledge, it is from the understanding of images that this ascent embarks. In Plato's philosophy, exact science never leaves the images behind. It translates the images into speech in order to pursue definition; but it trades on the same fount of memory and familiarity that binds the ordinary opinions together.

Descartes advances a theory of knowledge that not only leaves the perceptual images behind. Descartes' theory of knowledge regards the images of perception as fundamental obstacles to knowledge. This is indicative of the relationship between Descartes' new science of mind, which we call 'epistemology,' and the ordinary opinions. Plato's guardians can rely upon gentle persuasion with the people, largely because they share the same language and meanings. Descartes' scientific overseers, and any science which professes to know better than the ordinary opinions is such an overseer de facto—are not free to employ this kind of persuasion.

Descartes' model of philosophy insists that the evidence obtainable by sense perception is too tainted to be trusted; that if the senses can be seen to mislead us in even one instance, there is justification for excluding perception from the category of evidence altogether. "Sensory perception does not show us what really exists in things," Descartes writes, "but merely shows us what is beneficial or harmful to men's composite nature. . . . They normally tell us of the benefit or harm that external bodies may do to this combination, and do not, except occasionally and accidentally, show us what external bodies are like in themselves."[1] Hobbes maintains, like Descartes, that perception is equivalent merely to certain motions in the animal spirits of the human body; that it is mere delusion to think that the senses provide us with evidence as to external objects.[2] Spinoza for his part characterizes perception not merely as worthless in terms of evidence, but as morally perverse and contrary to intelligence.[3] This only brings us to the doorstep of Locke and Hume.

Descartes, for his part, offers us a very elaborate set of arguments in his *Meditations* as to why the senses cannot be trusted. In the *Meditations*, Descartes offers reasoning that suggests he is attempting to beat back a skeptical tide with his doctrines. However, in Descartes' early writings, there is a distinct affinity for the atomistic teaching that he learned from Isaac Beeckman. In his later writings, the atomistic philosophy is even more pronounced. During the *Meditations* themselves, it is possible to locate arguments which indicate that body, by its *nature*, is not divisible.

Descartes does not provide a formal political philosophy, such as Bacon does. He does not offer a definition of 'man congregate.' Yet it would be difficult to overlook the radical political and cultural implications of the model of mind that Descartes' natural philosophy is attempting to establish. Natural philosophy concerns itself with the domain of truth. Ultimately the truth concerns everyone; and therefore the philosophy that shapes a culture's definition of truth is a most formidable power. In the case of the Early Modern philosophers, the great thrust of their doctrines is to rebuke, upbraid, humiliate, denounce, and tweak the evidence that is obtainable by sense perception. The constellation of public opinion is the great meeting place where the laypersons and natural philosophy are destined to engage one another. Ordinary opinion does not possess any resources for this contest. As a result, it must either be suspended, or it must bend. Life cannot go on if ordinary opinion and belief are suspended; hence it bends.

In Locke, it is asserted that the human mind is a swarming hive of 'simple ideas' such as 'red,' 'round,' and 'sweet.' From Locke's perspective, this fragmented, broken-down, dislocated collection of perceptual fragments is what the mind originally obtains. In order to pronounce the name of 'apple,' to merely indicate its existence, Locke alleges, the mind must undertake to cobble together the 'simple ideas' into a compound. Thus our conception of an 'apple' is alleged to be the mind's artificial creation.[4] It cannot be truly known if that which we describe accurately indicates an external object or not, in Locke's teaching. Thus individuals are free to compile definitions for the names that they use, without any obligation to how other people use those same names.

One would have to look long and hard for a non-philosopher who is even familiar with these teaching of Locke's. For the class of natural philosophers, however, this 'liberty' opens up new vistas of extraordinary power and influence. People may define the name 'apple' in different ways. The simple ideas in their mental workshops might be slightly different. Different combinations might work their way into the individual's construction of the compound name. What to do then? In the case of apples, nobody much cares. Yet what about in the case of the name 'murder'? What about the case of the name 'justice'? What about the case of the name 'right'? In these instances, the weight of the new natural philosophy will descend with a seemingly unimpeachable severity: halting, harnessing, bridling the ordinary perception, pulling back harshly on the ordinary assignment of name to object, and leading it finally into the twilight of pliable uncertainty.

To know better than the others what the real and true bodies are is to know better than the others in every and any thing. The people's entire knowledge is anchored in the observations they make about the bodies they perceive; and thus to subvert the evidentiary status of perception is to undercut the authority of any other species of judgment the people may make. In the final analysis, Descartes' philosophers believe themselves unequal to the others, because they feel they know what is *right*. Yet they obtain this status by virtue of their privileged access to a knowledge that is allegedly clear, certain, and irrefutable. "In the same way, although the proofs I employ here are in my view as certain and evident as the truths of geometry, if not moreso, it will, I fear, be impossible for many people to achieve an adequate perception of them, both because they are rather long and some depend on others, and also, above all, because they require a mind which is completely free from preconceived opinions and which can easily detach itself from involvement with the senses."[5] The majority, in Descartes' view, do not possess this capacity, and need to be brought to a new set of opinions by the quasi-scientific political authority which consists in the power to *define*.

To examine the new natural philosophy's theory of perception and mind is to find a sufficiently elitist doctrine. Descartes' teachings on perception, to be discussed below, are too complicated for unsophisticated public opinion to navigate. By contrast, the technological works and promises of plenty which fuel the Early Modern teachings make quite the impression on the public mind. Descartes accordingly affects a democratic disposition, despite the severity of his indictment of the ordinary mental faculties.

What he will *do* for the human race, Descartes assures us, is the very height of public-spiritedness. He will organize the quest for material abundance. He will make war upon disease. He will increase the overall balance of pleasure, and reduce the overall amount of pain. Descartes has the audacity to claim that he does not affect to be more enlightened than the ordinary individual.[6] It is not a claim that bears close scrutiny. In addition, one can't underestimate the degree to which Descartes intends to obtain some egalitarian bona fides simply through caustic characterizations of Aristotle. "Indeed, the best way of proving the falsity of Aristotle's principles is to point out that they have not enabled any progress to be made in all the many centuries in which they have been followed."[7] The powerful elevation of philosophic authority and a corresponding depression of the authority of the ordinary opinions is not something Descartes is anxious to trumpet. However, it is the dominant pulse of his philosophy.

The grounds upon which Descartes commends his philosophy to us as a fountain of a modern, democratic sentiment is 'generosity': to have knowledge is to be obliged to use it for the common interest. "But as soon as I had acquired some general notions in physics and had noticed, as I began to test them in various particular problems, where they could lead and how much they differ from the principles used up to now, I believed that I could not keep them secret without sinning gravely against the law which obliges us to do all in our power to secure the general welfare of mankind."[8] Descartes' promised science is one that says it will deliver mastery over nature to the human race, with useful inventions and new conveniences of life. More, Descartes' philosophy promises to make an enemy of death: to elevate the art of medicine to the apex of the domain of wisdom.[9]

IS DESCARTES AN ATOMIST?

Most famously, Descartes in his *Meditations* characterizes the influence of skeptical philosophy upon his thoughts. Descartes claims, in the *Meditations*, to be ruling out the validity of sense perception due to the fact that it proves unreliable in at least some cases, and that our faith and trust must be placed in some knowledge that is absolutely certain.[10] Some writers, such as Edward Curley, have made a very great deal about the influence of skeptical philosophy upon Descartes.[11] However, more recent work by Stephen Gaukroger has supplied evidence to the effect that Descartes indicted the evidence of sense perception based upon an atomistic, corpuscular philosophy long before he ever wrote the *Meditations* or entertained the notions of Greek skepticism.[12] In fact, the origins of the indictment of sense perceptions in Descartes' body of work owe much more to the atomic theory than even Gaukroger has recognized.[13]

It should be observed of Descartes' theory that he officially disclaims any reliance upon the atomist arguments. "It is true that Democritus also imagined certain small bodies having various sizes, shapes and motions, and supposed that all bodies that can be perceived by the senses arose from the conglomeration and mutual interaction of these corpuscles, and yet his method of philosophizing generally meets with total rejection," Descartes writes. "This rejection, however, has never been based on the fact that his philosophy deals with certain particles so minute as to elude the senses, and assigns various sizes, shapes, and motions to them."[14] Descartes here speaks in a misleading fashion. That there are bodies too small to perceive is not the issue. That the larger bodies that we do perceive are merely epi-phenomena of such smaller bodies is the core of the atomist claim, and the basis for the indictment of perception. Both Plato and Aristotle deny that sense perceptions arise from invisible bodies. To Plato and Aristotle, the bodies that we ordinarily take ourselves to perceive are real ones.

Descartes is frequently identified as one of the founders of the revolution in modern science. The scholars who make that claim are focusing on Descartes' notions of physics, that is, of what constitutes a body and its nature.[15] In Descartes' philosophy, the skeptical attitude deployed towards the senses is still bashful in public. Descartes does not wish to call too much attention to it. However, the issue for the present moment is whether or not Descartes is an atomist. It is duly noted that Descartes professes, in several texts, to be no such thing.

> We know that it is impossible that there should exist atoms, that is, pieces Of matter that are by their very nature indivisible. For if there were any atoms, Then no matter how small we imagined them to be, they would necessarily have To be extended; and hence we could in our thought divide each of them into two Or more smaller parts, and hence recognize their divisibility. For anything we Can divide in our thought must, for that very reason, be known to be divisible; So if we were to judge it to be indivisible, our judgment would conflict with Our knowledge.[16]

The very first question we must ask is, what is an atomist? An atom is an indivisible unity. It has no parts. It is something that does not come into being, or pass out of being. It is incapable of undergoing alteration, or any other qualitative change. In the original atomist doctrine, the atoms have existed from eternity. They never had a beginning. They just always *were*. In this crucial respect, we must say that Descartes has introduced some innovations into the theory. According to Descartes, God created the atoms; and at His pleasure, God can destroy them.

The Almighty created the atoms, or body: and the Almighty can destroy body. Does this mean that Cartesian bodies are truly divisible? If the Cartesian bodies themselves are truly *divisible*, they *must* come-into-being and pass away. God is not necessary to either create or destroy body, if this body is truly divisible. For the moment, though, grant Descartes his hypothesis. What about the time in between God's creation and His destruction of 'matter' or bodies? Absent God's interference, can the bodies be divided? Do they come into being and pass away without God's interference? In the Cartesian teaching, it does not appear to be the case that they can.

Some things should be pointed out in Descartes' theory. In the first place, Descartes does not deign to give us a precise account of the origin of bodies. Instead of giving us an exact account of the origin of bodies, Descartes insists that he will provide us with a 'fable.'[17] In another place where Descartes discusses the origins of body, he claims to be taking our minds to 'imaginary spaces.' It is hardly possible to argue with a 'fable,' or to take the measure of 'imaginary spaces' with reason.[18] Yet this gives some indication of the original claim of this chapter: Descartes' model of philosophy is very much a model of authority, a model of authority that has no place for deliberation and disagreement.[19] If a subject gives rise to disagreement, Descartes reasons, this can only be due to the fact that we do not possess adequate knowledge of the subject; since adequate knowledge eliminates the possibility for disagreement.[20]

However, when a discussion is embarked upon something as remote and difficult as the discovery of the nature of body, or the origins of body, it is clearly too much to ask that there be a moratorium imposed on disagreement from the outset. Aristotle's *Physics*, towards which Descartes is quite critical, provides ample evidence as to how mistakes and disagreements are important to the progress of knowledge. Inadequate theories that conquer a small part of a subject matter give impetus to newer and stronger theories that seek to overcome the still weak and unsatisfactory parts.[21] Debate and discourse are fundamental to scientific investigation in Aristotle's view.

In the beginning, Descartes' fable informs us, God created all matter. 'Matter' is Descartes' term for body. God alone has the power to create this matter, Descartes insists; and God alone has the power to destroy this matter.[22]

> First, we need to know that absolutely all substances, or things that must Be created by God in order to exist, are by their nature incorruptible and cannot ever cease to exist unless they are reduced to nothingness by God's denying his concurrence to them. Secondly we need to recognize that body, taken in the general sense, is a substance, so that it too never perishes.[23]

There is another series of arguments in the *Principles* where Descartes indirectly advances the old Eleatic argument as to the indivisibility of body. In the relationship between cause and effect, Descartes argues, it is impossible for the effect to contain a greater amount of perfection than is present in the cause. Further, in the relation between cause and effect, it is not possible for the effect to possess anything that is not already evident in the cause. Descartes writes, "For it is very evident by the natural light not only that nothing comes from nothing but also that what is more perfect cannot be produced by—that is, cannot have as its efficient and total cause—what is less perfect."[24] If we apply this reasoning to the phenomena of coming into being, the implications become rather obvious.

> A stone, for example, which previously did not exist, cannot begin to exist unless it is produced by something which contains, either formally or eminently, everything to be found in the stone; similarly, heat cannot be produced in an

object which was not previously hot, except by something of at least the same order of perfection as the heat, and so on.[25]

Aristotle believes that it is nature's intention to bring about generation, and that generation aims at *fulfillment* of the objects brought into being. In Aristotle's physics, nature is like a craftsman. The Aristotelian theory of cause has four parts. First, there is the 'matter.' Matter, by itself, for Aristotle, has no being to speak of. It is limited to mere *potential* being. Secondly, there is the 'formal' cause, or the formula, which is meant to order the formless matter. By itself, once again, the formal cause is merely *potential* being. The third, or efficient cause is the agent that brings the form to the matter. The efficient cause itself accomplishes the bringing together of the matter and the form. The efficient cause, after it has accomplished its task, is no longer operative in the newly generated object. But the combination of matter and form, which comes after the merely potential existence of both in separation, is certainly a higher perfection than its causes. The formed matter possesses *actual* being; and it is thought to achieve its ultimate embodiment *at the end* of its process of development.[26] This entire process is prohibited by Descartes' argument that the cause must be at least as excellent as the effect.

Descartes speaks frequently about corporeal particles which are too small to be perceived, as the atomic theory insists.[27] Descartes' particles are also characterized by three qualities: size, shape and motion.[28] These happen to be the three properties allowed by Democritus and Leucippus to atoms.[29] Even in Descartes' account of sense perception, he ascribes the actual sensory experience to the single sense of touch in a way that mirrors the atomist theory and which rejects (without serious examination and debate) the Aristotelian theory of perception through a medium.[30]

Aside from the argument that God can divide particles of matter, Descartes really only has one other argument to separate him from the atomists of old. We cannot say that body is indivisible, Descartes says, because it is of the nature of body to be divisible. Yet what does Descartes mean by divisibility? Once again, he gives an answer that is quite empty. A human being, Descartes argues, can *imagine* that there are many divisions in body. Human beings will always possess this capacity to imagine divisions in body, Descartes argues; and therefore we cannot say that body is indivisible.

What does it have to do with the reality of body, if a human being should undertake to *imagine* something? Descartes is not anxious to be pinned down on the issue of body. It can be stated here that what a human being can imagine, as pertains to body, is irrelevant to one's doctrine of body. What matters is what one takes body to *be*. Descartes has a very definite account of what body must be taken to be. He calls it 'extension.' "By 'extension' we mean whatever has length, breadth and depth, leaving aside the questions as to whether it is a real body, or merely a space." "Let us now proceed to the sentence 'body possesses extension.' Here we understand the term extension to denote something other than body; yet we do not form two distinct ideas in our imagination, but just the single idea of extended body."[31] Extension is Descartes' definition of body. Is 'extension' divisible? In accordance with the Cartesian criteria for divisibility, which has not yet been adequately inspected, the answer is yes.[32]

Descartes' definition of body as extension reduces body to certain measurements. This is perfectly in accord with Descartes' attempt to achieve a mathematical physics. Mathematics is *about* measurement: of magnitudes, of numbers. Length, breadth and depth are measurements. This is Descartes' account of what extension is. "To each substance there belongs in principle one attribute; in the case of mind this is thought, and in the case of body it is extension."[33] However, this is not an adequate account of body. The essence of body cannot be 'extension.' For extension, measurable dimensions, cannot exist by itself. This takes us to Plato's *Parmenides,* and his investigation into the definition of body.

Parmenides defined body, or reality, as an indivisible unity. 'Being' has no divisions in it, Parmenides argues, because it is perfect. Plato argues, however, that unity cannot exist by itself. 'Unity' is a form, a nature, a part of reality. Yet it is indivisible. It cannot *include* 'being' as part of its own definition.[34] 'Being' or existence is a nature that is separate from that of unity. Being is a nature that is distinct. If 'unity' was to contain within itself, as a part of its own nature, the form of 'being,' then 'unity' would be *many.* The nature of unity cannot contain the nature of being, Plato argues.[35] What is unity itself cannot *be* many. If 'being' were derivative from unity somehow, then unity would be more than one. Yet this is not so.

Once it is conceded that unity and being are two coequal natures, then it is established: a body can only exist as a composite.[36] The unlike natures of unity and being are both necessary to a body. Body cannot exist without either of these natures. Fundamentally then, body is composed of unlike parts. This is something that the natural philosophy of the Enlightenment would rather not deal with. In the discussions of divisibility characteristic of Early Modern Philosophy, it is always assumed that body is somehow homogenous: and that the issue of divisibility or 'parts' in body can be reduced to cutting a body into pieces. This is not the case. The parts of 'unity' and 'being' are not themselves parts of magnitude. Infinite parts of unity and being exist in every body. Yet that does not interfere with the limits that are also characteristic of every body: coming into being and passing away.

To resume this explication of Plato's logic: any body whatsoever must be a whole of parts. 'Whole' and 'part' are therefore not mere contrivances or hypotheses of intellectuals. These are real patterns that exist in nature.[37]

Body is inseparable from the natures of whole and part. The issue of the whole is the next thing that Plato takes up. For in order to be a whole, a body must have *extremities*. Extremities include dimensions. This is where Descartes has located his own theory of body: in dimensions, as if the dimensions were the essence of body. Without the natures of 'unity' and 'being,' 'whole' and 'part,' there *is* no extension.

The proof that a body is a whole of parts that has extremities takes us to the very nerve of Plato's argument. A whole, in order to be a whole, must have limits. It must have a beginning, and it must have an end. In other words, in order to be a whole, a body must come into being and pass away. These are both forms of movement. Descartes insists that the only movement characteristic of nature is locomotion.[38] Not coincidentally, locomotion is the only movement that is conceded to the atoms in the classical theory.

In the case of the examination of time, for instance, we are once again met by Descartes' God who is able to perform any feat imaginable. The subject of our discussion must be a body. It is bodies that exist in time. Any body that you like, which is generated and which passes out of existence, must have the full complement of the parts of time. The object must have an existence in the present; it must have had some existence in the past; and it must have some existence to come in future time. Descartes, however, insists that we are free to divide time to infinity. We can isolate any moment that a body may be existing in, and from that vantage point, deny that the body has a past, or a future time. How can such a miracle be accomplished? By God's intervention.[39] God's supernatural power can cause the object to cease to exist at any moment, thereby depriving the body of a future time; just as Descartes' God can bring a body into existence at any moment, blurring any contemplation of a past time.

In the discussion of time, it is not hard to see Descartes attempting to implant the atomic postulates circuitously. If his fable is correct, in Descartes' view, we may speak of a body forever existing only in the present moment, and not possessing any parts of 'not-being.' The past exists in a certain sort of way, but it is not in existence *now*. Likewise Descartes believes that his fable enables him to deny the reality of future time. All of which would leave us with a body forever suspended in the present time, clutched tightly by God, with God alone determining its existence or its extinction. Yet if we remove the God fable, what we have is a changeless body that has no past and no future, which is exactly the atomist configuration.

One thing that can be observed is that the modern theories which employ a 'corpuscular' hypothesis have ceased to confess their dependence on the old Eleatic reasoning which is the true cause of the atomic theory. The modern corpuscular theories are folded into a new productive enterprise: the enterprise of causing effects, of producing 'utilities' in the effort to conquer nature. Descartes, as Bacon before him, can afford to be vague about his theory of the ultimate bodies in nature. The Early Modern philosophers do not want to present their theories of body in such a way as would lead scholars backwards in time towards the theoretical models of the ancients. For the critiques of those theories are also located in antiquity. As Richard Kennington has observed, Descartes can even afford not to have any final opinion about what the smallest or ultimate natural particles are like.[40] So long as the existence of the particles is stipulated in such a way as to remove authority from the ordinary perceptions, that is all the new utilitarian science appears to need from the atomist legacy.

Descartes is actually quite resolved to drive a wedge between the domains of intelligence and nature. Descartes feels determined to prove that the only intelligence in nature is contained in 'mind,' and that this in no way informs the nature of body. In his work *On Man*, Descartes asserts a radical distinction between body and mind. "First I must describe the body on its own," Descartes says, "then the soul, again on its own; and finally I must show how these two natures would have to be joined and united in order to constitute men who resemble us."[41] In his famous *Meditations*, Descartes undertakes to prove to us that we can only be certain of the existence of our mind, and that this does not require us to be able to prove that we even have bodies, or that any bodies exist at all.

It makes sense that this would be the object of Descartes' attack. In Descartes, as in Bacon, the thing that has disappeared is the *variety* of subject matter in the world. Descartes' method of reasoning ignores the differences between external objects, along with the diverse patterns and forms which characterize them. Looking to the universities, Descartes states that "Distinguishing the sciences by the differences in their objects, they think that each science should be studied separately." For this Descartes has no tolerance. "But here they are surely mistaken. For the sciences as a whole are nothing other than human wisdom, which always remains one and the same, however different the subjects to which it is applied, it being no more altered by them than sunlight is by the variety of the things it shines on."[42] The actual forms and patterns that characterize the real objects in the world have principles of movement inherent in themselves which are radically distinct. This is to say, that they have a condition of perfection or excellence which they will strive to achieve, unless they are otherwise impeded. It can come as no surprise that Descartes wishes to suppress the differences between the objects of study as separate sciences. His 'method' does certainly tend to erase from any and all subject matters the forms and qualities characteristic of them.

The rigid separation between soul and body that Descartes enforces is interesting on several levels. First, it is not clear why anyone would want to think of himself as a disembodied soul; or to think that the materiality in which his life is situated, can be coherently excluded from the attributes of soul which consist in virtue and vice, and thereby our prospects for happiness. Not many people, if any, labor for the knowledge of their mere existence. To labor after a knowledge of existence which so rigidly segregates the soul from the body, furthermore, does not merely delude the soul into thinking that its actions are somehow irrelevant to matter; or to the falsehood that our embodied choices and actions do not bear, most directly, on our happiness. The rigid separation between soul and body needs to be contemplated in terms of what it implies about body: that is, that body has no 'soul.' Another way of articulating Descartes' position is to indicate that body has no *form.*

Ah yes, form. That which Plato maintains is everywhere wrought into the fabric of material things; that which Plato and Aristotle both distinguish as the intelligible aspect of reality, of formed matter. Objects to Plato and Aristotle are forms, patterns, or kinds: men, trees, stones; but also courage, justice, virtue. One century after Descartes, David Hume will argue that there is nothing morally foul in the actual actions of a murderer; that the malevolence of such a deed, can only arise within our own breasts, as our own productions and subjective preferences, for it is not allowed to inhere in the deed that was itself done. In Hume's teaching, we cannot know the objects external to us with any kind of certainty, where matters of fact and truth are concerned. All we can know is our own *feeling*. Thus the 'ought' cannot be derived from the 'is.' For Plato and Aristotle, however, the common names indicate patterns inherent in the perceived actions and bodies themselves. Common names indicate 'universals' for Plato and Aristotle, that is, patterns that are distributed to innumerable perishable objects.

Not surprisingly, in the Cartesian mantra, 'universals' are categorized as mere fictions created by the mind. Since we first of all perceive forms, in the Platonic canon, scientific knowledge must build upon the information obtained by the senses. In Descartes, sense perception is relegated to the defamatory category of 'imagination.' 'Understanding,' the home of the intellect for Descartes, abruptly turns its back on perceptual evidence. This not only foreshadows an antagonism between public scientific authority and ordinary intelligence. It threatens to suffocate intellect, to cut off its access to reality. It threatens to make us dumb.

It is thus not necessarily appropriate to laud Descartes for his 'mathematization of reality.' Descartes, when he banishes from the essence of body all but allegedly innate mathematical information ('extension'), can be seen to repudiate the forms otherwise conveyed to us by our perceptual experience. To dismiss perceptual evidence to the precinct of 'imagination' is to impose upon ordinary perception the onus of brute insensibility; whereas in the Peripatetic philosophy that Descartes so loathes, the insight packed into perception and memory is respected as the most preliminary criteria of intelligence. The severe construction of Cartesian conceptions of body is directly tied to his atomistic definition of what perception is. Descartes' bodies are said to pursue, not fulfillment, but 'preservation.'[43] Rest, in Descartes' physics, can only be discussed in terms of motion that is being forcibly restrained. We are not to be allowed to think about natural objects as possessing any trajectory of motion of their own. For related reasons, Descartes frankly denies the distinction between movements in natural bodies that are in accordance with the nature of said bodies, and those which seek to force a body against its natural principles of movement, or violent motion. "You ask whether I think that water is in its natural state when it is liquid or when it is frozen. I reply that I do not regard anything in nature as violent, except in relation to the human intellect, which calls 'violent' anything which is not in accordance with its will, or not in accordance with what it judges ought to be the case."[44]

DESCARTES ON PERCEPTION

The political dimension of Descartes' work is powerfully on display in his presentation of theories of perception. A very sharp dichotomy is opened up between the perception that is known by the commonality of the human race, and those who are initiated into Descartes' philosophy. Descartes chooses to employ the term 'perception' for his own theory, so as to blur the differences that he enacts from what is ordinarily understood by the term. Yet the chasm is gaping.[45]

It should be noted, too, that Descartes correctly recognizes Aristotle as a representative of the ordinary notion of perception. Descartes pointedly avoids a direct analysis of Aristotle's theory of perception. The implications of Descartes' new theory of perception can be documented rather easily by referring to Descartes' conception of a body. It is certain that none except those initiated into his theory, or some variety of atomism, would recognize what they perceive in Descartes' conception of the term.

Descartes' indictment of sense perception occurs in his earliest works and in his latest works. In the earlier works, Descartes' indictment of the sense perceptions does not appear to come from skeptical postulates at all, but from the quasi-atomist theory that he himself has chosen to work with. "For although everyone is commonly convinced that the ideas we have in our mind are wholly similar to the objects from which they proceed," Descartes writes, "nevertheless I cannot see any reason which assures us that this is so."[46] Aspects of bodies which the ordinary thinkers take to be quite real, such as hot and cold, moist and dry, are in Descartes' argument rejected as things that cannot really be known. Descartes insists that these qualities must be effects from more subtle, invisible causes, that is, from minute bodies that cannot be perceived. "If you find it strange that in explaining these elements I do not use the qualities called 'heat,' 'cold,' 'moisture' and 'dryness'—as the philosophers do—I shall say to you that these qualities themselves seem to me to need explanation. Indeed, unless I am mistaken, not only these four qualities but all the others as well, including even the forms of inanimate bodies, can be explained without the need to suppose anything in their matter other than the motion, size, shape and arrangement of its parts."[47]

In the late *Principles of Philosophy*, which Descartes intended to serve as a college textbook, he is making arguments to the effect that in our childhood, we got into the habit of allowing our minds to be infected by our bodies. Color, shape, hardness, smell, and all other qualities are not really in the bodies external to us, Descartes insists. "And when nothing very beneficial or harmful was happening to the body, the mind had various sensations corresponding to the different areas where, and ways in which, the body was being stimulated, namely what we call the sensations of tastes, smells, sounds, heat, cold, light, colors and so on—sensations which do not represent anything located outside our thought." "Sensory perception does not show us what really exists in things, but merely shows us what is beneficial or harmful to man's composite nature."[48]

According to Descartes, we must not either think that the bodies external to us can be made known to us by our senses; only intellectual perception or intuition is allowed to count. What our perceptions are, in the neo-atomist language, is motions in our own bodies, which are either pleasurable or not pleasurable to us. This almost exactly replicates the Epicurean argument. Descartes says that there is no 'compelling reason' for us to believe that our perceptions convey information about the external objects that actually exist. Supposedly, since we have not yet proved the existence of the external bodies, and only have the proof of the existence of our minds, we need not even engage the question at this point. Yet there is more of the God fable from Descartes in book two of the *Principles*.

In *The World*, Descartes makes it very plain to us that he has never found himself fighting against skeptical postulates. Gaukroger is correct. In *the World*, Descartes is determined to indicate that there is a difference between our sensations and the objects that we experience. He indicates that when we utter names, these names do not resemble the objects that they refer to (Plato thought that names try to do exactly this, imitate the natures of the objects to which they refer).[49] How can a name bring before our mind an object that is not resembling the name?[50] The name is an instrument of memory, we should reply to Descartes. Moreover, the names in question are names of *forms*. For all common names indicate a form. In the memory, as Plato argues in his *Theaetetus*, there is a memorial seal of the kind of object with the name.[51] It is so effortless for human beings to recall the forms as they come across them that they indicate to one another without duress the huge variety of forms that they encounter in everyday life. Everybody knows to which forms the names refer: water, house, quickness, courage, justice, cow, heat, cold, so on and so forth. But Descartes is attempting to make the argument that the name is unlike the form. If the name does not imitate the form, then why does it make us think of the form? Descartes wants to make the argument that our sense perception, in a like manner, need not give us a true indication of what the external objects are. Both things, names and sense perceptions, could just be referred to as 'signs' of some sort. This is the kind of theory that Descartes would like to work with.

What the name brings before our mind, however, is a certain object in our memory. That object is a form. It is the forms in the world that are in need of names; names are instruments to enable us to communicate to one another about those forms and the bodies in which they are enmattered. If our sense perceptions did not reach to the forms or natures of the external objects, names would be impossible for us to formulate.

Indeed, in Aristotle's view, there is no such thing as imageless thought. Descartes would deny this proposition. Yet the fact of the matter is that Descartes certainly does have images which guide his thoughts. They are simply not shared images, that is, images shared with the audience for his science. Again, Descartes thinks that he knows better for the human race from the vantage point of 'generosity' and allegedly irrefutable knowledge. The images which lead his thought are not directly shared with the public; but they are certainly there.

Aristotle's theory of form is a bit more complicated than that of Plato. For Aristotle, a form must have some work to do: that work is to order matter, so that the matter follows a certain schedule of development (the basis for a distinction between natural and violent motion). Aristotle calls the form a 'formula' in his theory of causes, but he is as one with Plato in affirming that the ordinary use of names is intelligent: it is the form of an object that is the basis of our perception, not the matter of the object that we perceive.[52] This is precisely the argument that Descartes is determined to reverse. By adopting a quasi-atomistic theory of what perception is, that is, invisible corpuscles literally colliding with our own sense organs, the evidentiary value of the ordinary perception is suspended; but the way is opened up for the Cartesian theory of 'perception,' with its alien notions of body.

Descartes deploys more rhetoric in his claim that Aristotle is in agreement with him "and all the other philosophers" to the effect that perception is a species of 'touch.' The contact that is part of the Aristotelian perception is however not like the touch that is characteristic of the atomistic Cartesian theory. For Aristotle, perception occurs through a medium: the object affects the medium (usually air), and the air conveys the substantial image of the object to the sense organ.[53] In Cartesian theory, imperceptible bodies are said to assail the sense organs directly, in which case no image is conveyed, and no form is communicated.

Descartes' early work with Isaac Beeckman, as Gaukroger has pointed out, supplied him with a very typical atomistic conception of the sensory organs.[54] The sensory organs in Descartes' theory are entirely bodily. This is at variance with the arguments made by both Plato and Aristotle. For Plato and Aristotle both, perception is a function of the soul, not the body. As Plato describes it in the *Theaetetus* and in his *Phaedo,* perception is a case where the soul *uses* the sense faculties to acquire information about the objects in the external world.[55] Perception, in Plato and Aristotle both, recognizes form, that is, *the intelligible*. It is in Descartes' theory where perception is reduced to an entirely gross type of touch, an entirely bodily process, which is said to deliver no information about external reality whatsoever. "But if I may express myself more freely, I will not hide the fact that I am convinced that what affects our senses is simply and solely the surface that constitutes the limit of the dimension of the body which is perceived by the senses," Descartes argues. "For contact with an object takes place only on the surface, and nothing can have an effect on any of our senses except through contact, as not just I but all philosophers, including even Aristotle, maintain."[56]

Gaukroger makes the argument, and it is not lacking in persuasiveness, that Descartes' later work in the *Meditations* seeks for a less controversial means to introduce the atomist embargo upon perceptual knowledge.[57] In the *Meditations,* Descartes goes to great lengths to appear not to be in direct conflict with Aristotle's teachings. There, Descartes simply argues that all thoughts that have been infected with the bodily influence are inimical to true reason; and that the search for true reason, for absolute certainty, must commence with the resolution that thought itself must be stripped clean of every influence from the sense faculties.

Descartes examines the perceptions as so many physiological processes. He is always tracing out nerves, and the flow of blood, and he talks about the objects of perception 'striking' the bodily organs. This is hardly a new argument. That there are physiological processes that enable and correspond with perceptual experience is not at issue: what is at issue is Descartes' determination to exclude the familiar, form-based ordinary recognition of objects from the explication of what sense perception entails. What he is trying to do away with is the forms, that is, the intelligible aspects of objects which is what people know them (the objects) by. It is noted that Descartes uses the terms 'common sense' and 'imagination' very loosely, and situates them very closely to one another.

Descartes' theory of sense perception is manifest in the way that he conflates the terms of sense perception and 'imagination.' Descartes does not formally characterize the relationship between these two domains, except to make it very difficult to tell the difference between them in his descriptions. It needs to be pointed out that for Aristotle, sense perception and imagination are radically distinct things.

> For imagination is different from either perceiving or discursive thinking,
> Though it is not found without sensation, or judgment without it. That this activity
> Is not the same kind of thinking as judgment is obvious. For imagining lies within
> Our own power whenever we wish (e.g. we can call up a picture, as in the practice
> Of mnemonics by the use of mental images), but in forming opinions we are not
> Free: we cannot escape the alternative of truth or falsehood. Further, when we
> Think something to be fearful or threatening, emotion is immediately produced,
> And so too with what is encouraging; but when we merely imagine we remain as
> Unaffected as persons who are looking at a painting or some dreadful or
> Encouraging new scene . . .[58]

Descartes, because he is resolved that sense perception itself must be discredited, establishes a firm conceptual link in his argument between sense perception and imagination. This is as much as to defame sense perception through a false association. In the Early Modern Period, this association is made with ferocity and frequency, and it is a piece of intellectual territory that we need to return to, evaluate, and re-argue. Descartes says that:

> the intellect can either be stimulated by the imagination or act upon it. Likewise the imagination can act upon the senses through the motive force, by directing them to objects, while the senses in their turn can act upon the imagination, by depicting the images of bodies upon it. But memory is no different from imagination—at least the memory which is corporeal and similar to the ones that animals possess. So we can conclude with certainty that when the intellect is concerned with matters in which there is nothing corporeal or similar to the corporeal, it cannot receive any help from those faculties; on the contrary, if it is not to be hampered by them, the senses must be kept back and the imagination must, as far as possible, be divested of every distinct impression.[59]

Descartes affirms, in the *Meditations*, that he wants to keep as far from error as possible. Yet he offers a rather bizarre analytic. Descartes insists that he must abstract himself most deliberately from the objects of sense perception, because these are wrapped up in matter, and are not purified of involvement with it. This is why Descartes insists that his metaphysics or first philosophy acquaints him with nothing but his own incorporeal mind. The incorporeal mind has nothing to do with the body, according to Descartes.[60] Descartes alleges that the incorporeal mind is infinitely more clear and distinct than our ideas of any body can be.[61]

CLEAR AND DISTINCT IDEAS

The domain of the clear and distinct ideas is the core of Descartes' sophisticated teaching. This is not presented as a fable. Nor does God simply bestow this knowledge upon individuals, in Descartes' account. To the contrary: the mind, in order to approach clear and distinct ideas, must first of all purge itself of everything it is capable of doubting. This includes the existence of God. The existence of God is something that stands in need of proof, the proof that can only come from an absolute certitude, the content of the clear and distinct ideas.[62]

Descartes is very emphatic about how his clear and distinct ideas must be conceived of. They must be conceived of as entirely independent of sense perception. This is the very source of their authority. By virtue of the clear and distinct ideas, science ascends above sense perception in authority. In both Plato and Aristotle, intellect depends upon the sense perceptions for its initial evidence of what exists. But Descartes' question does not concern what exists. It concerns that which he can be absolutely certain of and which therefore does not depend on anything except his own mind. He is looking for a point of leverage intellectually, from which his arguments cannot be resisted.[63] Once Descartes has achieved a single clear and distinct cognition, an absolute certainty upon which he can rest all of his psychological weight—and from which all dependence on sense perception has been forbidden—then from this single Archimedean point, absolutely all other knowledge can allegedly be deduced.

Descartes insists that the mind and the body are distinct substances. It has been indicated above that Descartes is determined to view body as bereft of intelligence (form), as a means to convert the notion of body into something that cannot really be known for itself. When it is established that body cannot really be known with certainty even to exist, as Descartes maintains, then this makes it possible for the mind to theorize body in a way fitted to dominate it. The theory of body employed by Descartes for a prototype is not recognizable.

Descartes' conception of the clear and distinct ideas is evident in his very early writings. "We should attend only to those objects of which our minds seem capable of having certain and indubitable ideas," Descartes writes.[64] 'Intuition' is a name that Descartes used, early on, to denote the 'clear and distinct ideas.' 'Intuition,' in this sense, indicates cognitive operations that do not have, allegedly, any dependence on sense perceptions. "By 'intuition' I do not mean the fluctuating testimony of the senses or the deceptive judgment of the imagination as it botches things together, but the conception of a clear and attentive mind, which is so easy and distinct that there can be no room for doubt about what we are understanding. Alternatively, and this comes to the same thing, intuition is the indubitable conception of a clear and attentive mind which proceeds solely from the light of reason."[65]

The clear and distinct ideas cannot have any relationship to body, for Descartes. They must not be linked to the patterns that exist in the actual perishable bodies, for this would reveal the existence of the bodies in their own right and with their own natures. The clear and distinct ideas, then, must emerge as if out of the mind itself. They must be sui generis: they must be ascribed to something called 'logic' or to the 'innateness' of the mind. Above all the clear and distinct ideas must be characterized, for Descartes, as ideas that the mind can grasp entirely without any aid or dependence on sense perception; for in this arises their independence from body, and subsequently their superiority to it.[66]

Descartes refers to his clear and distinct idea, at least in its most famous manifestation, as a *proposition.* Something is known, and from that something another thing is inferred or deduced. "I think, therefore I am." In his earliest writings Descartes refers to this formulation of a clear and distinct idea; except that in the early writings, before the *Meditations*, Descartes directs us to the "I think, I exist" formulation as simply the *easiest* and most convenient, clear, and distinct idea. He recommends it to us because it is obviously easy to access and to test. Yet later on in the *Meditations*, when Descartes seems to introduce for the first time the appearance of a great threat from skeptical teachings, the pursuit of the "I think, therefore I exist" formulation is made to seem infinitely more complex and mysterious.

Descartes, in his writings, makes the occasional reference to Plato's Socrates and his famous profession of ignorance. Descartes misrepresents it, however. Socrates' profession of ignorance is meant to be a litmus test for philosophers: it is designed to evoke the requisite humility from interlocutors at the outset of an investigation. According to the profession of ignorance, the philosopher must agree that he does not begin his enterprise knowing anything more than the ordinary folk around him. This is Plato's position. Descartes refuses to accept this position. For he is interested in something else. He is interested in certainty, for the sake of certainty, as a property of knowledge which he shares with no one else. He insists that the mind turn towards, not the objects of the common world, such as the sense perceptions open up to us; but rather inward upon itself. "So the difference between this mode of thinking and pure understanding may simply be this," Descartes writes: "when the mind understands, it in some way turns towards itself and inspects one of the ideas which are within it; but when it imagines, it turns towards the body and looks at something in the body which conforms to an idea understood by the mind or perceived by the senses."[67] Choose the simplest possible thing that can be known, Descartes tells us, such that this knowledge cannot be doubted or gotten wrong.

The thing to which Descartes urges us to turn is the fact that we are thinking, that we exist. Descartes again, in *The Rules for the Direction of the Mind*, one of his earliest writings, indicates no role at all for the skeptical postulates that he will treat with such high status in his *Meditations* more than a decade later. But note also, that Descartes is looking for something that nobody can possibly deny, this being the common ground. In short, Descartes is making his own mind into the object of knowledge. This is not done through perception. Descartes insists that it is intellect which 'perceives,' not the senses. The mind is certain that it is thinking, and it is certain that it exists because it is thinking. Nobody can take this away from Descartes. All will have to agree to it, and this is certitude: and it is only simple things like this, irreducible things like this, that Descartes will allow to qualify as the universals and simples of thought. Only things upon which there is no possibility of disagreement, and where no possible discord can ensue, will be allowed to count as clear and distinct ideas.

> I will now shut my eyes, stop my ears, and withdraw all my senses. I will
> Eliminate from my thought all images of bodily things, or rather, since this is
> Hardly possible, I will regard all such images as vacuous, false, and worthless.
> I will converse with myself and scrutinize myself more deeply, and in this way
> I will attempt to achieve, little by little, a more intimate knowledge of myself.
> I am a thing that thinks.[68]

Descartes has taken a very abstract and eccentric object for his original knowledge. Descartes allows that the awareness he points to is the easiest, and seems silly for even enumerating because it is the easiest. It will indeed be obvious to everyone that they both think and exist; they had never doubted it. But the point is that Descartes' principles of clear and distinct ideas, that is, that he thinks and therefore that he exists, qualify as these universal and 'simple' things, out of which everything else more complicated can allegedly be obtained.[69]

It is absolutely imperative for Descartes' argument that the 'I think' not be construed in any way as a product of what is ordinarily called perception. This cannot mingle with evidence obtained by the senses. Memory, infected by imagination, is ruled out. The clear and distinct ideas, of which the 'I think' is the first, must belong to a different realm than that of sense perception, in Descartes' argument. For this is where the source of authority, including political and moral authority, must have its origin: not in something that is shared with others through common evidence, but which each individual who is going to be so initiated must undergo alone without common evidence.

After all, in the Cogito argument, what the individual is certain of is *his* thought. He does *not* possess this certainty about the existence of his body. He will never possess this certainty about the existence of body per se. Mind and body are to be considered as separate substances, in Descartes, due to the fact, and only due to the fact, that he is capable of separating the two things in thought. It has been illustrated above how it is not really possible to separate body from form, (i.e., matter from form), in Aristotle's and Aquinas's proofs. Yet in Aristotle and Aquinas, the aim is truth. In Descartes, the aim is being impregnable somehow; it is a personal condition of mind that he is after, as well as a pivot of authority and persuasion. "And my purpose is not to explain, as they do, the things which are in fact in the real world,"

Descartes writes, "but only to make up, as I please, a world in which there is nothing that the dullest minds are incapable of conceiving, and which nevertheless could be created exactly as I have imagined it."[70]

In the *Principles*, he makes a brief reference to things that the mind must of course know in order to be able to undertake the propositional thought process of "I think, therefore I exist." One must know what 'existence' means, Descartes allows; one must know what 'think' means. Descartes, however, indicates that it is not necessary for us to know anything about the existence of the things that these names indicate, and so he is not obliged to take the matter up seriously.[71] In this Descartes is surely wrong. His dependence on ordinary language, as antecedent to the thought process that he wishes to enshrine as 'innate' to the mind, does indeed oblige his entire theory to the sense perceptions which are inextricable from the matrix of language in which the distinct objects obtain their names. If memory is really confounded with imagination, as Descartes has insisted, then he cannot access the simplest word without tumbling his entire edifice of philosophy into a jumbled heap of unreliable nonsense.

Note now, that the *implication* of Descartes' claim here is that perception, through the senses, is not the medium through which the mind is enabled to think, and to know itself. How can the mind know itself absent its possession of a language, through which *it can* think? Descartes gets very testy about questions like this, but such things cannot be allowed to interfere with our investigation. It is clearly true, that the human being is not born self-aware. It takes years of life for the human infant to discriminate between its own wants and needs and the external world. Just as at a certain stage of life every man and woman is to the child 'momma' and 'poppa,' somewhat earlier the entire world is merely an extension of the child's needs. The infant recognizes no distinction between its needs and the external world which must meet those needs. It takes a while for the child to begin to recognize the difference between himself and other objects. It is only the emotional and psychological development which his learning of language makes possible, that enables his capacity of communication, even when this communication is merely between himself and his thought. All of these developmental steps must be accomplished before it becomes possible, down the road, to undertake such an abstract proposition as "I think, therefore I am."

Descartes speaks prematurely when he says that his 'intellectual' perception or intuition, that he both thinks and therefore exists, is not dependent on *sense* perception. These thoughts have arisen in no other way. Descartes could not have an 'I' nor conceive of an 'I' if he did not know the distinction between himself and the other objects in the world; and he could not have learned the signification of the name existence, except and in proportion as he has experience of non-existence, that is, of things deteriorating or dying or having died.[72]

These facts threaten to subvert Descartes' model of 'innate' ideas.[73] For in Descartes' model, the pursuit of the clear and distinct ideas, allegedly purified of dependence on sense perception, enable the mind to arrive at its own isla bonita; at a paradise where the existence of other people, or even of one's own body, are external to the mind's true or necessary knowledge. One's own body and the other bodies all fall merely into the realm of the probable: and this does indeed point up the infantile aspect of Descartes' philosophy, the furious egotism, which becomes more clear in his discussion of the passions.

The blades of Descartes' clear and distinct ideas exclude from the domain of original knowledge everything except knowledge of self-consciousness. Nor will Descartes ever tolerate the emergence of a domain of common evidence to which his own mind, along with the others, must submit as a premise of making investigations. It is thus the communal basis of evidence that Descartes' model is striking at: and in so doing, it is a *political* philosophy.

Descartes loosely identifies the "I think, therefore I exist" proposition as the cradle for the rest of the mind's clear and distinct ideas. For the other pivotal names that he adduces as evidence of clear and distinct ideas, they are simply names that cannot be analyzed or divided into smaller elements. When Descartes talks about body, he will employ the names of size, shape, and movement. He will not employ the terms of unity and being. Descartes reduces the category of body to the hypothetical, by denying that it can be certainly known to exist—since clear and distinct ideas, the cradle of evidence, arise from the isolated consciousness which to this degree is said to be independent of the body. Yet this does not appear to be true.

The instruments with which the mind thinks, in Descartes' model, have cast off their moorings from the domain of common evidence, perception of what the bodies and beings actually are. Whether the issue concerns number, or duration, or order of any kind, these terms to Descartes simply indicate the attitude that the mind prefers to take as regards these objects. The doctrine of the clear and distinct ideas is the mind's proclamation of its own power, of being effectively self-referential; that is, thought is now to be determined in accordance with the rules that the mind is subjected to by Descartes' model.[74] All of the categories through which body has traditionally been evaluated are now reduced to mere modes of the mind, that is, 'attitudes' or postures of the mind through which it contemplates the object in question.

DESCARTES' CLAIM TO MORAL AUTHORITY

One needs to bear in mind that when Descartes starts to make arguments about the nature of the soul, about the nature of the passions, and about the nature of virtues, he is not in his view obligated to begin with the rest of us and a common domain of evidence. The soul, in Descartes' view, has *no parts*.[75] It is possible to begin to feel the authoritarian dimension of Descartes' attitude towards disagreement with a bit more profundity in this area. Even Descartes himself finds a need to employ different names when he discusses the powers of the soul. Yet his peculiar mathematics nevertheless collapses the different names into a unity. Descartes' machinations become a bit plainer when he briefly takes up the Aristotelian theory of the soul to illustrate his own argument.

For Aristotle, appetite and intellect are opposite domains. The intellect strives for the truth of 'what is,' whereas the appetites concern mental powers which involve seizing or driving away, self-referential calculations of pleasure or pain. In Aristotle's view, the appetites are properly the slaves of the intellect: truth is the proper master of the desires and aversions.[76] Descartes, however, erases the distinction between the two. Descartes' representation of the true essence of sense perception plays a role in this discussion.

For Plato and Aristotle, sense perception is the opposite of imagination. For Plato and Aristotle, sense perception concerns the truth of what an object is, and the function of the perceptual organs, at the service of the soul, is to convey true information about those objects. For Descartes, as we have seen, the clear and distinct ideas have wholly repudiated their dependence on sense perceptions. Truth is the province of what the soul can take out of itself, via its 'innate ideas.' What then is the purpose of sense perceptions, and how does Descartes propose to construe their function?

In fact, Descartes has adopted the Epicurean view of sense perceptions. The sense perceptions are fitted to do one thing for Descartes, and one thing only: to determine what is pleasant or good for the body, and to determine what is bad or evil for the body. "The function of all the passions consists solely in this, that they dispose our soul to want the things nature deems useful for us, and to persist in this volition," Descartes writes.[77] In other words, Descartes has subordinated sense perception to the appetites. It is no wonder, then, that Descartes proposes to abolish the distinction between the intellect and the appetites that Plato and Aristotle theorized. "From what has been said it is easy to recognize that the utility of all the passions consists simply in the fact that they strengthen and prolong thoughts in the soul which it is good for the soul to preserve and which otherwise might easily be erased from it."[78]

Once the doctrine of clear and distinct ideas has been formulated in Descartes' theory, the radical estrangement of the mind from the body has come to its end. Now, supplied with superior authority independent of the sense perceptions or 'what is' type knowledge, Descartes unveils the mind as the proposed master of the passions.

For Aristotle, the human being's feelings are deeply involved in the very assertion of a claim to justice. The emotions are therefore among the most important things to educate and train in Aristotelian moral theory. In that theory, character is the decisive thing. In what does character consist, in the Aristotelian view? Suppressing appetites, perceiving facts of the situation, and preparing an action that is neither too little nor too much to properly respond to those facts. Indeed, to advance one's own claim to justice while respecting the like claim to justice among other citizens is the political education of ultimate consequence for Aristotle; but this all goes down the drain in Descartes' new disposition towards the passions. Here the intellect strives to repress and deny the interpersonal origins of the passions that are evoked, and to categorize them instead as mere epiphenomena of biology and the circuitry of a mechanistic body.

In two ways, then, Descartes is asserting moral authority for the way that his allegedly infallible science of clear and distinct ideas proposes to construe and culturally condition the souls of the populace. All perceptions will be regarded as actions of self-interest, by Descartes, as the body's defense of its own. Secondly, Descartes proposes that it is the intellect's function to deny or suppress any recognition that the passions themselves are interpersonal or political in origin. In *The Passions of the Soul,* on page after page, Descartes reduces human passions to physiological processes, as if their intelligible context were nothing but a dream.

Descartes often likes to affect a pose towards the rest of the human race, such that he only wants to look after himself, engage in intellectual pursuits for his own private interest, and therefore would not like to meddle with the lives of the others. This pose is more than offset by the pretentious claims that Descartes makes in his attitude towards the accumulated knowledge of philosophy. "On the other hand, I compared the moral writings of the ancient pagans to very proud and magnificent palaces built only on sand and mud. They extol the virtues, and make them appear more estimable than anything else in the world," Descartes writes, "but they do not adequately explain how to recognize a virtue, and often what they call by this fine name is nothing but a case of callousness, or vanity, or desperation, or parricide."[79] The power to determine the signification of names is a mighty power indeed.

Descartes has admitted, in between those moments where he affects to be doing nothing more than personal investigations, that he seeks to establish a new foundation for the sciences. It becomes clearer and clearer that the foundations of the sciences are slotted for transformation, by Descartes, in order that he may prescribe new moral directions for the culture. Those who adhere to the moral virtues taught by Plato and Aristotle, namely justice, courage, wisdom, and temperance—are in the way of the new utilitarian ethos.

Bacon has made the point that it is human affections that stand in the way of reason, or the new conception of reason; that the utility which will be for the sake of the human race can only be approached in proportion as the affections and moral emotions of the people are diverted, or stalemated. Descartes' definition of the virtues appears to follow in league with that of Bacon. Which brings up the question: are not the relationships that people have with one another the ultimate basis of what is truly useful to them? In other words, is it not true that it is relationships which people value most; and that therefore utility ought to march in step with those public measures which strengthen, and magnify the importance of these emotions and attachments?

When Descartes defines the utility of perception in terms of pleasure or pain for the individual body, this means to act as a corrective to ordinary perception. Ordinary perception believes that it perceives mutual facts. The new science intends to educate people to believe that their perceptions can teach them about nothing more than their own preferences. Since this issue bears so heavily on human happiness, however considered, it seems that the truth of Descartes' doctrine of perception is indeed of paramount importance.

'Virtue' itself, for Descartes, is infinitely private. It can only concern that over which the individual has absolute and solitary control: anything that the individual does not possess absolute and solitary control over, in Descartes' view, cannot be said to be linked to his free will; and therefore virtue can only apply where these conditions hold. "We must, then, utterly reject the common opinion that there is a Fortune outside us which causes things to happen or not to happen, just as it pleases. And we must recognize that everything is guided by divine Providence, whose eternal decree is infallible and immutable to such an extent that, except for matters it has determined to be dependent on our free will, we must consider everything that affects us to occur of necessity and as it were by fate, so that it would be wrong for us to desire things to happen in any other way."[80]

Let us recur to Aristotle's formulation of virtue. The very feelings and actions which are called for, in Aristotle's estimation, are summoned from us by the situation which we encounter. The facts of the case, always interpersonal, are the determinants of the proper feeling, the proper degree of feeling, and the end which ought to be brought about in that situation by the moral actor.[81] Aristotle's entire conception of virtue and vice, which professes to be, and indeed in large measure is, but an explication of the functioning of ordinary praise and blame, is all ruled out of Descartes' theory of 'virtue' from the outset.

The goal of 'virtue' in Descartes' theory is always the same: tranquility, contentment of the soul. In other words, the self is the constant object of Descartes' conception of virtue. This certainly squares with his *prima philosophia* of the clear and distinct ideas, which after all only leave a man all locked up in his own mind. Nor however is this all. Where the individual finds himself in circumstances where he is not in absolute and solitary control of the outcome, Descartes insists to us that we must regard, not fortune, but irresistible or absolute fate as the cause of whatever happens. To suppose otherwise would after all disturb Descartes' tranquility of mind.

As for 'generosity,' Descartes' conception of virtue par excellence, what is it? It is a condition of aloofness from the rest of the human race; a position from which one has rendered oneself closed off and immune to any emotions which either their actions or their sufferings could otherwise evoke from us. It is, one could say, an idealization of a false superiority, a conceit.[82] Descartes' virtue is to love himself.

It is quite clear that Descartes regards the entirety of the ancient philosophy as a pretentious failure, and this must include the four classical virtues recognized by Plato and Aristotle: justice, temperance, courage, and wisdom. Wisdom is the only one of the four virtues that Descartes talks very much about. Wisdom, in the philosophy of Plato, cannot exist except in company with the other three virtues. In Descartes, however, wisdom seems to qualify as an all-encompassing virtue; and wisdom in Descartes' philosophy is said to be the absolute certainty of knowledge that his clear and distinct ideas represent. All other knowledge for Descartes is derived from his clear and distinct ideas.

Descartes begins to talk about his doctrines of utility and clear and distinct ideas as a new criteria for what shall count as *civilized*. To really be virtuous, in Descartes' analysis, indicates that one must climb on board the new practical philosophy of invention, the one first formulated by Bacon. "Through this philosophy we could know the power and action of fire, water, air, the stars, the heavens," Descartes promises, "and we could use this knowledge—as the artisans use theirs—for all the purposes for which it is appropriate, and thus make ourselves, as it were, the lords and masters of nature."[83] The harsh indictments that Descartes makes against Aristotle and Plato begin to point up the older civilization as a barbarous one, especially in view of the virtues that they understood and prized. "This would oblige all those who desire the general well being of mankind—that is, all who are really virtuous, not virtuous only in appearance or merely in repute, both to communicate to me the observations they have already made and to assist me in seeking those which remain to be made."[84]

In his *Passions of the Soul*, Descartes takes a further step in the direction that Machiavelli beckoned in chapter 15 of his *Prince*. In chapter 15 of the *Prince*, Machiavelli asserted that 'what appears to be virtue' must be evaluated and judged in accordance strictly with its effects on the fortunes of the individual who is the moral actor.[85] There, Machiavelli substitutes the old sophist conception of pleasure and pain for the criteria of goodness and badness which Plato and Aristotle distilled, not from antique deities principally, but from the *substance* of ordinary praise and blame.

In Descartes' long discussion of the human passions, both those passions said to be of the body and of those passions said to be of the soul, what is obliterated is the interpersonal nature of morality itself. The obliteration of the interpersonal nature of morality could never advance, in western philosophy, absent the revival of the Eleatic based physics which attempts to give the lie to the perceptions that the human race obtains through its unaided faculties. Descartes' very conception of virtue and truth is consecrated for a moral isolate: and in truth, there is very little difference between the passions of the soul and the passions of the body that Descartes details. If we remember, it is Descartes who has decided that the soul itself has no parts; that the classical distinction between appetite and intellect is but a trope to be swept away in the new scheme of classifications.

Just consider what becomes of the object to be indicated by the name of 'love.' For Descartes, this name merely indicates some variety of pleasure. Ordinary opinion is always attempting to sanctify that object which is designated by this name; but in Descartes, we are informed that we must use the name love for any strong desire whatsoever. "Nor do we need to distinguish as many kinds of love as there are different possible objects of love. Consider, for example, the passions which an ambitious man has for glory, a miser for money, a drunkard for wine, a brutish man for a woman he wants to violate, an honorable man for his friend or mistress, and a good father for his children. Although very different from one another," Descartes argues, "these passions are similar in so far as they all partake of love."[86]

When the public signification of the name 'love' becomes disseminated after Descartes' fashion, it is certain that the ordinary meaning of the name will be scattered to the winds. When love is the name used to identify lust, ambition, or inebriation, it is going to confuse people. The new theory of nature is still only nascent in Descartes' philosophy, but the estrangement he enacts between the ordinary use of words and external reality is prescient. These are the early footsteps which lead to the emergence of the world as meaningless; the world where science redefines all human perception as the epiphenomena of bodies and movements that they have no access to in their experience. These are the early footsteps, at least, for the popular culture, in which it is fated to become lost.

Descartes' conception of the deity, for that matter, resembles nothing so much as the divinity of Epicurus and Lucretius. What are the obligations of this divinity? What are the chief characteristics of this deity? To turn a cold indifference upon the world of human suffering. Descartes himself, for all his professed generosity, is not moved to action by the sight of suffering and injustice. To the degree that the object of his observation is even minutely independent of his own personal will and control, Descartes insists that we regard such facts as absolute fate, as irresistible divine providence. Why must this be so? It must be so in order that we will not be bothered by these scenes, that is, in order that we may retreat to private citadels of self-satisfaction and complacency.

Descartes' god is repelled by the existence of sadness and misery only because it would mean that God himself would have to endure unpleasantness. Descartes refuses to allow any such perceptions or thoughts to reside in his own mind. To the contrary, he reinforces the legitimacy of any such facts, the better in order to make himself comfortable in their presence. Which leaves us, as we indicated before, with the narrow role of 'free will' that Descartes consecrates in the larger system of Providential determinism.

> Thus I saw that doubt, inconstancy, sadness and the like could not be in God, since I myself would have been very glad to be free from them . . . But since I had already recognized very clearly from my own case that the intellectual nature is distinct from the corporeal, and as I observed that all composition is evidence of dependence and that dependence is manifestly a defect, I concluded that it could not be a perfection in God to be composed of these two natures, and consequently that he was not composed of them.[87]

The reader will wonder: is there no contradiction to be found in Descartes' assertion of both free will and the argument that everything not absolutely dependent on the will is surrendered to irresistible fate? There really is no contradiction, if one examines with greater care the doctrine of 'free will' that Descartes prepares. In that domain of absolutely personal control, where alone Descartes defends the existence of free will, he celebrates this as the bounty of the clear and distinct ideas, that is, of the absolute intellectual sovereignty of the individual over the contours of his own mind. It is these clear and distinct ideas from which every other form of knowledge is alleged to be 'deduced.' It has been demonstrated that what Descartes 'deduces' from his clear and distinct ideas is the condition of moral isolation: the repudiation of the common facts.

From *prima philosophia*, Descartes proceeds to redefine sense perception as limited to one thing and one thing only: the body's capacity to know what is good or bad *for itself*. The will may be said to be free, for Descartes, only insofar as the individual is free to act in such a way as to ensure that the body's determination of the good and the bad for itself may constitute the ultimate criteria of rightness and wrongness, virtue and vice. In both the providentially ordered world, and also in the actions ordained by 'free will' in Descartes, the goal and the aim is the same: to achieve insulation against the disturbing or inconvenient facts of the world, the interpersonal facts; and where the individual is not free to so act on his own behalf with absolute certainty of success, then he will exert all of his intellectual power (if he is Descartes) to insist on the rightness of any facts that he is himself unable to master. In both cases, the goal is to shield his soul from any interpersonal context that might obligate him to act in a way that is not at the service of his own bodily pleasures and pains, or his soul's chances for tranquility.

The reader who spends any significant amount of time with Descartes, and who measures the diverse threads of his argument, cannot help but feel, after a while, that this man is one of the poorest logicians to ever receive so much admiration and praise in the tradition of western philosophy. Let us pick up once more the case of the deity. Descartes, after his own brand of wisdom, professes that the deity must be 'infinite,' in that the power of God must be beyond all human reckoning. It must be within the power of such a God to divide otherwise indivisible matter; to make it so that the foundations of arithmetic are cast into confusion and irrelevance; and even to make it possible that the human race have perfectly free will while actually operating under the guidance of irresistible divine decree and fate. Yet let us stay with the notion of God as infinite.

Aristotle and Thomas Aquinas both take up the nature of the infinite for discussion. Both thinkers, led by Aristotle obviously, come to the conclusion that it is the nature of the infinite never to be present all at one time.[88] All of which is to say, that the divinity who is 'infinite' must be divided. To be divided, in Descartes' own epistemology, is to be dependent: to be dependent is to be imperfect. Descartes' clear and distinct ideas can entertain no such divisibility, from his point of view: the consciousness must be able to have absolute certainty as regards them. Body is defective precisely because it is divisible, for Descartes.

If the deity is 'infinite,' then the deity can never be a unity. The deity must have parts. Maimonides could have told Descartes that the deity must be simple, in order to be perfect. Parmenides could have told Descartes that that which lacks nothing is perfect, but that that which lacks anything is imperfect. Parmenides' being is without divisions. It is a finite whole, a globe. But Descartes' divinity is a scattered collection of pieces which can never exist together all at the same time: for it is the nature of the infinite that there is always another part to be counted.

Descartes makes some telling analogies in his *Principles of Philosophy*, which give us some better instruction as to how he envisions his own project. Descartes characterizes his project in one of two ways: it is either like the person who has discovered how to make massive profits with very little effort; or it is like a commander who is taking towns and cities.[89] 'Battles' are what Descartes is fighting. Except that he had rather not openly confront his adversaries.

Aristotle's code of morality, which happens to be ordinary praise and blame—is scorned as lacking in true virtue. That *morality*, at least, didn't inspire new inventions. Thus the indictment of Aristotle's logic and the syllogism: they are guilty because they only tell us what we 'already know,' rather than helping us to invent new knowledge. What is taking place during the seventeenth century is a sweeping redefinition of culture, but the really problematic part of it is moral. The 'practical' science is a dubious one: for it is terrified of *truth*. Descartes certainly has not proved his assertions, but it is his intention to indicate that his 'practical' science is the definition of a civilized society. The strong implication is that the influence of Aristotle is tantamount to the influence of the 'barbarous' and uncivilized.[90]

NOTES

1. *Principles of Philosophy*, part 2, article 3, p. 224. In *The Philosophical Writings of Descartes*, volume I. Translated by John Cottingham, Robert Stoothoff, and Dugald Murdoch. Cambridge University Press, 1985 (Henceforth *CSM* I). "When I said that the entire testimony of the senses should be regarded as uncertain and even as false, I was quite serious; indeed this point is so necessary for an understanding of my *Meditations* that if anyone is unable or unwilling to accept it, he will be incapable of producing any objection that deserves reply" (243). From the "Fifth Set of Replies," *CSM II*.

2. *The Elements of Law Natural and Politic*. Edited by J.C.A. Gaskin. Oxford University Press, 1994. Part I, chapter 2, article 10, p. 26.

3. *Treatise on the Emendation of the Intellect*, p. 23 near the bottom. In *The Complete Works*. Translated by Samuel Shirley. Indianapolis: Hackett, 2002.

4. *An Essay Concerning Human Understanding*. Edited by Peter Nidditch. Oxford: Clarendon Press, 1979. Book III, chapter ix, articles 7–8, pp. 478–79.

5. Dedicatory Letter to the Sorbonne, *the Meditations*. CSM II, 5.

6. *Discourse on the Method*, 111. "For my part I have never presumed my mind to be in any way more perfect than that of the ordinary man; indeed, I have often wished to have as quick a wit, or as sharp and distinct an imagination, or as ample or prompt a memory as some others. And apart from these, I know of no other qualities which serve to perfect the mind . . ." In *The Philosophical Writings of Descartes*, volume 1. Translated by John Cottingham, Robert Stoothoff, and Dugald Murdoch. Cambridge: Cambridge University Press, 1985.

7. *Principles of Philosophy*, 189. In CSM I.

8. *Discourse on the Method*, CSM I, 142.

9. *Discourse on the M*ethod, CSM 1, 143.

10. *First Meditation*, 12. In CSM.

11. *Descartes Against the Skeptics*. Cambridge: Harvard University Press, 1978.

12. *Descartes: An Intellectual Biography*. Oxford: Clarendon, 1995.

13. Daniel Garber (*Descartes' Metaphysical Physics*. Chicago: University of Chicago Press, 1992; and "Descartes' Physics," in J.C. Cottingham, editor. *The Cambridge Companion to Descartes*. Cambridge University Press, 1992) notes the considerable influence that atomism had on Descartes. Garber makes some important arguments as regards Descartes' theory, including the view that Descartes' early writings do not reflect the substance of the later writings such as the *Meditations*, the *Principles of Philosophy*, and *The Passions of the Soul* (DMP, 48). Garber thinks that there are distinct limits to the atomistic influence that Descartes picked up from the corpuscularian theories (117). I must disagree with Garber in his characterization of Descartes as a writer more or less evenly divided in his sympathies as regards atomism. The characteristic of body as 'extension' (length, breadth and depth) depends on the atomist or corpuscularian physics. The theoretical basis for the atomist argument (developed by Leucippus and Democritus, Epicurus and Lucretius) is that the perceptible bodies cannot be the real ones, because the perceptible bodies are impermanent. It is this atomist conviction, as I will demonstrate in this chapter, that makes it possible for Descartes to deny that we can truly know the actual bodies in the world as they are. It is this passage that makes it possible for Descartes to conceive of a theorem of body which, as Garber admits, borrows heavily from the atomist philosophy. Moreover, the size, shape, position, and movement that Descartes' bodies are allocated are not themselves accessible via sense perception either. These qualities can only

exist in the theorized particles for Descartes, especially seeing as how the bodies we observe do often move themselves, do often truly rest, neither of which is compatible with Descartes' 'extensions.' It is true that Descartes rejects the atomist conception of the void or vacuum (117). However, it is not true ultimately that Descartes sincerely disavows the indivisibility of what he calls 'matter.' As I will point out in the chapter text, Descartes insists that true body (the particles) cannot be destroyed. Since Descartes rejects the prime matter theory of Aristotle, he also rejects the division of body as involving distinctive parts such as unity and being, whole and part, along with the extremities of coming into being and passing away that a whole must possess in order to be a whole. The only divisibility of body that Descartes entertains is a gross mental division that a mind could imagine about a body, along the lines of Zeno's arguments. The particles of Descartes' physics have their origin, as Descartes says, in God's creation; and it is only God who can reduce them to nothing. Where coming into being and passing away as parts of body are proscribed, and I will show how Descartes does proscribe them—then *authentic* divisibility is not a property of said 'body.' Inauthentic divisibility, such as Zeno's model, originates in the mind or instrument of the investigator; but authentic divisibility originates in the diversity of parts characteristic of the body in question. This is to say that in authentic divisibility, diverse natures are located among the parts of a body: unity, being, whole, part, shape, size, times past, present, and future, and so on.

14. *Principles of Philosophy*, CSM 287, volume 1.

15. Tom Sorell's essay on the subject of Descartes' modernity is an interesting one ("Descartes' Modernity." In *Reason, Will and Sensation: Studies in Descartes' Metaphysics*. John C. Cottingham, editor. Oxford: Clarendon, 1994). In Sorell's opinion, there is some reason to justify the view, held by other scholars, that Descartes remains enmeshed in the Scholastic tradition. As Sorell notes (40), this could be explained by Descartes' desire to appear less controversial. Sorell argues, however, that Descartes is doing something new; opening a new path which at least gave impetus to the new set of scientific approaches designated modern. Sorell's analysis is insufficient, however. What is new about Descartes' science, is not the atomism, of which he is a true adherent—but which is as old as Democritus. What is new is the application of the quasi-atomistic doctrine in a more narrowly political form. As Sorell notes, it is practical utility that the modern atomists advertise. This practical utility, which subordinates (or suppresses) the pursuit of truth in order to advance the pursuit of inventions and power (not the same thing), indicates a new *political* relationship between the rising practical science and the people. The Aristotelian philosophy of metaphysics anchors its evidence in the knowledge characteristic of ordinary thinkers, whereas the Cartesian *prima philosophia* begins its proofs with the wholesale dismissal of said evidence (dependent on perceptual faculties). John Cottingham's analysis of Descartes puts the latter squarely in the Scholastic camp for which the latter evinces no sympathy whatsoever (*Cartesian Reflections.* Oxford University Press, 2008). "What I shall argue, in brief, is that what began as a seemingly straight-forward exercise, that of supporting the trunk of his physics by unearthing its metaphysical roots, gradually overwhelmed Descartes by its complexity; and that in attempting to complete the task, he was drawn, little by little, to fall back on the very Scholastic apparatus that he so derided in his scientific work" (58). Cottingham thinks that the prominent role assigned to the deity in Descartes' philosophy gives the game away. This would be a better argument if Descartes' deity did not so much resemble the Epicurean, as I explore in the chapter's concluding section. Cottingham wants to make the argument that Cartesian ideas more closely resemble ordinary public ideas (116). Like Plato, Cottingham suggests, Descartes seeks to separate the ideas from the physical things. With due respect, this does not reflect a very good understanding of Plato. For Plato, we have *no* access to forms or patterns except through perception initially. The form of courage, to Plato, is actually partly constitutive of the brave soldier, patient, or lover. When Plato abstracts a form or idea, it is the evidence from the sensibles that guides the process. In Descartes all this has been reversed. Not even the so-called mathematical attributes of shape, size, position, and movement are accessible to a Cartesian through perception. Sensory perception is an epistemological null set for Descartes. His conception of ideas, though doubtless 'borrowed' from ordinary experience, fully proclaims its independence from the sense perceptual. One need only consult Descartes' notion of 'place' or 'alteration' (to Descartes there is no such thing) to verify what I say. It is certainly not inconceivable that a program for mechanical

arts like Descartes' could occur under the umbrella of Aristotelian metaphysics: but in that event, truth would not be reduced to questions of utility, and the reality of forms would remain the metaphysician's most significant knowledge.

16. *Principles of Philosophy, CSM* I, p. 231. To continue with the quote: "Even if we imagine that God has chosen to bring it about that some particle of matter is incapable of being divided into smaller particles, it will still not be correct, strictly speaking, to call this particle indivisible. For, by making it indivisible by any of his creatures, God certainly could not thereby take away his own power of dividing it." Cf. *Letter to Plempius for Fromondus, 3 October 1637,* CSM III, p. 61. There Descartes dismisses the work of Democritus and Epicurus as "heories which have nothing to do with me." Cf. *Letter to Mersenne, 30 September 1640*, CSM III, 154: "It is a serious mistake to believe that one can conceive distinctly that an atom, or even a part of matter, can occupy indifferently a larger or smaller space. First of all, an atom can never be conceived distinctly, since the very meaning of the word implies a contradiction, that of being a body and being indivisible."

17. *Discourse on the Method*, CSM I, 112.

18. *The World*, CSM I, 90.

19. This is what makes the relationship between science and perception quite so urgent a matter in the Cartesian Oeuvre. In Aristotelian or Platonic philosophy, perception and the ordinary assignment of name to object serve to anchor the discussion in common facts. On the road towards science, however, discursive analysis of the subject matter requires, for the sake of progress, a rich and diverse debate. In Cartesian analysis, which as I have already indicated borrows considerably from the ancient arsenal of Eleatic rhetorical style, the simple possibility of disagreement at any stage of the process is said to be indicative of an argument that lacks the requisite Cartesian *authority*. He who knows the Cartesian truth must be able to compel assent, which is not quite the same thing as engaging in rational persuasion. In the early *Rules for the Direction of the Human Mind*, Descartes writes that "we should attend only to those objects of which we seem capable of having certain and indubitable cognitions" (CSM I, 10). "But whenever two persons make opposite judgments about the same thing, it is certain that at least one of them is mistaken, and neither, it seems, has knowledge. For if the reasoning of one of them were certain and evident, he would be able to lay it before the other in such a way as to eventually convince his intellect as well" (11). Descartes proceeds to indicate that arithmetic and geometry are the only models for such science before the proposed Cartesian clear and distinct ideas. It will pay for a moment to note how Descartes does not model his science upon arithmetic and geometry in their *natures*. Arithmetic depends upon the origin of number: in truth, the origin of number is the unlikeness of the objects that we perceive. Geometry depends upon the origins of shape and magnitude: these two things depend upon abstraction from perceived objects. In Descartes' mathematics, number and shape/magnitude profess no debt to what is perceived. For Descartes, mathematics begins with mental power, power to 'order' and 'measure.' The mind in Cartesian studies reaches for unjustifiable authority to use names from this vantage point that is not dependent on evidence external to the 'clear and distinct ideas.' Descartes' home-style arithmetic can be observed in the way he counts the divisions in his own model of mind. Perceptual sensation, intellectual thinking, the experience of the passions, the use of a rationally determined will are indicated by Descartes as parts of the mental operation. However, Descartes dissolves imagination, sense perception and the cogitation of clear and distinct ideas under *one* name, 'thinking'; and he reduces character, diverse passions and judgment to one name also: 'willing.' "By the term 'thought,' I understand everything which we are aware of as happening within us, in so far as we have awareness of it. Hence, thinking is to be identified here not merely with understanding, willing, and imagining, but also with sensory awareness." *Principles of Philosophy*, CSM I, 195. The link that binds mathematics or geometry to *truth* is the one Descartes snaps. John A. Schuster ("Descartes' Mathesis Universalis: 1619–28," in Stephen Gaukroger, editor. *Descartes: Philosophy, Mathematics and Physics*. Totowa: Barnes and Noble, 1980) correctly indicates that Descartes and Beeckman sought to mathematize corpuscles, which of course depends on the reality of corpuscles (48). Yet Schuster characterizes Descartes' attempt to banish form from physics as a struggle against the 'unintelligible' (76). There is hardly anything unintelligible about the forms of Aristotle and Aquinas. These just are what all people effortlessly perceive about objects in nature. Cf.

Stephen Gaukroger, "Descartes' Project for a Mathematical Physics" (214) in the same volume. In my view Gaukroger overstates the debt that Cartesian science pays to 'common sense.'

20. *Rules for the Direction of the Human Mind,* CSM I, 10–11.

21. *Physics* 191a25.

22. *The World,* CSM I, 90: "let us suppose that God creates anew so much matter all around us that in whatever direction our imagination may extend, it no longer perceives any place which is empty."

23. *First Meditation,* CSM II, 10.

24. *Principles of Philosophy,* CSM I, #18, 199.

25. *Third Meditation,* CSM II, 28.

26. *Metaphysics* 983a24. Cf. Thomas Aquinas. *Commentary on Aristotle's Metaphysics.* Translated by John P. Rowan. Notre Dame: Dumb Ox books, 1995, 25.

27. *The World,* CSM I, 83, 85, 87, 89, 95; *Principles of Philosophy* CSM I 226, 239; *Letter to Mersenne* 15 April 1630, CSM III, 21–22.

28. *The World,* CSM I, 89.

29. Aristotle's *Physics* 188a22.

30. For Descartes, see *The Passions of the Soul,* CSM I, 332, 333, 337, 339; *On Man,* CSM I, 101. For Aristotle, see *De Anima* 424b17: "Generally, about all perception, we can say that a sense is what has the power of receiving into itself the sensible forms of things without the matter, in the way in which the piece of wax takes on the impress of a signet ring without the iron or gold; what produces the impression is the iron or gold, but not qua iron or gold: in a similar way the sense is affected by what is colored or flavored or sounding not insofar as each is what it is, but insofar as it is of such and such a sort and according to its form."

31. *Rules for the Direction of the Human Mind, CSM I,* p.59.

32. See *Letter to Mersenne 28 October 1640, CSM III,* p.155: "Two indivisible things could only make a single thing divisible into two parts at most; but before saying that they could make up a body, you must know what is meant by the term 'body.' In fact it means a thing which has length breadth and extension, and so cannot be composed of indivisible things, since an indivisible thing cannot have any length or breadth or depth. If it had, we could divide it at least in our imagination, which would suffice to guarantee that it was not indivisible; for if we could divide it in imagination, an angel could divide it in reality."

33. *Principles of Philosophy, CSM* I, 210: "A substance may indeed be known through any attribute at all; but each substance has one principal property, which constitutes its nature and essence, and to which all of its other properties are preferred. This extension in length, breadth and depth constitutes the nature of corporeal substance . . ."

34. *Parmenides* 141e.

35. *Parmenides* 129e.

36. *Parmenides* 142d.

37. Ibid.

38. *Principles of Philosophy, CSM* I, 233: "By 'motion,' I mean local motion; for my thought encompasses no other kind, and hence I do not think that any other kind should be imagined to exist in nature."

39. *Third Meditation,* CSM II, 33: "I do not escape the force of these arguments by supposing that I have always existed as I do now, as if it followed from this that there was no need to look for any author of my existence. For a lifespan can be divided into countless parts, each completely independent of the others, so that it does not follow from the fact that I existed a little while ago that I must exist now, unless there is some cause which as it were creates me afresh at this moment—that is, which preserves me. For it is quite clear to anyone who attentively considers the nature of time that the same power and action are needed to preserve anything at each individual moment of its duration as would be required to create that thing anew if it were not yet in existence. Hence the distinction between preservation and creation is only a conceptual one. And this is one of the things that are evident by the natural light."

40. Richard Kennington's analysis of Descartes' theory of reason seems much closer to the mark. Reason in Descartes, Kennington argues, is 'an instrument' (*On Modern Origins: Essays in Early Modern Philosophy*. Edited by Pamela Kraus and Frank Hunt. Lanham: Lexington Books, 2004, 110). Kennington observes of Descartes that his "reason serves the passions, and

doesn't naturally seek knowledge" (110). Kennington correctly traces how Descartes seeks to exclude the intelligible from the perceptual evidence: that pain is bad and that pleasure is good, that is, the claims of the ancient sophists, is the new function which Descartes seeks to assign to sense perception. Kennington is also correct to observe that Descartes simply withholds from us his views as to what the ultimate parts of matter are (120). For Descartes, as for Newton after him, professed ignorance as to the ultimate parts of matter does not interfere in the least with his purpose: which is to subject 'nature' to experiments that will produce beneficial effects. Knowing bodies for what they are could well interfere with converting them into objects of utility, and this goes for human bodies as well. I think Kennington is further correct in his estimation that Descartes participates in the new 'moral realism' of Machiavelli and Francis Bacon. Cf. Mike Marlies. "Doubt, Reason and Cartesian Therapy." In Hooker, editor. *Descartes: Critical and Interpretive Essays*). Marlies uncritically accepts Descartes' proposition that ordinary perception amounts to 'prejudice,' and that only by banishing 'prejudice' from reason's presence can the 'intelligible' be pursued (95–6). Marlies correctly observes that sensory objects are not perfect: but his claim that the road to perfect intelligibility requires the suppression or denial of the validity of perceived images is to convert the 'intelligible' into a mere assertion of power, one that has forsaken the quest for truth.

41. *On Man,* CSM I, 99.

42. *Rules for the Direction of the Human Mind,* CSM I, 9.

43. *The World,* CSM I, 89, 94, 95.

44. *Letter to Mersenne* January 1638, CSM III, 79.

45. The fact remains that the indictment of sense perception was not occasioned by the experience of the human race. This indictment has its origin in theory, as even Descartes can do little more than trump up ancient and long ridiculed examples to justify his own disdain. Margaret Wilson for instance writes that, in Descartes' view, sense experience does not involve the 'incorporeal,' but only the corporeal and physical ("Descartes on Sense and Resemblance," 210. In John Cottingham, editor. *Reason, Will and Sensation: Studies in Descartes' Metaphysics*. Oxford: Clarendon, 1994). Sense perception, however, does involve the incorporeal. To perceive a piece of wax, for instance, is to perceive formed matter. The form of wax, inescapably a part of any piece of perishable wax, refers to a pattern or *nature.* This pattern or nature exists in many other enmattered objects. Descartes' attack on the forms of Aristotle and the Scholastics are superficial, rhetorical ones. Not only do they not persuade, but they are reluctant to even enter into a serious investigation of the issue. Rhetoric, not argument, is Descartes' response to Aristotle. Lili Alanen ("Sensory Ideas, Objective Reality and Material Falsity" in the same volume) indicates that Descartes has delivered final proofs of the distinction between primary and secondary qualities. "Descartes' mechanistic account of sense perception in combination with his metaphysics has, it seems, once and for all rendered any attempts to build the relation between the mental (including the sensory) representations and physical reality on likenesses vain" (232). Such, at least, may be the fate of those who are theoretically prostrate before the atomist doctrines that Descartes can be seen to rely upon. This is certainly no desideratum of truth.

46. *The World,* CSM I, 81.

47. *The World,* CSM I, 89.

48. *Principles of Philosophy*, CSM I, 218, 224.

49. Descartes fully understands the relationship between perception and language. He offers a misdirected argument in order to try to illustrate his point that sense perception does not provide us with an actual likeness of the object perceived. Descartes wants to subject names to a critique, precisely insofar as names indicate the *form* (the universal) aspect of objects. "Finally, because of the use of language, we tie all our concepts to the words used to express them; and when we store the concepts in our memory we always simultaneously store the corresponding words. Later on we find the words easier to recall than the things; and because of this, it is very seldom that our concept of a thing is so distinct that we can separate it totally from our concept of the words involved" (*Principles of Philosophy* CSM I, 220). Descartes' concept of 'distinctness' is not comfortable with the link between common names and universal forms: this is why Descartes wishes to deny that the name adequately indicates the object. "The fourth cause of error is that we attach our concepts to words which do not precisely correspond to real

things" (220). Descartes elsewhere argues that names are entirely arbitrary: the names don't resemble things, so why must sense perceptions do so? Once again, the common names, though they vary from language to language, indicate forms that cannot be confined to one single object, since the form exists in many objects. Names indicate forms, which thereby *do* give us knowledge of the nature of the objects. In Aristotle's account of perception, it is not an account of matter in contact; "we can say that a sense is what has the power of receiving into itself the sensible forms of things without the matter" (*De Anima* 424b17). Descartes insists that "words, as you well know, bear no resemblance to the things they signify and yet they make us think of those things. . . . Now if words, which signify nothing except human convention, suffice to make us think of things to which they bear no resemblance, then why could nature not also have established some sign which would make us have the sensation of light, even if the sign contained nothing in itself which is similar to this sensation?" (*The World*, CSM I, 81). Descartes errs when he says that words are 'merely' human convention: it is the task of words to indicate that which is *not* our contrivance, that is, the natures and forms *in the objects*. It is the forms in need of names that designate the instrumental function of words: that which needs to be indicated, is the form we know through sense perception, not some dummy object of artificial concoction.

50. *The World*, CSM I, 81.

51. *Theaetetus* 191c.

52. *De Anima* 424b17. Descartes, when he refers to sense perception, seems to use the word almost interchangeably with 'imagination.' In Aristotle's discussion of the soul, sense perception and imagination are strictly separated. The latter does not inform us about reality, while the former does. See Thomas Aquinas, *Commentary on Aristotle's De Anima*. Translated by Kenelm Foster and Silvester Humphries. Notre Dame: Dumb Ox Books, 1994, 195.

53. *De Anima* 419a12.

54. *Descartes, an Intellectual Biography*, 70.

55. *Theaetetus* 184d, *Phaedo* 75b.

56. *Fourth Set of Replies*, CSM II, 173.

57. Gaukroger persuasively makes the case that Descartes' intellectual career never included a serious crisis provoked by an encounter with skeptical philosophy. Skepticism is obviously present in the *Meditations*, "and it is generally assumed that skepticism made Descartes' epistemology. But I have been unable to find any concern with skepticism before the 1630s, and the kind of epistemology that is pursued in the treatment of cognition in the Regulae and in L'Homme is naturalistically inclined, showing no concern at all with skeptical issues. Moreover, I have found no evidence that Descartes ever went through a crisis of a skeptical kind. . . ." (*Descartes: An Intellectual Biography*. Oxford: Clarendon, 1995, 11). In Gaukroger's view, it was the condemnation of Galileo which provoked Descartes' interest in skepticism: by representing his work as a response to skepticism, Descartes thought he might avoid a similar collision with Church authorities. Gaukroger's argument is largely aimed at E.M. Curley (*Descartes Against the Skeptics*. Cambridge: Harvard University Press, 1978, ix, 38, 188).

58. *De Anima* 427b7.

59. *Rules for the Direction of the Human Mind*, CSM I, 43.

60. *First Meditation*, CSM II, 17.

61. *Second Meditation*, CSM II, 16.

62. One must always appraise a philosophy in terms of its relation to the ordinary human experience. It is necessary to do this, because philosophies, especially philosophers that aim to guide the formal knowledge of a civilization (its sciences), would and will enact postulates *as if* they were so. If it were the customary experience of the human race that its sense perceptions led it into disastrous errors (mistaking the motor oil for the milk, walking into walls instead of through doors), then Descartes' indictment of the senses would not provoke our attention. Yet this is not the case, and never has been the case. Just in proportion as the alleged crisis besetting sense perception is a gross exaggeration to say the very least, so too may we shine a critical light upon the alleged 'natural light' of the deliverance in so-called rational method that the philosopher proposes to us, as the remedy for the supposed crisis. Lest we speak incompletely here, let it furthermore be noted that simply because philosophers know how to confound ordinary opinion with perplexing questions about the validity of their perceptions, need not

indicate (and never has indicated) that sense perception has misled human beings about the true facts of the world. Peter Schouls ("Human Nature, Reason, Will," in Cottingham, edited. *Reason, Will and Sensation*. Oxford: Clarendon, 1994) well notes that Descartes' philosophy, especially his clear and distinct ideas, hinge upon 'an act of will' (173). Philosophy which seeks to substitute an act of will for 'passively' received sense perception indicates a philosopher who refuses to abide Plato's Socrates' profession of ignorance, the admission that he must begin with the same knowledge that the others have.

63. Descartes claims that he is doubting everything that it is humanly possible for him to doubt. That he is henceforth forswearing belief and confidence in any the last bit of knowledge which has been derived from sense perception; which would then leave him in a position to 'intuit' what he could still nevertheless know. To accomplish this task, Descartes decides, requires an active concentration of the will to literally oppose all of the opinion and information based on sense perception. Because his habits keep reviving the knowledge based on sense perception, "I think it will be a good plan to turn my will in completely the opposite direction and deceive myself, by pretending for a time that these former opinions of mine are false and imaginary" (*FM*, CSM I, 15). As I have indicated in the text of this chapter, to really undertake this enterprise would deprive Descartes of his very access to language. It would literally render him mute. Yet Descartes does not aspire to this intellectual integrity. Instead, he proposes to pretend too, that he can formulate propositions which do not depend on the indicted sense perceptual knowledge, beginning with the proposition that, he alleges, he may still formulate: "But I have convinced myself that there is absolutely nothing in the world, no sky, no earth, no minds, no bodies. Does it now follow that I too do not exist? No" (*SM*, CSM I, 16). "If I convinced myself of something then I certainly existed." This feat enables Descartes to claim that what he is alone sure of is his own mind and its existence; and from this vantage point, he now professes, he may draw from a purified well of absolutely irrefutable ideas. As we have indicated, Descartes' claim is false. Yet it needs to be noted what extraordinary psychological leverage Descartes is able to draw from this empty well. For he is now in a position to claim, now that all evidence external to us has been suspended in its probative value, that the formulations of his mind are irresistible; and that if he can imagine it, from this vantage point, it must be possible that it is true. I could not imagine a more disastrous mental destination for the political dimension of civilization.

64. *Rules for the Direction of the Human Mind*, CSM I, 11.

65. *Rules for the Direction of the Human Mind*, CSM I, 14.

66. *Discourse on the Method of Rightly Conducting One's Reason*, CSM I, 127.

67. *Sixth Meditation*, CSM II, 51.

68. *Third Meditation*, CSM II, 24.

69. I have made the point earlier that Descartes' claim to found his clear and distinct ideas on a mathematical foundation is a baseless claim. From the "I think, therefore I exist" proposition, the first allegedly mathematical deductions that Descartes makes concern the natures of the bodies which he will now hypothesize as the real and true ones: and these are nothing other than the corpuscles, the 'particles,' which are descended from the atomist theory, which Descartes, his protestations notwithstanding, sustains in his own physics. I have addressed this directly in the first part of this chapter. The size, shape, position and motion of the alleged real bodies are nothing other than the characteristics that Leucippus and Democritus allowed to their atoms. Plato proved that there are no shape, size, position, or movement to talk about at all, unless and until we have conceded the first two parts of any object that has existence: namely, unity and being. Descartes allows that all things he examines must participate in unity, but he does not deal with the issue of being. The combination of unity and being constitutes the whole of parts, the whole of parts which alone possesses the extremities of shape, size, position, and movement.

70. *The World*, CSM I, 92.

71. *Principles of Philosophy*, CSM I, 196.

72. I have noted in this chapter that Descartes' God resembles the Epicurean God. This god is indifferent to the universe of human suffering and injustice, deaf and dumb to the desire for justice. Descartes' unique combination of a limited domain of 'free will' and a much larger sphere of absolute fate and divine necessity has another cause which can be traced to the

Epicurean tradition. The Epicureans were among the earliest devotees of philosophies of will, picking up Parmenides' legacy with vigor and devotion. Anxiety is foremost among the Epicurean evils, something which they are determined to drive away at all cost. The prospect of free will is very threatening to the temperament which puts will before facts. Anxiety is produced by the refusal to abide by common evidence, anxiety that concerns the prospect of being judged. This anxiety extends to the belief in a divinity which judges, and which the Epicurean divinity accordingly banishes through its doctrine that all things occur by 'necessity.' Descartes shares this Epicurean legacy as well, and it is to this end that his free will is limited to that over which he has absolute and solitary control. All else, for the sake of tranquility of mind, must be surrendered to blind fate, and ferociously affirmed as simply the way things have to be. "We must then utterly reject the common opinion that there is a Fortune outside us which causes things to happen or not to happen, just as it pleases. And we must recognize that everything is guided by divine Providence, whose eternal decree is infallible and immutable to such an extent that, except for matters it has determined to be dependent on our free will, we must consider everything that affects us to occur of necessity and as it were by fate, so that it would be wrong for us to desire things to happen in any other way." (*The Passions of the Soul*, CSM I, 380).

73. *Third Meditation*, CSM II, 26.

74. According to Descartes, the "exclusive concern of mathematics is with questions of order or measure and that it is irrelevant whether the measure in question involves numbers, shapes, stars, sounds, or any other object whatever." *Rules for the Direction of the Human Mind*, CSM I, 19. Number and magnitude are first of all properties of objects, not powers derived from the human *inclination* to 'order or measure.'

75. *The Passions of the Soul*, CSM I, 345–46.

76. *Nichomachean Ethics* 1102b13, 1103b3.

77. *The Passions of the Soul*, CSM I, 349.

78. *The Passions of the Soul*, CSM I, 354.

79. *Discourse on the Method*, CSM I, 114.

80. *The Passions of the Soul*, CSM I, 380.

81. *Nichomachean Ethics* 1131a10 illustrates Aristotle's conception of the virtue of justice.

82. *The Passions of the Soul*, CSM I, 384–85.

83. *Discourse on the Method*, CSM I, 143.

84. *Discourse on the Method*, CSM I, 144.

85. *The Prince*. Translated by Harvey C. Mansfield Jr. Chicago: University of Chicago Press, 1985, 62.

86. *The Passions of the Soul*, CSM I, 356–7.

87. *Discourse on the Method*, CSM I, 128 Cf. *The Essential Epicurus*. Translated by Eugene O'Connor. Buffalo: Prometheus books, 1993, 62.

88. *Physics* 203b1.

89. *Discourse on the Method*, CSM I, 145: "Those who gradually discover the truth in the sciences are like people who become rich and find they have less trouble making large profits than they had in making much smaller ones when they were poorer. Or they may be compared to military commanders, whose forces tend to grow in proportion to their victories, but who need more skill to maintain their position after losing a battle than they do to take towns and provinces after winning one. For attempting to overcome all the difficulties and errors that prevent our arriving at knowledge of the truth is indeed a matter of fighting battles: we lose a battle wherever we accept some false opinion concerning an important question of general significance . . ."

90. *Principles of Philosophy*, CSM I, 180.

Chapter Three

Hobbes's Natural Science

What concerns us in this book is Hobbes's political and moral philosophy, his civil science. In my view, it is not possible to comprehend Hobbes's civil science absent a thorough analysis of his natural science. The roots of Hobbes's natural science and the foundation of Hobbes's civil science converge in the domain of definitions, the significations ascribed to names. The authority for creating definitions in Hobbes resides in the natural science, in its pretensions to supersede the customary authority of perception and communal naming. In this particular way, therefore, it is my view that Hobbes's political science fully depends on the authority obtained in his doctrines of physics and perception respectively.

Civil science, famously, is something Hobbes claims to have founded and discovered. However, this is not the only science that Hobbes claims to have originated. He also claims this title for his *Optics*, that is, for his science of perception.[1] The relationship between Hobbes's theory of perception and politics, between physics and political science, is therefore of the utmost importance for this inquiry.

Hobbes, in his *De Homine*, insists that it is not necessary for a person to understand the physics, in order to understand the civil science. "Having completed this section, *On Man*, I have finally fulfilled my promise. For you now possess the prime Elements of my philosophy in all its divisions and subdivisions. Moreover, it happens that the two parts whereof this section consists are very dissimilar. One is very difficult, the other very easy; one consists of demonstration, the other of experience; one can be understood by few, the other by all."[2]

It is not usually regarded by the human race to be an extraordinary feat, to know what are the meanings of names. This is to say that to assign a name properly to an object is something that the human race manages to do quite easily, as Plato observes in his *Theaetetus* (192ac). It is, however, the belief of the non-philosophers that the names they articulate are indicating common objects for the most part external to them. Hobbes does not intend to base his use of names, or his definitions of them, upon such common usage. In the domain of language Hobbes affects to be rather an inventor.

> By this it appears how necessary it is for any man that aspires to true knowledge, to examine the definitions of former authors, and either to correct them, where they are negligently set down, or to make them himself . . . So that in the right definition of names, lies the first use of speech; which is the acquisition of science: and in wrong, or no definitions, lies the first abuse; from which proceed all false and senseless tenets; which make those men who take their instruction from the authority of books, and not from their own meditation, to be as much below the condition of ignorant men, as men endued with true science are above it.[3]

Hobbes's confident assertion that it is not necessary to know his natural science in order to understand the definitions of his civil philosophy is not so easily accepted by scholars. There does not seem to be a bigger issue in Hobbes studies than the relationship (or lack of one) thought to exist between his physics and civil philosophy. Does the civil and moral philosophy depend on the physics, in such a way that the former cannot be understood absent a correct understanding of the latter?[4] Hobbes's civil science is, in *De Cive* and other works, a chain of arguments concerning human beings, their natures and feelings, their rights and obligations. In *De Cive*, Hobbes explains to us how we are to understand the names 'good' and 'evil,' 'just' and 'unjust,' 'right' and 'wrong,' in accordance with his civil science.

It is true that the ordinary speaker can receive instruction from Hobbes's use of the names 'good' and 'evil,' 'just' and 'unjust,' 'right' and 'wrong,' as they are retailed in *De Cive*, *Leviathan*, and *The Elements of Law*. However, the ordinary speaker does not therefore understand whether these names are being used in a true or false fashion. For Hobbes famously makes a break with the ordinary speakers, not merely insofar as they are speakers of moral words; but insofar as they are the utterers of any names whatever.

The name 'accident' is used by ordinary speakers to refer to things that happen out of turn, as exceptions to the customary order of events and objects. Hobbes, however, employs the name 'accident' in a vast manner. Many scholars resist the proposition that Hobbes is a natural philosopher in the atomist lineage. Yet that is the lineage which defines all of perceptual experience as a product of 'accident.'[5] It is also the lineage, as I will attempt to demonstrate in this chapter, that best squares with Hobbes's teaching as to what 'body' is. The atomists wish to indicate, by this use of language, that the perception we have is not indicative of the objects external to us. Instead, for the atomist, the perceptual experience is the product of collision, when invisible (because too small) bodies collide with our own. Epicurus established a name for the science of definition that attends to this atomist theory of perception. He calls it 'prolepsis.'[6] However, if the Epicurean prolepsis is descriptive of the nomenclature that Hobbes employs, our scholarship does not begin to so indicate. As regards perception, Hobbes says:

> And to proceed to the rest of the senses, it is apparent enough that the smell and taste of the same thing, are not the same to every man, and therefore are not in the thing smelt or tasted, but in the men . . . That as in conception by vision, so also in the conceptions that arise from other senses, the subject of their inherence is not the object, but the sentient. And from thence also it followeth, that whatsoever accidents or qualities our senses make us think there be in the world, they are not there, but are seemings and apparitions only. The things that really are in this world without us, are those motions by which these seemings are caused.[7]

All original information comes to us through what Hobbes calls 'sense,' and *all* of this original information is said to be a false representation of what exists external to us. Scholars have kept their distance from the part of Hobbes's science that he advertises as difficult, or at least most political theorists have. It is nevertheless surprising that perception has not arisen as more of a concern in Hobbes scholarship. It also bears pointing out that Hobbes follows both Bacon and Descartes with his doctrine of perception. They are all based on a proto-atomistic theory. As discussed in the first chapter, Bacon indicts ordinary perception and attempts to render it dependent upon scientific experiment in order to function.

There is no doubt some confusion about as well. For Hobbes insists that sense is the origin of all human thoughts, and that sounds rather democratic. Yet the argument that our sense perceptions do not obtain accurate information about external objects is not democratic. It is not tolerant of the faculties with which the majority of human beings must obtain knowledge about the world around them.[8]

> Concerning the thoughts of man, I will consider them first singly, and afterwards in train, or dependence upon one another. Singly, they are every one a representation or appearance, of some quality, or other accident of a body without us; which is commonly called an object. Which object worketh on the eyes, ears, and other parts of a man's body; and by diversity of working, produceth diversity of appearances. The original of them all is that which we call SENSE, for there is no conception in a man's mind, which hath not at first, totally, or by parts, been begotten upon the organs of sense. The rest are derived from this original.[9]

Hobbes's promise, therefore, that the reader unacquainted with his physics may nevertheless fully comprehend the civil science laid down in *De Cive* is not justified by the facts. All of the subject matter discussed in *De Cive*, and in the *Leviathan*'s political parts for that matter, is dependent on the authority of sense perception from the ordinary *reader*'s point of view. The authority which Hobbes espouses for his use of key names is not perception, or understanding based upon perception.

The philosophical foundations for the above doctrine of perception are not discussed by Hobbes. Of course no philosopher has ever seriously maintained to the public that it could not rely upon its perceptions in order to live. What they have often taught is that these perceptions cannot be offered in the way of evidence as to the truth of fact. This is where the political issues arise: for such philosophers, whether Parmenides, Protagoras, Democritus or Francis Bacon and Descartes, envision a unique position of authority for their trained philosophers, for the science that they practice. The political conceptions of authority in Early Modern Philosophy therefore have their original roots in the doctrines of physics, which challenge the competence or perception and which seek to establish deference to the practitioners of the new science.

Scholars have been rather determined, however, to downplay the significance of Hobbes's theory of perception. It is true that Hobbes reduces all original experience to matters of 'sense.' It does not therefore follow, as Oakeshott thinks it does, that Hobbes regards sense perception as the authoritative source on matters of fact. In his commentary on the *Leviathan*, Oakeshott says that Hobbes respects the authority of sense perception for non-scientists; that Hobbes places sense perception outside of the realm of science. Yet we have already seen that this is not so. "This is the end and crown of experience," Oakeshott states. "In the mind of the prudent or sagacious man, experience appears as a kind of knowledge. Governed by sense, it is necessarily individual; a particular knowledge of particulars. But, within its limits, it is 'absolute knowledge'; there is no ground upon which it can be doubted, and the categories of truth and falsehood do not apply to it."[10]

This discussion of perception is not prompted by the question as to whether human beings can be entrusted with their sense faculties when they are walking through doors, or handling dangerous tools, or chatting casually with friends. It is concerned with perception insofar as it bears upon some knowledge of truth. The domain of perception that Oakeshott recognizes in Hobbes does not have competence to distinguish false from true. That is the problem.

Oakeshott argues that Hobbes simply has a philosophical nature, and that this leads him to behold the evidence of sense perception in a different light from most people. Oakeshott insists that philosophers' orientations towards matters such as sense perception are not generated by bookish learning. Oakeshott does not give any evidence to suggest that he himself is familiar with the history of Eleatic atomism. Hobbes by contrast provides considerable evidence to the effect that he is all too familiar with it. The key doctrine articulated by Hobbes in this particular is the indestructibility of matter, which Hobbes calls 'body.' Simply because Hobbes chooses not to indicate his predecessors does not therefore prove that he was able to come upon these thoughts in his daydreams.[11]

Since the authority by which Hobbes's philosopher assigns names to objects is inconsistent with the authority by which ordinary speakers assign names to objects, things are definitely at an impasse. For Hobbes does not merely assert the independence of his own philosophy of naming: he fully indicts and attempts to convict that authority by which ordinary speakers assign names to things. "But seeing names ordered in speech (as is defined) are signs of our conceptions, it is manifest that they are not signs of the things themselves; for that the sound of the word stone should be the sign of a stone, cannot be understood in any sense but this, that he that hears it, collects that he that pronounces it thinks of a stone."[12]

In the above quotation, Hobbes takes a swipe at the reputation of Aristotle. Aristotle, however, is not such a difficult author to understand. Is it not possible for Hobbes to actually explain Aristotle's error, so that the reader could be enlightened? Aristotle after all has many things to say about those physicists, such as Hobbes himself, who would reduce everything in the knowable universe to one thing called 'body.' Perhaps Hobbes does not have the time or patience to conduct such comparative reasoning. This study will undertake to fill in some of these gaps, however, and hopefully for the benefit of all.

The sort of scientific authority that Hobbes envisages for society is aimed at a new, emerging conception of culture. On matters of truth of fact, in this new culture, science will be asserting its prerogatives. That which can be known by ordinary speakers will not be rudely interfered with. Yet on the other hand, facts which would be settled in accordance with the testimony of untutored perception will be treated by the new science as naïve, as mere raw material for the actual agencies that determine truth. There are profound political implications in this arrangement. This is doubly the case for a science that would serve as the foundation for a social contract model of government.

For if the people are not deemed qualified to pronounce upon the truth of matters of fact by Hobbes's science, how much freedom and authority can their actual opinions and decisions hold in the new model of government? It is well known of course that freedom of opinion is not an attribute of the commonwealth that Hobbes proposes to build; yet he certainly makes the argument that the state itself must be derived from contract or consent. The predicament of the Hobbesian notion of conception therefore afflicts his political theory at its roots.

We will later examine the strong theory of individual reason that many scholars ascribe to Hobbes. The nub of the matter concerns the nature and origins of this theory of sense perception that Hobbes upholds. Oakeshott's answer, that Hobbes's science simply leaves the terrain of sense perception to itself, is demonstrated to be untenable. "And in all its characteristics, (sense perception) is distinguished from philosophical knowledge, which (because it is reasoned) is general and not particular, a knowledge of consequences and not of facts, and conditional and not absolute," Oakeshott states.[13]

In part, Oakeshott's point is well taken. Philosophy is concerned with universals or general knowledge. Yet, at least according to the philosophies of Plato and Aristotle, which respect the evidence obtained by perception, the element of the universal is present in the *particular* facts. Plato and Aristotle would both argue that a tree contains not merely the perishable matter of a singular object, but *partakes* of a universal form. The particular tree is not the only tree. Tree is a kind, a nature, a form, a pattern. Hobbes bitterly opposes this type of argument. This point illustrates the manner in which science, at least a science of definition, is properly *obliged* to the ordinary use of names, and the perception which makes it possible. Ordinary opinion is based on familiarity with images, science with definitions in speech. Yet in the philosophy that Hobbes attacks, the definition in speech must obtain its original evidence from the images known to all. Converting the images into definitions is a hard task; but Hobbes proposes to do away with the role that images play in this equation of the development of definitions. The will of the philosopher is the driving force of definition in the Hobbesian model. Oakeshott is therefore wrong to think that the issue of perception is not relevant to Hobbes's philosophy. In fact, Hobbes's theory of perception brazenly confronts and rebukes the authority of ordinary perception, with the power of his new definitions.

This raises the question as to the origins of this theory of philosophy that Hobbes is working with. Upon what grounds does Hobbes possess this alleged knowledge, that our sense perceptions are indicative merely of what is present *in us,* but not in the objects themselves? By what evidence will Hobbes ultimately seek to prove that the operation of perception cannot acquaint us with the true natures of objects, as it were, the real objects being too miniscule for our perceptions to handle?

Frithiof Brandt tells us the following: "Regarding the first point we find, in a letter from the stay in Paris in 1636, a pronouncement which is thoroughly typical of Hobbes throughout the whole of his development. . . . Hobbes writes to Newcastle: 'In things that are not demonstrable, of which kind is the greatest part of natural philosophy, as depending upon the motion of bodies so subtile as they are invisible, such as are ayre and spirits, the most that can be attained unto is to have such opinions, as no certayne experience can confute, and from which can be deduced by lawful argumentation to no absurdity, and such are your Lordship's opinions in your letter . . . That the variety of things is but variety of local motion in the spirits or invisible parts of bodies.' When we said that the pronouncement made by Hobbes is thoroughly typical of him, it is inasmuch as it regards the possibility of a verification of the physical hypothesis. In order to understand Hobbes on this point we must carefully remember his conception that the majority of physical phenomena, viz. all the sensible qualities, depended upon motions in the parts of the body so small that they were invisible . . . this conception, Hobbes has, as we shall subsequently see when treating *de Corpore*, always retained" (197).[14]

Hobbes's natural philosophy depends upon a *theory* as to the nature of what is real. Hobbes's theory pushes aside the authority of what we are able to perceive with our faculties of sense perception. Though Hobbes is frequently cast as the enemy of 'metaphysics,' it is proposed here that Hobbes depends on *his own* metaphysics. One of the great problems in our more modern discussions of metaphysics is that the name 'metaphysics' has become associated with moral teachings of a highly abstract nature. In Aristotle, metaphysics is a science concerning, not morals, but *being*. In this sense, Hobbes's own philosophy is indeed a metaphysical theory, because these 'subtile' bodies which he regards as real, and which he cannot demonstrate, depend upon theory and argument to substantiate.

The scientific principle upon which Hobbes's theory rests, is the shibboleth of *indestructible body*: that is, of indivisible body. This principle, which was originated by the philosopher Parmenides, insists that coming into being and passing out of existence are false appearances. "There is only one other description of the way remaining, namely, that what is, is. To this way there are many signposts: that Being has no coming-into-being and no destruction, that it is whole of limb, without motion, and without end."[15]

Epicurus's atomism preserves the same fundamental premise set forth by Parmenides, along with several innovations. Epicurus indicates that the fundamental bodies of the universe are atoms, and that these can be neither generated nor destroyed. "Having made this distinction, we must now consider what is not evident to our senses: first of all, that nothing is created from what does not exist. For everything would be born from everything without the need for seed. And if that which is destroyed were dissolved into what does not exist, everything would be destroyed, since that into which they were dissolved does not exist."[16]

An investigation into Hobbes's natural science must be undertaken merely in order to understand what is going on in Hobbes's civil science. In Hobbes's civil science, he claims that the ordinary understanding of right and wrong is the very cause of disorder, the very source of violence and mayhem. It is Hobbes's attack on the authority of sense perception in general which makes this broader scathing criticism of the customary culture possible. "For were the nature of human actions as distinctly known as the nature of quantity in geometrical figures, the strength of avarice and ambition, which is sustained by the erroneous opinions of the vulgar as touching the nature of right and wrong, would presently faint and languish; and mankind should enjoy such an immortal peace, that unless it were for habitation, on supposition that the earth should grow too narrow for her inhabitants, there would hardly be left any pretence for war."[17] This argument is borne from Hobbes's science of definition; which itself rests upon the authority of his indictment of sense perception; which lastly depends squarely upon the authority of his conception of what is body to begin with.

THE IDEA OF BODY IN THE LITTLE TREATISE

Most scholars agree that the *Little Treatise* or *Short Tract* discovered among Hobbes's papers at Chatsworth by Ferdinand Tonnies is a significant exposition of Hobbes's youthful philosophy. Tonnies places the dating of this manuscript between 1630–1636. It is thought that the date is closer to 1630. This *Little Treatise* constitutes Hobbes's first adventure into natural philosophy.[18] One must say that the *Little Treatise* is the precursor of Hobbes's *prima philosophia* that is only finally developed in *Thomas White's De Mundo Examined* and *De Corpore*.

Frithiof Brandt, still the most celebrated investigator into Hobbes's natural science, characterizes the *Little Treatise* as Hobbes's first exploration into the nature of 'motion.' It is motion, in Brandt's view, that is the focus of Hobbes's physics. Brandt, however, is not so good a philosopher. Hobbes's *Little Treatise* is principally a treatise about *body*. It is a treatise about *the nature* of body, the one that Hobbes believes to exist.[19] This is the indestructible body of atomist heritage, which some scholars have vehemently denied plays a serious role in Hobbes's natural science.[20]

To start very simply with the *Little Treatise*, we must inquire as to its fundamental categories. 'Substance' and 'accident' would be those divisions. 'Substance' and 'accident' are a terminology that Aristotle developed *first*. Like Bacon and Descartes, Hobbes appears to wish to blur the distinction between his own theory and that of Aristotle in this early work.[21]

Socrates is a substance. Socrates wearing a hat is an accident; because the hat does not make Socrates Socrates; and removing the hat from Socrates does not affect the being of Socrates. This, in any event, would be Aristotle's enumeration of the two terms. "And since natural science, like other sciences, confines itself to one class of beings, that is, to that sort of substance which has the principle of its movement and rest present in itself, evidently it is neither practical nor productive" (*Metaphysics* 1025b18). The 'accidental,' for Aristotle, indicates those qualities that usually do not belong to a substance, that is, qualities which are not 'necessary' to it, and which therefore do not describe the object's essence. "Since, among things which are, some are always in the same state and are of necessity (not necessity in the sense of compulsion but that which means the impossibility of being otherwise), and some are not of necessity nor always, but for the most part, this is the principle and this the cause of the existence of the accidental; for that which is neither always nor for the most part, we call accidental" (*Metaphysics* 1026b27).

Socrates's capacity to reason is, like the hat, something that can be predicated of Socrates. However, the capacity to reason is part of the essence of a human being. The capacity to reason cannot be removed from the human being without severely damaging that being. Socrates's capacity to speak is likewise part of his essence; as is Socrates's mortality. One cannot be immortal and still be Socrates. All men are mortal. Mortality is a quality universally true of human beings. The essence of mortality is part of the human essence. There is nothing *accidental* about Socrates being mortal. When Hobbes chooses to define a human being as 'a living creature,' he is flexing the muscles of his science of definition. Do not human beings die? Is it impossible to speak of a dead human being, a dead man? "From hence also this may be deduced, that the first truths were arbitrarily made by those that first of all imposed names upon things, or received them from the imposition of others. For it is true (for example) that *man is a living creature*, but it is for this reason, that it pleased men to impose both those names on the same thing."[22]

Hobbes loosely embraces Aristotle's definition of a substance, but he replaces Aristotle's conception of accident with the Epicurean atomistic variation. "Substance is that which hath being not in another, so as it may be of itself, as aire, or gold," Hobbes writes (*Little Treatise* section 1, principle 15). "Accident is that which hath being in another, so as, without that other it could not be, as color cannot be, but in somewhat colored" (*LT*, section 1, principle 16). For Hobbes, substances are those things that can exist of themselves, such that they do not need another to exist *in*. Accidents are qualities that depend upon substances in order to exist in them. *Hobbes does not develop the distinction that Aristotle makes between essence and accidents.* Aristotle regards the distinction between essence and accident as pivotal for knowledge: "And it is an accident that a man is white (for this is neither always nor for the most part so) but it is not by accident that he is an animal" (*Metaphysics* 1026b34).

The very first principle of the first section of Hobbes's *Little Treatise* reads as follows: "That, whereto nothing is added, and from which nothing is taken, remains in the same state it was." The second principle is as follows: "That which is no way touch'd by another, hath nothing added to nor taken from it." In these two postulates, we behold the entire foundation for Hobbes's mechanistic physics. No body or substance is capable of moving itself. Any body that is moved, therefore, must be caused to move by some other body. It is hard not to notice, that Hobbes here tosses Aristotle's entire conception of natural substance out the window; for in Aristotle's conception of natural substance, it is that which has its principle of motion within *itself.* "Therefore in one sense nothing will have a definition and nothing will have an essence, except substances, but in another sense other things will have them. Clearly, then, definition is the formula of the essence, and the essence must belong to substances either alone or chiefly and primarily in the unqualified sense" (*Metaphysics* 1031a10). "For those things are natural which, by a continuous movement originated from an internal principle, arrive at some end: the same end is not reached from every principle, nor any chance end, but always the tendency in each is towards the same end, if there is no impediment" (*Physics* 199b16).

In the conclusions to section 1 of the *Little Treatise*, Hobbes reiterates that "Nothing can move itself" (#10). Hobbes's reasoning for his conclusion is as follows. If an object had the power to move itself, it must always have the power to move itself. If an object has the power to move itself, it must in fact move itself always. Therefore, if an object be at rest, then it must not have the power to move itself. It must remain at rest always unless and except as some other body touches it (Ibid.).

Hobbes admits, in Conclusion #4, that any accident depends on its substance in order to have being. "For seeing that which is moved must have being while it is moved; and no accident can have being without that subject in which it is (by 16 Prin.): it follows, that no accident can be locally moved out of his own subject." This is to say, Hobbes admits that the *substance* must have the part *of being*.

Hobbes, however, is leaving something out of the equation. For anything to exist, for being to be possible at all, there must be some *unity* to possess this part of being. Hobbes's substance, in other words, cannot be *merely* being. It must have the part of unity. Unity and being are *unlike natures*. In other words, Hobbes's substance must be a whole of parts: a unity that has the part of being. In just these few sentences, we have seen that Hobbes's substance cannot lack the natures of unity and being, part, or whole: no being can exist, unless that being is paired with the part of unity; and any unity that has the part of being, constitutes a whole of parts. The natures of whole and part are therefore proved.[23]

It is the nature of a whole that it must have *extremities*. As a whole, the object must have a beginning, and it must have an end. It must come into being, and it must pass away. We have already disproved the very first thesis that Hobbes advances in his *Little Treatise*. Hobbes has said that "whereto nothing is added, and from which nothing is taken, remains in the same state it was." And the second principle, to repeat: "That which is no way touch'd by another, hath nothing added to nor taken from it." To come into being and to pass away concern the object *by itself*. Something comes into being, and that object itself passes out of being. It acquires the part of being at one time, and loses the part of being at another time. Since a real object must possess the three parts of time—past, present, and future, it will also undergo *alterations*. It will undergo alterations of size and shape, of texture and color. It will undergo all of these motions irrespective of the movements of any external objects.

Plato describes the nature of an object in the following manner: "Further, since its parts are parts of a whole, the one, in respect of its wholeness, will be limited. For the parts are contained by the whole, and the container must be a limit. Therefore, a 'one which is' is both one and many, whole and parts, limited as well as indefinitely numerous. Since it is limited, it will have extremities, and if a whole, it will have a beginning, middle, and end. A thing cannot be a whole without all these three: if any one of them is lacking it will no longer be a whole. Thus the one will have beginning and end and middle. But the middle can only be what is equidistant from the extremities" (*Parmenides* 145ab).

Nothing is added to the object, because the object does not exist yet, before the acquisition of being; and nothing is taken away from the object, insofar as this movement is concerned, since the object does *not remain* when the part of being is lost. In its extremities, the object undergoes the most radical motion. Hobbes's definition of a substance is therefore disproved. These moments of acquiring and losing being cannot be, or ought not to be confused with *alteration*. That which begins obviously constitutes a beginning, a new thing; whereas the object that loses being is not therefore altered, but ceases to have the part of being.

In the second part of the *Little Treatise*, Hobbes has begun to talk about the causes of perception. Indeed, in Brandt's analysis, it is Hobbes's interest in the phenomenon of perception that is the driving force of the entire *Little Treatise*.[24] Hobbes, in this early work, adopts a theory of perception which is effectively identical to that of Democritus, Epicurus, and Lucretius. Hobbes says that the substance emits 'species': 'species' are small bodies, bodies too small to be perceived. Thus if a body is going to affect another body, if *A* is going to affect *B*, this must be accomplished by *A*'s emitting these tiny objects called 'species.' "Every agent that worketh on a distant patient, toucheth it, eyther by the medium, or by somewhat issueing from itself, which thing so issuing lett be call'd species" (Section 2, principle 1).

It is true that Hobbes believes that some substances can work upon other substances, through "actions on the parts of the corporall medium" (Section 2, conclusion 1). At a later date in his career, Hobbes will dispense with the thesis of species emission. Yet that is nothing to the present purpose, for a 'species' in Hobbes' definition is a *substance*, a body which can itself cause motion in other substances. The act of perception in a human being, for example, can only, in this Hobbes's earliest theory, be produced by motion, which is itself caused by the species when it touches the patient.

If the above is true, the 'species' itself offers us a second opportunity to look at Hobbes's definition of body, in this his earliest known work of natural philosophy. Hobbes spends some time, in section 2, examining the local motion of a species. And indeed, if a body is to travel from one body *A* to a second body *B*, it must indeed move or travel locally, that is, it must locomote somehow. Only that can locomote which is a body, which is a substance. Thus the species is, like the Epicurean image, a thin film of body which is emitted 'infinitely' by the agent (Section 2, conclusion 4). Let us consult Epicurus' account of the 'image,' which Hobbes's account of the 'species' so resembles:

> Furthermore, there are images having the same shape as the solid objects, but far removed from objects apparent to the senses owing to their subtlety. For it is not impossible that such emanations arise in the air surrounding the objects, nor that there arise tendencies toward the manufacture of hollow spaces and subtleties, nor that such emanations maintain the constant position and order they had in the solid bodies. We call these images films that are given off by the object and that convey an impression to the eyes. (*LH* 46a)

Hobbes makes the argument that as regards substances, in particular as regards species, that there is "no minimum in eyther," that there is no "least part in line and time" (section 2, conclusion 8). Hobbes speaks incorrectly here. The extremities of a whole cannot be infinitely divided. The species is a body, a substance; and it therefore must have the nature of a *whole*, as proved above. It must possess the parts of being and unity; if it lacks either one, it is nothing. If it possesses both, it is a whole of parts. If the extremities could be infinitely divided, they would not be extremities. The extremities of a whole must include time past, present, and future. The past time of the object is bounded by the moment when it acquired being; the future time of the object is bounded by the moment when it loses the part of being. The present is the part of time which is neither past nor future. These parts are not subject to infinite division. There is a least part of time, and a greatest part of time: the least part of time consisting in the moment when the object acquires being, and the greatest part of time recorded at that moment when the object loses the part of being (for that is when it is oldest).

When Hobbes is speaking about 'species,' he makes an observation that is characteristic of his natural philosophy, and indeed, of the natural philosophy of his predecessors. These alleged workings are offered to us on the premise that such things "as all the wayes of nature, be to us not so perceptible" (*Little Treatise*, section 2, conclusion 8).

When Plato refutes the several varieties of atomistic philosophy in his dialogue the *Parmenides*, he does so with the express purpose of evaluating the challenge that these theories make to the authority of ordinary sense perception. When Plato disproves the possibility of indivisible body, when Plato proves that any object in nature must come into being and pass away—he is in that moment disproving the theories that allege that the bodies ordinarily perceived are not the real ones. The small species that Hobbes characterizes do not have the part of being. They 'are not.' They cannot be perceived nor correctly spoken of. "And if a thing is not, you cannot say that it 'has' a name or that there is anything 'of' it. Consequently, it cannot have a name or be spoken of, nor can there be any knowledge or perception or opinion of it. It is not named or spoken of, not an object of opinion or of knowledge, not perceived by any creature" (*Parmenides* 142a).

THOMAS WHITE'S DE MUNDO EXAMINED

Hobbes composed *Thomas White's De Mundo Examined* at the request of Marin Mersenne. Mersenne was a Catholic priest who was at the cutting edge of the new scientific developments in France. Mersenne served as intermediary between Hobbes and Descartes. It was Mersenne, in fact, who arranged for Hobbes to be one of the respondents to Descartes' *Meditations*.[25]

White was a philosopher whose book *De Mundo* addressed a wide variety of concerns that were central to Hobbes's project. Mersenne asked Hobbes to undertake the study, in order the better to introduce the Mersenne circle to the full scope of Hobbes's arguments in science and *prima philosophia*. In *Thomas White's De Mundo Examined*, Hobbes continues to feign a loose allegiance to Aristotelian principles. The Aristotelian principle of central importance for Hobbes's argument about body, is the former's doctrine of 'prime matter.'

Aristotle studied long and hard the question that had perplexed the early Milesian scientists, and which had been aggressively addressed by Democritus and Leucippus. It is clear that the objects that we witness, coming into being and passing away, must come into being out of something, these earlier thinkers reasoned. It is also clear that they must be returned into something, these perishable objects, when they pass away. How this question is answered makes a great deal of difference.

Parmenides had made the argument that "nothing comes from nothing," and that therefore the true reality in nature should be consecrated as 'being,' which itself neither ever comes into being nor passes away. Parmenides adds on to this observation the wrinkle that something cannot really come to exist out of something either: because it *is* already. In Parmenidean philosophy, coming into being is an illusion. The atomist philosophy aspires to offer a theory of coming into being, but ultimately the atoms are incapable of being either wholes of parts themselves, or parts in some larger whole. This reasoning has been discussed elsewhere.[26]

Empedocles and Anaxagoras, the natural philosophers who first labored from beneath the heavy shadow of Parmenides' argument, reasoned that the true 'roots' or 'seeds' of nature must be indestructible, ungenerated things. Democritus and Leucippus also reasoned in this vein, producing their theory of the atoms. In the atomist theory, the true bodies are ungenerated and indestructible.[27] Perishable bodies are caused by the collision of the atoms with one another, in the atomist theory; and when a perishable body is seen to pass away, this is thought by those atomists to merely represent the reallocation of the true bodies, the atoms, in different configurations.

Aristotle thought Empedocles, Anaxagoras, and Democritus were incorrect. Coming into being and passing away are real and true things, Aristotle insists. Nature intends for generation to come to pass, in Aristotle's view; and nature has plans for that which comes into being. We here begin to touch upon Aristotle's famous theory of 'final causes': that is, that there is an excellence, a potential fulfillment, different for the various kinds of natural object, towards which they strive upon coming into being—unless otherwise interfered with. For Aristotle, accordingly, coming into being and passing away are not only real, they are nature's *aims* and limits. Motion, which to Hobbes is but an 'accident,' is for Aristotle the very essence of nature's intention. "Nature is a principle of motion and change, and it is the subject of our inquiry. We must therefore see that we understand what motion is; for if it were unknown, nature too would be unknown" (*Physics* 200b13). "It is the fulfillment of what is potential when it is already fulfilled and operates not as itself but as movable, that is motion" (*Physics* 201a29). "The solution of the difficulty is plain: motion is in the movable" (*Physics* 202a12).

Aristotle it is who authored the famous distinction between natural and violent movements in nature. Quite obviously, bodies which are indestructible have no nature. It is impossible to treat a body violently, if it is impossible to *destroy or fulfill* said body. Democritus's and Leucippus's atoms are such indestructible bodies; and it is not therefore a surprise that both of these thinkers do away with the principle of natural movement that Aristotle finds inherent in every natural being. "If then it is both by nature and for an end that the swallow makes its nest and the spider its web, and plants grow leaves for the sake of the fruit and send their roots down (not up) for the sake of nourishment, it is plain that this kind of cause is operative in things which come to be and are by nature. And since nature is twofold, both the matter and the form, of which the latter is the end, and since all the rest is for the sake of the end, the form must be the cause in the sense of that for the sake of which" (*Physics* 199a26).

We have seen, already in this work, that both Francis Bacon and Descartes do away with the distinction between natural and violent motion. Despite the disguises which they employ to adorn their versions of the atomist theory, atomist philosophers is what these thinkers are. Hobbes follows in the footsteps, not of Aristotle, but of Bacon and Descartes, Epicurus and Lucretius, Democritus and Leucippus.

According to Aristotle, 'prime matter' is something that *does not have being*. Prime matter is matter in its eternal aspect. This is to say that for Aristotle, 'matter' deprived of form is only *potential* being, not actual being. "By matter I mean that which in itself is neither a particular thing nor of a certain quantity nor assigned to any of the other categories by which being is determined" (*Metaphysics* 1029a20). In order for prime matter to become actual being, *it must be married to a form.* "That form indeed is nature rather than the matter; for a thing is more properly said to be what it is when it exists in actuality than when it exists potentially" (*Physics* 193b7). It is the form alone which gives identity, order, and purpose to prime matter: as it is form alone which carries matter into a mortal career, towards a path of potential development towards some fulfilled natural state, a 'final cause.' Nature intends to realize something through generation, Aristotle reasons; and therefore Aristotle's theory of prime matter is designed to illuminate the reality of genesis and destruction, while also providing a coherent explanation for matter in its eternal state.

'Matter' by itself does not exist, in Aristotle's view. There is no such thing as 'body' which is merely matter. There is only formed matter. Human beings have no experience of anything other than formed matter. All perishable objects are formed matter, or *substances*. It is the nature of substances to be perishable, in Aristotle's theory. We have already seen, in Hobbes's *Little Treatise*, that he denies all sorts of self-movement to *his* substances. Coming into being and passing away are the alpha and omega of movements that all substances know, in Aristotle's philosophy. When a perishable object ceases to be, it *really* perishes: and what is left no longer fits into the category of 'being.' For Hobbes, things are quite otherwise: for Hobbes, when an object perishes, this is merely a deception of the human senses. 'Matter,' for Hobbes, not only has being by itself; but it is indestructible. Aristotle's prime matter does not have being. Hobbes's version of prime matter does. Hobbes, again, is a philosopher in the atomist tradition.

Hobbes tells us, in *Thomas White's De Mundo Examined*, that 'Philosophia Prima' deals with "essence, matter, form, quantity, the finite, the infinite, quality, cause, effect, motion, space, time, place, vacuum, unity, number, and all the notions which Aristotle discusses" (23–24). "Birth and death are sometimes taken in the respective senses of 'creation out of nothing' and 'destruction' or 'reduction to nothingness.' These processes, however, can be carried out and understood by divine omnipotence alone, for we cannot in any way comprehend in our imagination how something can be created from nothing or how nothing can result from something" (63).

> Sometimes, indeed frequently, the terms are taken to mean 'generation' and 'corruption,' that is, change. This is not, however, 'any change you like,' but only a change which makes us assign to, or remove from a thing the name that answers the question, 'what is the thing?' Thus we do not say that man is produced because the material of which he is composed takes on a new color or the characteristics of flesh or bones, or is subjected to another, minor change—we say it only when the material undergoes so great a change that, on the change's manifesting itself to us, it creates a picture such as leads us habitually to call any thing 'man.' (63)

According to Hobbes, "the material in which consists the nature of bodies does not perish, however" (63). "Again, the premise 'what can perish is mortal' is not a correct use of terms, for the death of any substance is not perishing but a certain prescribed change of percipient bodies into non-percipient. Nor is perishing, what we call reduction to nothingness, death" (63).

Hobbes brazenly and falsely conflates Aristotle's definition of prime matter, which we have described above, with the atomist philosophy. Again, for Aristotle, substances really do perish, really are mortal, really *do* lose the part of being. For Aristotle, nature is the principle of movement which is working towards a state of fulfillment. It is an *end*. "Therefore action for an end is present in things which come to be and are by nature" (*Metaphysics* 199b8). Not so for the atomists. What is at stake in this debate is the absolutely vital distinction between natural and violent motion, for one thing; and in the second case, what is at stake is the care and caution ordinarily felt for the perishable bodies, among communities of human beings. The theory of indestructible body that Hobbes proposes makes possible a certain kind of detachment for the political philosopher, as regards the lives and hopes and feelings of the members of the polity. This detachment is usually conveyed in a language of deliberate abstraction; but its impulse is aggressive. For this new science is, from the start, committed to imposing force on the perishable bodies, on imposing artificial forms upon them. The chief characteristic of this science is its focus on will, its own will. "The end of knowledge is power," Hobbes writes, "and the use of theorems (which among geometricians, serve for the finding out of properties) is for the construction of problems; and lastly, the scope of all speculation is the performing of some action, or thing to be done" (*EW* I.7).

It is the Eleatic atomist heritage which best serves Hobbes's project of philosophy. Knowing what objects are in their essence would impose constraints upon the philosophy of mastering and ordering said bodies. The atomist doctrine provides a sort of veil of ignorance for the natural philosopher: he claims not to be able to know the objects in their essences, which in turn enables him to limit his speculations to things that he can *do* to the objects, whatever they prove to be. Hobbes would prefer not to have to directly confront Aristotle's theories as to what objects in nature *are*. In misrepresenting Aristotle's theory of prime or eternal matter, Hobbes can suggest that the two projects, atomist and Aristotelian, more or less converge. "Recognizing this, Aristotle introduced as the principle of things a materia prima which could take on the form of everything perceptible and which could be changed in any fashion whatever, that is, in an unlimited number of ways; hence the variety of things is infinite. This is accepted by those philosophers who believe that everything consists of atoms, for they make those atoms homogeneous, differing from one another in shape alone, as if to say that those atoms were that unique matter which Aristotle called prima materia" (80).

Coming into being is a real thing for Aristotle, as is ceasing to be. This is because there is a definite point at which a form is joined to matter. A form by itself does not qualify as being for Aristotle; matter by itself does not qualify as being for Aristotle. It is only the combination that qualifies as being; and when the movement imparted by the form reaches its limit in the appropriate matter, a limited endeavor for Aristotle, the limits of the object's being are reached. Neither matter or form can be characterized as 'accidents' in Aristotle's philosophy; for the only thing thought to be characteristic of matter, is that it desires form.

> The matter comes to be and ceases to be in one sense, while in another sense it does not. As that which contains the privation, it ceases to be in its own nature; for what ceases to be—the privation—is contained within it. But as potentiality it does not cease to be in its own nature, but is necessarily outside of the sphere of becoming and ceasing to be . . . (For my definition of matter is just this—the primary substratum of each thing, from which it comes to be, and which persists in the result, not accidentally) (*Physics* 192a25).

For Aristotle, contra Hobbes, the very subject matter of physics is delimited by perception. "Our inquiry is limited to our special subject matter, the objects of sense" (*Physics* 204b2). It is Hobbes's account of body which is a single thing; and in Aristotle's view, a theory such as Hobbes is compelled to regard all coming into being as mere alteration. This is not however Aristotle's view. "For if the change is alteration, then the substratum is a single element; that is, all things which admit of change into one another have a single matter" (*On Generation and Corruption* 314b30). "In one sense things come-to-be out of that which has no being without qualification; yet in another sense they come-to-be out of what is. For there must pre-exist something which *potentially* is, but actually is not." (*On Generation and Corruption* 317b15). "Matter, in the most proper sense of the term, is to be identified with the substratum which is receptive of coming-to-be and passing away"

(Ibid., 320a3). "Now cause, in the sense of matter, for the things which are such as to come-to-be is that which can be and not be; and this is identical with that which can come to be and pass away, since the latter, while it is at one time, at another time is not" (*On Generation and Corruption* 335a23).

For Aristotle, one form cannot change into another form. This is where the two traditions of physics finally lock horns. In the world of Plato and Aristotle, notwithstanding the differences in their respective theories of form, all of the forms are distinct and none of them are interchangeable. In the materialistic theory of nature which excludes form from the equation, all change is merely alteration. Everything is basically the same.

Later on in his work examining Thomas White's *De Mundo*, Hobbes gives us his definition of body. This definition of body echoes the theory of Descartes. That is body, Hobbes argues, which takes up space, which has dimensions, or which is to say the same thing, has extension. "The essence of body, or corporeity, consists (as has been said earlier) in the body's possessing dimensions or (which is the same thing) in its occupying place" (320). In this definition of body that Hobbes provides us with, time is not allowed to qualify as a motion in the things themselves. Time is merely ascribed to our imaginations: "It remains, then, that time is not a motion in things themselves, outside the mind, but is mere imagination" (339).

It can be seen very easily that Brandt is incorrect to christen Hobbes as the philosopher of 'motion.' Hobbes is not the, but *a* philosopher of body. From Hobbes's first ruminations in his *Little Treatise* to his account of Thomas White's work and through to his *De Corpore*, the 'body' that Hobbes enumerates as the true one is absolutely unrecognizable to human experience. Hobbes's bodies cannot move themselves. They cannot, in reality, come into being, or be destroyed into not-being. Not only this: but they cannot have the quality of movement, the quality of time, in themselves.

Aristotle would certainly not regard Hobbes as the philosopher of 'motion.' Hobbes does away with the very source of movement in nature, the formal cause. In Aristotle's view, "the form indeed is nature rather than the matter; for a thing is more properly said to be what it is when it exists in actuality than when it exists potentially" (*Physics* 193a7). "What grows qua growing grows from something into something. Into what then does it grow? Not into that from which it arose but into that to which it tends" (*Physics* 193b16).

It would be useful to remind the reader at this point of the stipulation offered at the beginning of this chapter. It is a great debate in Hobbes scholarship, whether his civil science or political and moral philosophy can be studied competently in separation from, and in exclusion of, Hobbes's natural science. I answered that these studies cannot be severed. Here, at this juncture in our discussion, we may begin to appreciate the reasons for this argument afresh.

Hobbes's civil science, as we will later examine, rests upon the premise of a nature that is at best indifferent to man. It is more truly based, Hobbes's civil science, on a theory of nature that is *incomprehensible* to man. The 'definitions' and configuration of names that Hobbes employs as the foundation of his civil science, obtain their principal justification and bearings from Hobbes's *natural* philosophy. It is only based upon the premise that nature is disordered, or without *any* order of its own, that Hobbes eventually becomes free to assign definitions to human beings and human things which deny the order to which they adhere, or *would seek* to adhere, if left unimpeded by civil philosophy and the institutions to which it seeks to give birth.

Hobbes does allow us, in *De Corpore*, to set aside the language of 'accidents' such as must be employed, in his view, whenever we are talking about perceptible bodies. That is, in *De Corpore*, Hobbes speaks about body as it must really and truly exist in nature. It is this latter sort of characterization that we are interested in. For both Plato and Aristotle, perception does give us accurate information about the true objects that exist in nature. Not so for Hobbes.

Hobbes writes, "The extension of a body, is the same thing with the magnitude of it, or that which some call real space. But this magnitude does not depend upon our cogitation, as imaginary space doth; for this is an effect of our imagination, but magnitude is the cause of it; this is an accident of the mind, that of a body existing out of the mind."[28] It is important to take note of the fact that Hobbes does consider himself able to talk about authentic body, that is, about a body which is not the subject of mere 'phantasms' and 'imaginations.' The criticisms that we propose to aim at Hobbes's description of body involves this one that exists apart from our minds. In this definition of actual body, Hobbes is omitting to characterize the parts of unity and being without which no object in nature can exist.

Extension concerns the extremities of an object. Shape concerns the extremities of an object. Color concerns the extremities of an object. Place concerns the extremities of an object. Time concerns the extremities of an object. What has extension, shape, color, place and time can only be a whole of parts. These parts are not figments of the imagination, nor are they creations of the mind. Nor are the parts *alike*. Being is not unity; the nature of a whole is not the same as the nature of a part. The nature of a whole, as we have traced out, must involve extremities in time, not merely of dimension. The extension of a body is not eternal, nor can it be. Again, we here speak of a body as it exists 'out of the mind.' Extension is something that a body acquires, when it comes into being; and it is indeed something that is lost, when the body perishes.

This disproves Hobbes's argument that time is merely a part of the human imagination. This disproves Hobbes' argument that bodies have no movement inherent in themselves. That body which lacks coming into being and passing away as true movements in itself, and lacks the parts of time so acquired, is no body at all. It is but a phantom, a delusion, a sophism.

Frithiof Brandt, among others, makes the point that Hobbes is simply not much remembered by historians of natural science. Hobbes, in Brandt's analysis, is quite detached from the experimental tradition of science that Bacon launched.[29] Yet natural science is, for Hobbes, but a premise, a foundation, for the true goal of his intellectual labors. That goal is cultural and political. It is to gain control over the dispensation of naming, that Hobbes labors so hard in the fields of natural philosophy. It is in his account of perception, in the domain of physics, where Hobbes attacks the authority of the ordinary assignment of name to object. The failure of Hobbes's account of perception has now been demonstrated, as the failure of the theory of body upon which said account rests, has been demonstrated.[30]

DE CORPORE

De Corpore is perhaps Hobbes's most difficult work. In this effort Hobbes lays down his arguments, on metaphysics, physics, names, and logic, in a final form. The design of Hobbesian argument, as I comprehend it, involves a three-fold formulation. At the foundation is a theory of body. This theory is present in all three of Hobbes's works that seriously treat of body: the *Little Treatise*, *Thomas White's De Mundo Examined*, and *De Corpore*. This theory which Hobbes unfolds, as it pertains to body, is *metaphysical*: it is the thesis that true body is ungenerated and indestructible. This is the legacy of ancient atomism, and I have distinguished it sharply from Aristotle's theory of materia prima, which it in no way resembles.

Hobbes's theory of body is the absolute and indispensable antecedent of his critique of ordinary human perceptual faculties. Hobbes's metaphysical theory of body reduces to the omnibus category of 'accident' all of the perceptions that ordinary human beings have: and it is, lastly, this attack on the authority of perception which enables Hobbes to lay claim, for his science, to the authority to define names anew. Hobbes's philosopher regards himself as an inventor of language. Indeed, he has insisted that the philosophers who follow him regard themselves as like unto God at the time when the original chaos was brought into the order of creation:

> Philosophy, therefore, the child of the world and your own mind, is within yourself; perhaps not fashioned yet, but like the world its father, as it was in the beginning, a thing confused. Do, therefore, as the statuaries do, who, by hewing off that which is superfluous, do not make but find the image. Or imitate the creation: if you will be a philosopher in good earnest, let your reason move upon the deep of your own cogitations and experience; those things that lie in confusion must be set asunder, distinguished, and every one

> stamped with its own name set in order; that is to say, your method must resemble that of the creation.[31]

In this case Hobbes is the creator, or would-be such; and as chancellor of what names shall be construed to indicate in discourses committed to the ascertainment of truth, he undertakes to unite the project of philosophy with rhetoric. In this particular Hobbes can be seen to follow Bacon. It was Bacon who assigned the category of logic and truth so narrowly, to those trained in his experimental philosophy; as it was Bacon who conceived of the multitude as limited to the domain of 'imagination.' Yet Bacon does not possess the blazing rhetoric or logic that Hobbes possesses. Hobbes's rhetoric is intended to be socially transformative, as we will see.

In *De Corpore*, Hobbes once again makes it all too clear that he clutches, close to his bosom, the shibboleth of the atomist faith: that there is indestructible body, and that we are incorrect to say of any perishable body that the true body is destroyed *ever*. "But that magnitude for which we give to any thing the name of body is neither generated nor destroyed. For though we may feign in the mind that a point may swell to a huge bulk, and that this may again contract itself to a point; that is, though we may imagine something to arise where before was nothing, nothing to be there where before was something, yet we cannot comprehend in our mind how this may possibly be done in nature. And therefore philosophers, who tie themselves to natural reason, suppose that a body can neither be generated nor destroyed, but only that it may appear otherwise than it did to us, that is, under different species, and consequently be called by other and other names; so that which is now called a man, may have at another time the name of not-man; but that which is once called a body, can never be called not-body" (*De Corpore*, *Works*, I.116).

Hobbes attempts here to link his theory of body to Aristotle's idea of 'materia prima.' Yet this proposed connection cannot bear scrutiny. For Aristotle's materia prima cannot have quantity, it cannot have extension; it cannot have the part of *being*. Hobbes's conception of body, however, cannot be separated from magnitude, number, or being. "And as for that matter which is common to all things, following Aristotle, usually call material prima, that is, first matter, it is not any body distinct from all other bodies, nor is it one of them. What then is it? A mere name; yet a name which is not of vain use; for it signifies a conception of body without the consideration of any form or other accident except only magnitude and extension, and aptness to receive form and other accident. So that whensoever we have the use of the name body in general, if we use that of material prima, we do well" (Ibid., 118). "Materia prima, therefore, is body in general, that is, body considered universally, not as having neither form nor any accident, but in which no form nor any other accident but quantity are at all considered, that is, they are not drawn into argumentation" (119).

As indicated, there is no such thing as an *enumerable* formless matter in Aristotle's metaphysics or physics. Coming into being is a real thing in Aristotle's nature; so is ceasing to be, passing out of being, or destruction. Nor is it possible that any object whatsoever, can possess the attribute of extension, or dimensions, without at the same time possessing the extremities of coming into being and passing out of being, the very things that Hobbes is denying.

Hobbes speaks very harshly about things incorporeal. It is not inconvenient that we should be talking about Plato's proof of the realities of objects in nature here; for in those proofs, Plato also demonstrates the reality of the incorporeal things. Plato has proved, in the above cited passages, the reality of the natures of unity, being, whole, and part. Plato has also proved that anything whatsoever that can be said to exist in nature, must possess the coequal natures of unity and being. The whole, therefore, must have the coequal natures of unity and being; but so too must each part in the whole, possess its *own* natures of unity and being. The whole possesses the natures of unity and being; and so does the part possess the natures of unity and being. Unity and being, therefore, exist in more than one place. These natures, make no mistake, are *in* the objects so denominated.

Plato writes: "Again, among all these parts there cannot be any which is part of being and yet not a (one) part. If it is, then, so long as it is, it must always be some *one* part; it cannot be *no* (not one) part. Consequently, unity must belong to every part of being, and be lacking to none, smaller or greater. . . . It appears, then, that unity itself is parceled out by being, and is not only many but indefinitely numerous." (*Parmenides* 144c). Unity is in all of the parts, but it is always the same as itself. This is, incidentally, the characteristic, or one of the characteristics, of what Aristotle means by 'form.'

Hobbes disputes this roundly, that there is any such reality in nature that can be in more than one place at a time. Hobbes denies that there is an essence of a human being, that is in each and every human being. Hobbes denies that there is an essence of a tree, such that it is present in each and every tree. Universality, such as Plato is proving to exist across the many parts of even one single object, is in Hobbes's view merely and solely the creation of grammarians. Names alone, can be characterized as universal for Hobbes; not *natures*. Yet 'being' and 'unity' are natures that are not created by grammarians. Plato has proved above that they are indeed present, these two natures, in every whole, and in every part of every whole, that exists in nature.

Unity, being, whole, part: these are the incorporeal forms which we have just, following Plato, proved to exist in any body whatsoever. From this foundation, we could make arguments extending to Aristotle's distinction between matter and form: reviving his doctrine of the formal cause, or formula, which orders a specific matter towards a particular path of development appropriate to its kind. We could plant our feet in this discussion of the proofs of the incorporeal forms, demonstrating that there is such a thing as the essence of a human being, the essence of a tree, that 'quiddity' is not a stupid or barbarous term, as Hobbes would have us believe.[32]

HOBBES ON NAMES AND DEFINITIONS

In *Leviathan*, Hobbes makes some sober comments on the degree of authority that the new science, especially the new natural science, aims to wield. The authority of this natural science is to be imposed with some intensity upon the ordinary use of names. Hobbes's denunciation of the perceptual faculties of human beings is preparatory to this project of language, that is, of laying a new, near impregnable foundation for the implementation of government by rhetoric. It is clear enough that the individual who does not have the slightest familiarity with the new science's representation of the simplest bodies available in nature, is going to be in no position to rely upon his own faculties any longer, when it comes time to determining what shall be construed as reality, as fact.

"Want of science," Hobbes writes, "that is, ignorance of causes, disposeth, or rather constraineth a man to rely on the advice, and authority of others. For all men whom the truth concerns, if they rely not on their own, must rely on the opinion of some other, whom they think wiser than themselves, and see not why he should deceive them" (*Leviathan* chapter 11. #17, 69).

It is not necessary that we resort to any crude anticipation as to how Hobbes's sciences of rhetoric and its related use of names would play out in any society. It is certainly not necessary that the new rhetoric fully supplant the ordinary use of names. It would be quite a sufficient and significant feat if the new science of rhetoric and naming could flummox or question or otherwise challenge the ordinary use of names. This indeed would be a far more effective exercise of scientific authority. For how could people navigate it? For the public to experience discrepancy and frustration in its own use of names, as compared to the authoritative pronouncements upon truths of fact, is quite potent. It is certain to raise the temperature of the passions in society, which is one of Hobbes's principal goals. It makes people anxious when things that they take to be obvious facts are hoist up in public, even under authoritative colors, under quite different descriptions and characterizations. As indicated above, society must always have order, and it always will succeed in obtaining order. The materials available to the achievement of order will therefore be decisive as regards the nature of the order that is finally obtained. Hobbes's science of rhetoric wholly dispenses with the cultivation of deliberation. Hobbes's mere definitions of human nature, let alone his theory of what authority names are obliged to, quite effectively see to that. At the very least one can observe that Hobbes's doctrines are attempting to dispose society towards a form of order that does not have access to peaceful deliberation. In and of itself, that is a very powerful accomplishment, if indeed it can be accomplished.

In every society, the people must be dependent upon those who are *reputed* to know best. In the history of philosophy, as Socrates was among the first to have proved, those are reputed to know best who know best how to win public arguments. Hobbes's science of speech seems to focus on a more indirect angle of the human psyche of societal members. To bewilder human beings is far easier to do than to subject them to commandments. Bewildered human beings must seek for guidance. Hobbes's civil science is prepared to offer such guidance. "Ignorance of the signification of words," Hobbes states, "which is, want of understanding, disposeth men to take on trust, not only the truth they know not; but also the errors; and which is more, the nonsense of them they trust: for neither error, nor nonsense, can without a perfect understanding of words be detected" (Ibid., #18).

There is no way to get at the root of Hobbes's theory of language which does not keep front and center his theory of perception. Hobbes famously begins his reasoning with a tip of the hat to Descartes' radical embrace of the skeptical postulates. A man, Hobbes reasons, when he begins his first investigations, should pretend to himself that the entire world, excepting himself, has been annihilated. This would leave the Hobbist philosopher in a situation ideally suited to his nature: he is the only man in existence, as it were; he has borrowed the language from the human race but no longer owes them allegiance to the common significations, seeing as how they have been 'annihilated.' Now he may refashion the language as he pleases.

"In the teaching of natural philosophy, I cannot begin better (as I have already shewn) than from privation; that is, from feigning the world to be annihilated. But if such annihilation of all things be supposed, it may perhaps be asked what would remain for any man (whom only I except from this universal annihilation of all things) to consider as the subject of philosophy, or at all to reason upon, or what to give names unto for ratiocination's sake" (*Works* I.91). It is from this psychological point of departure that Hobbes will undertake his project of assigning names their significations in this theory of definition. He will first suppose the external world, of things and natures and the ordinary indications thereof in speech, as non-entities.

> I say, therefore, there would remain to that man ideas of the world, and of all such bodies, as he had before their annihilation, seen with his eyes, or perceived by any other sense; that is to say, the memory and imagination of

> magnitudes, motions, sounds, colors, as also of their order and parts. All which things, though they be nothing but ideas and phantasms happening internally to him that imagineth; yet they will appear as if they were external, and not at all depending upon any power of the mind. And these are the things to which he would give names, and subtract them from, and compound them one with another. (*EW* I.92)

Hobbes's rivalry with Descartes is apparent here. For Descartes, the decision to regard the evidence from the senses as defective plunges the individual into an abyss of anxiety which finally only results in the proof that his *consciousness* exists. Hobbes is a materialist. Thus Hobbes insists that when he performs such a mental experiment upon himself, he is left with the assurance that both his body and mind endure.

So much of the authority implicit in the development of Early Modern Philosophy concerns the new attitude towards nature that is being developed. The atomist philosophy conjures up a certain estrangement between the human consciousness and the natural world. The attempt to remove the category of 'final causes' from physics plays a central role in this new attitude. To ordinary opinion, the expectations that they harbor as regards the various sorts of natural bodies does not involve any prejudices on their part. They do not impose human expectations upon their understanding of natural objects; yet the atomist philosophy suggests that they do. Bacon and Descartes helped to lay the foundations for this attitude, but the rhetorical powers of Hobbes and Spinoza dramatically altered the cultural landscape in unprecedented ways.

The reader should bear in mind that the first place in which this new authority to assign definitions is to be applied, in Hobbes, will be in the department of *manners* and *dispositions*. If the degree of distance between Hobbes's theory of perception and the objects as ordinarily perceived is vast, even unbridgeable, this is only because the alterations that Hobbes proposes to make in the cultural vocabulary of the society are so immediate, so omnipresent, so intimate. It is indeed impossible to understand Hobbes's attempt to make the people feel and inhabit the realm of *disordered nature,* which is his metaphysics, without understanding that it is in the domain of his definitions as pertaining to human nature where the shoe is intended to pinch the most sharply.

Yet it needs to be understood that in Hobbes's view, he is under no obligation whatsoever to dispute with us. In his theory, he is as free to assign names and what definitions he pleases, as any other natural philosopher is. We, his readers, are free to either accept his definitions of words, and thereby be 'taught' by his 'inventions,' or to take our business elsewhere. "And seeing teaching is nothing but leading the mind of him we teach, to the knowledge of our inventions, in that track by which we attained the same with our own mind; therefore the same method that served for our invention, will serve also for the demonstration to others, saving that we omit the first part of our method which proceeded from the sense of things to universal principles, which, because they are principles, cannot be demonstrated" (*EW* I.80).

In Aristotle, for example, it is the absolute test of universals that they can be demonstrated. To define an object in its universal aspect, is to be able to test that definition by examining any particular case of that object. This is indeed the structure of Aristotle's syllogistic logic: All men are mortal. Socrates is a man. Socrates is mortal.

In Hobbes, however, the universals are not derived from the particulars as ordinarily understood. Hobbes's universal principles are rather applied to individuals. The categories of *prima philosophia* predetermine what individuals can be from Hobbes's vantage point; and Hobbes's *prima philosophia* is, like Bacon's, a theory as to what body is.

It is in this sense that we must appraise Hobbes's definition of the 'mark,' which is the word that Hobbes uses to indicate the philosopher's imposing his own definition on a name. It matters not what a word or name indicates in the common domain of meaning, from this point of view. Natural philosophers, Hobbes insists, have the right to employ any name, with scientific authority, for the meaning which the philosophers prefer.

Hobbes pushes the fight over the nature of 'body' directly into a wrestling match for sovereignty over the signification of names themselves, broadly considered. Hobbes here insists that there is no common essence or *nature* that is the same in one human being, and another human being; that there is no such thing as an incorporeal form, such that 'man' indicates the same *nature* in both Peter and John. In Hobbes, the only universal things are those names created by men, to group, as it were, some objects, by virtue of some similitude between the things. It is not the case, in Hobbes's argument, that the objects themselves simply possess natures, irreducible natures, of the same pattern or form.

> Secondly, of names, some are common to many things, as a man, a tree; others proper to one thing, as he that writ the *Iliad*, Homer, this man, that man. And a common name, being the name of many things severally taken, but not collectively of all together (as man is not the name of all mankind, but of every one, as of Peter, John, and the rest severally) is therefore called an universal name; and therefore this word universal is never the name of anything existent in nature, nor of any idea or phantasm formed in the mind, but always the name of some word or name; so that when a living creature, a stone, a spirit, or any other thing, is said to be universal, it is not to be understood, but only that these words, living creature, stone, are universal names; that is, names common to many things.[33]

It is clear that when ordinary speakers use the names 'man,' 'stone,' 'courage,' they do not take themselves to be referring essentially to *names*. When an individual refers to 'men,' he does not seek to indicate that 'men' is basically a name. The ordinary speaker recognizes objects by the pattern or form which they partake of; the forms and patterns are in and of the objects. Hobbes's philosopher does not confess dependence on the common language in which he has been raised. Rather, Hobbes has omitted to discuss the genesis of the philosopher among the constellation of ordinary meanings. The very possibility of philosophy would not exist absent the cradle of language usage that permits the communication of individual with individual. Yet in the portrait of language that Hobbes delivers to us, it is as if the philosophers merely emerge out of the thin air, equipped with the authority to redefine common words and meanings for them. "From which it follows that, for the acquiring of philosophy, some sensible monuments are necessary, by which our past thoughts may be not only reduced, but also registered everyone in its own order. These monuments I call marks, namely, sensible things taken at pleasure, that, by the sense of them, such thoughts may be recalled to our mind as are like those thoughts for which we took them" (*EW* I.14). Hobbes's philosopher is not dependent on the common use of names. It will be his project to render the rest of the community dependent rather on *his* use of names: "Words so connected as that they become signs of our thoughts, are called speech, of which every part is a name. But seeing (as is said) both marks and signs are necessary for the acquiring of philosophy, (marks by which we may remember our own thoughts and signs by which we may make our thoughts known to others), names to both these offices; but they serve for marks before they be used as signs" (*EW* I.15). "Again, though some one man, of how excellent a wit soever, should spend all his time partly in reasoning and partly in inventing marks for the help of his memory, and advancing himself in learning; who sees not that the benefit he reaps to himself will be not much, and to others none at all? For unless he communicate his notes with others, his science will perish with him. But if the same notes be made common to many, and so one man's inventions be taught to others, sciences will thereby be increased to the general good of mankind."[34]

Just as Hobbes's philosophy claims for itself the right to be a founder of the signification of words, and to this degree distinct from, and superior to, the domain of ordinary usage, so Hobbes unveils two names to distinguish the one sort of locution from the other. We have indicated the characteristics of the 'mark' above. 'Marks' are names 'taken at pleasure' by the man who seeks to philosophize. Nor is he bound in any way by the common use of names. Quite to the contrary: he is free to originate his marks as he pleases, even if this be in contravention of the ordinary usage of the community. "The difference therefore, betwixt marks and signs is this, that we make those for our own use, but these for the use of others" (*Works* I.15).

In Hobbes's view, not even the 'signs' allocated to the community for its own use, shall be conceded to indicate the actual objects or natures referred to. The entire domain of naming is subordinated by Hobbes to his regulations as pertain to perception: to whit, in proportion as we are capable of perceiving nothing but certain 'accidents,' therefore our use of names shall be limited to what we conceive of in our minds. "But seeing names ordered in speech (as is defined) are signs of our conceptions, it is manifest that they are not the signs of the things themselves; for that the sound of this word stone should be the sign of a stone, cannot be understood in any sense but this, that he that hears it collects that he that pronounces it thinks of a stone. And therefore that disputation, whether names signify the matter or form, or something compounded of both, and other like subtleties of the metaphysics, is kept up by erring men, and such as understand not the words they dispute about" (*EW* I.17).

Hobbes is sometimes said to harbor an affection for Plato, but there could not be any more dramatic difference between them than in their attitudes towards language.[35] For Plato, the first among all things are the patterns *out there* in nature, in objects, which stand in need of names. These are common objects; and no human writer has ever been able to communicate in a language, which he did not in considerable measure share with the others. For Plato, accordingly, it is the *ordinary* assignment of name to object which serves as the philosopher's original evidence. Plato's philosophy, to be clear, enforces the authority of the ordinary assignment of name to object, as the original kernel of evidence to which all true philosophy is subject. Such is the meaning of Socrates's famous profession of ignorance: that the philosopher knows no better, the original evidence, than the meanest of his community.[36]

Plato, in the *Parmenides*, after he has proved the reality of the incorporeal natures of unity, being, whole and part, leads the reader directly to the issue of language. Plato's theory of language is as the reverse of Hobbes's. Common names are used an infinite number of times, Plato allows, precisely because the very same incorporeal nature is present an infinite amount of times in diverse objects. "What I mean is this. When you use any word, you use it to stand for something. You can use it once or many times, but in either case you are speaking of the thing whose name it is. However many times you utter the same word, you must always mean the same thing" (*Parmenides* 147d). In the *Republic* Plato argues for the immaterial substances in the following way: "And in respect of the just and the unjust, the good and the bad, and all the ideas or forms, the same statement holds, that in itself each is one, but that by virtue of their communion with actions and bodies and with one another, they present themselves everywhere, each as a multiplicity of aspects" (*Republic* 476a).

Names, for Plato, have nothing to do with 'invention.' For, while the original founders of languages must be innovators, nevertheless their task is circumscribed by what the founders are *given to name*. It is the objects about us that stand *in need* of names; and it is the repetition of patterns in the diverse objects that give rise to the common names. For Hobbes, it is the philosopher's *thoughts* which are the *source* of the meaning of names. Hobbes does not regard himself as beholden to the patterns in the external objects and the names they have received in popular usage; for these have been defined as mere 'phantasms.'

What is universal to Plato, however, is not the names, but the *natures* that the names indicate. The reason why people use the name 'courage' many times, and to designate different situations, is because we recognize the same pattern of courage *in* all of those diverse objects. The individuals who display courage differ; the circumstances in which courage is manifest differ; but courage always indicates the same form, and it is this which ordinary speakers recognize *and name* in the objects about them. "I was asking about courage and cowardice in general. And I will begin with courage, and once more ask what is that common quality, which is the same in all these cases, and which is called courage? Do you now understand what I mean?" (*Laches* 191e).

Hobbes seeks to make a huge distinction between the objects that are without us, and our 'conceptions' of the same. We have traced out in considerable detail just how fervently Hobbes denounces the authority inherent in the human being's *per*ception of an external object. Now we behold a correspondent liberty that Hobbes insists must be made available to the natural philosopher, for his '*con*ceptions.' Hobbes says that his natural philosophers are free to appropriate and bend the language ordinarily used; and that what the natural philosophers do with this language, shall be regarded as their 'invention,' which the others are not at liberty to contest, but only to follow or be silent about: for who is in a position to dispute a man's conceptions? Things are much different, however, if we are talking about the objects external to us, which is of course what we *mostly are* talking about. "Moreover," Hobbes says, "whatsoever the common use of words be, yet philosophers, who were to teach their knowledge to others, had always the liberty, and sometimes they both had and will have a necessity, of taking to themselves such names as they please for the signifying of their meaning, if they would have it understood" (*EW* I.16).

It does not matter one whit, to this end, that different communities of people, in different geographic regions, generate different languages. In the Spanish, the name 'agua' signifies that pattern, which in English we denominate 'water.' In neither the case of the Spanish, nor in the case of the English, is it properly the will of the philosopher which determines the signification of the name. The name indicates what exists, not what is 'conceived.'

In the *Cratylus*, the very crafting of the alphabet is an exercise in imitation, of sounds. Names themselves, in the *Cratylus*, attempt to indicate the nature of things. This squares very well with the ordinary use of names, which is not, first of all, to bend objects to a use so much as it is to indicate the 'whatness' of the objects. This is a theory that Hobbes derides. "What others say—however—that names have been imposed on single things according to the nature of those things—is childish. For who could have it so when the nature of things is everywhere the same while languages are diverse?"[37]

Plato's Socrates, in his discussions of non-philosophers, relentlessly turns our attention to the way that names indicate patterns and forms. Ordinary speakers recognize images, and name them accordingly; Platonic philosophers attempt to translate images into logos, into definitions in speech, driving out the contradictory elements. In *the Euthyphro*, Socrates says: "So, in the name of heaven, tell me now about the matter you just felt sure you knew quite thoroughly. State what you take piety and impiety to be with reference to murder and all the other cases. Is not the holy always one and the same thing in every action, and again, is not the unholy always opposite to the holy, and like itself? And as unholiness does it not always have its one essential form, which will be found in everything that is unholy?" (5d)

Readers may think that it is easier to recognize the form of a man than it is to recognize the form of holiness or unholiness. Yet even in this case, Hobbes throws up his hands, and insists to us that there is no such thing as a nature that is common to all human beings. "It is obvious that any individual thing is one, and singular. Like Peter and John, each man is an individual, and because there exists no man who is not one of a number of individuals, it follows that no man is universal. In the same way we prove that no stone, no tree, and in sum, no thing is universal. What, then, is 'universal' if a thing is not universal?"[38] "One universal name is imposed on many things, for their similitude in some quality or other accident; and whereas a proper name bringeth to mind one thing only, universals recall any of those many" (*Leviathan*, 22).

Aristotle's discussion of form in his *Metaphysics*, it is true, aims some criticisms in Plato's direction. This is because Plato expresses belief in the isolated existence of the forms in nature, something that Aristotle cannot understand. Aristotle nevertheless has a theory of form—one which can only be said to have more than potential being when it is married to matter. Aristotle's form indicates the pattern, the matter that which is ordered and governed in terms of the form or formula. In both the discussions of Plato and Aristotle, 'form' indicates something immaterial, which is nevertheless the ordering principle in the perishable objects. "Each thing then and its essence are one and the same in no merely accidental way, as is evident both from the preceding arguments and because to know each thing, at least, is to know its essence, so that even by the exhibition of instances it becomes clear that both must be one" (*Metaphysics* 1031b18). "It is obvious then from what has been said that the thing, in the sense of form or substance, is not produced, but the concrete thing which gets its name from this is produced, and that in everything which comes to be matter is present, and one part of the thing is matter and the other form" (*Metaphysics* 1034b16). In Aristotle's view, 'matter' is not intelligible by itself; only formed matter is intelligible; and the form or essence is common to many objects, precisely the principle that Hobbes is determined to dispute. "For each thing must be referred to by naming its form," Aristotle says, "and as having form, but never by naming its material aspect as such" (*Metaphysics* 1035a7).

NOTES

1. *English Works* (*EW*) VII, p. 470–471: "How doe I fear that ye attentive reader will find that which I have delivered concerning ye Optiques fit to bee cast out as rubbish among the rest. If he do, he will recede from ye authoritie of experience, which confirmeth all I have said. Butt if it bee found true doctrine, (though yet it wanteth polishing), I shall deserve the reputation of having beene ye first to lay the grounds of two sciences: this of Optiques, ye most curious, and ye other of natural justice, which I have done in my booke *De Cive*, ye most profitable of all other." Cf. *EW* I, ix: "Natural philosophy is therefore but young; but Civil Philosophy is yet much younger, as being no older (I say it provoked, and that my detractors may know how little they have wrought upon me) than my own book *De Cive*."

2. Epistle Dedicatory, *On Man*, 35. In Thomas Hobbes, *Man and Citizen*. Edited by Bernard Gert. Indianapolis: Hackett, 1991.

3. *Leviathan*. Chapter 4, #13. Oxford: Oxford University Press, 1998.

4. Some scholars of more recent times have rushed in to defend Hobbes's claim. Leo Strauss argues that Hobbes's civil science possesses its own independent basis in the common-sense understanding, one for which knowledge of physics is by no means necessary. "In the case of Hobbes, the attempt to base political philosophy on modern science led to the consequence that the fundamental differences between natural 'right' and natural appetite could not be consistently maintained. If the significance of Hobbes's principle of 'right' was to be duly recognized, it had, therefore, first to be shown that the real basis of his political philosophy is not modern science. To show this, is the particular object of the present study" (ix). *The Political Philosophy of Hobbes: Its Basis and Its Genesis*. Translated by Elsa M. Sinclair. Chicago: University of Chicago Press, 1952 (Midway Reprint, 1984). Tom Sorrell argues the same point: "What is the relation between the project in morals and politics and the project he carried out in composing the trilogy? The question has been raised often, but it has not, I believe, received a satisfactory answer. Hobbes's work on the 'doctrine of justice and policy' is usually interpreted in the light of his exposition of a unified science. Its claim to have put politics on a scientific footing is standardly understood as a claim to have grafted politics on to the body of natural science sketched in *De Corpore* and *De Homine*. Up to a point, the interpretation is correct. Hobbes was concerned to show that natural and civil philosophy were parts of a single subject. But he did not think his politics derived its scientific status from the possibility of being located within a unified science. And although he acknowledged connections between natural philosophy and politics, it was not on account of these links that politics was supposed to be a science. He thought politics had an independent claim to be a science, indeed a better claim to be a science than physics" (3 4). *Hobbes*. London: Routlege and Kegan Paul, 1988. There are representative thinkers on the other side, including C.B. MacPherson. In MacPherson's view, "We know that Hobbes wished his political philosophy to be regarded as the final part of a consistent system the first part of which was a mechanical materialist statement of laws of motion. He argues that the sciences of human nature and of politics could be deduced from 'the first part of philosophy, namely geometry and physics.' He never attempted this task for he discovered a more feasible way. The psychological principles from

which a political science could be deduced need not, he said, themselves be deduced from the laws of motion of material but could be obtained directly by self-examination. This did not mean that his materialist philosophy was cut off from his political thought. Nor does it indicate, as some have suggested, that his materialism was never vitally connected with his theory of human nature and politics. In making a fresh start from observations he did not forget the scientific concepts he had developed in thinking out the outlines of his whole materialist system" (177–78). From "Hobbes's Bourgeois Man," in *Hobbes Studies*. Edited by Keith Brown. Cambridge: Harvard University Press, 1965. J.W.N. Watkins likewise resists Strauss's attempt to sever Hobbes's natural science from his political and moral philosophy. "This book is about Hobbes's ideas and certain logical connections between them. It is not biographical. The question it answers is, How much of Hobbes's political theory is implied by his philosophical ideas? The conclusion it reaches is that the essentials of his political theory are so implied" (9). *Hobbes's System of Ideas*. London: Hutchinson University Library, 1965. Thus it is instructive, that a great deal of significant Hobbes scholarship flocks to this pivotal issue. Strauss, in his considerable work on Hobbes, goes so far as to attempt to shut down the scholarly attention being focused on Hobbes' natural philosophy. Macpherson, in his analysis of Hobbes, argues in a contrary direction. According to Macpherson, Hobbes's work is a unity, a systematic vision of the human being in nature, that must be addressed whole. Michael Oakeshott appears to attempt to walk a narrow path in between the two thinkers, but he is clearly closer to Strauss. Hobbes, in Oakeshott's view, does not engage in metaphysics. Hobbes begins with sense experience just like the ordinary person, Oakeshott maintains; the only difference is that Hobbes possesses an innately philosophic mind, Oakeshott continues, so that he is led to make inferences from his perception that ordinary thinkers would not arrive at. "The coherence of his philosophy, the system of it, lies not in an architectonic structure, but in a single passionate thought that pervades its parts. The system is not the plan or key of the labyrinth of the philosophy; it is rather, a guiding clue, like the thread of Ariadne. It is like the music that gives meaning to the movement of dancers, or the law of evidence that gives coherence to the practice of a court. And the thread, the hidden thought, is the continuous application of a doctrine about the nature of philosophy. Hobbes's philosophy is the world reflected in the mirror of the philosophic eye, each image the representation of a fresh object, but each determined by the character of the mirror itself. In short, the civil philosophy belongs to a philosophical system, not because it is materialistic but because it is philosophical; and an enquiry into the character of the system and the place of politics in it resolves itself into an enquiry into what Hobbes considered to be the nature of philosophy" (xx). Introduction to *Leviathan.* Oxford: Basil Blackwell, 1957. Gary Herbert (*Thomas Hobbes: The Unity of Scientific and Moral Wisdom*. Vancouver: University Press of British Columbia, 1989) argues for the unity of Hobbes's system. "The intention is to restore Hobbes's thought to its more traditional philosophical location, first, by taking seriously his clearly stated intention to produce a 'system' of philosophy grounded upon physics, working through an egoistic account of human nature to a realistic account of civil association" (x). "However difficult the task may be, one cannot arrive at an adequate interpretation of Hobbes's thought without taking seriously his intention to be systematic, that is, his intention to ground the first principles of human nature and political association on the principles of natural philosophy" (9). Herbert is interested in Hobbes's claim that he is a systematic philosopher. Hobbes does say that his *De Corpore*, *De Homine*, and *De Cive*, published 1655, 1657, 1642 respectively, were meant to be parts of one system: body, man, civil body. Most scholars balk at the prospect of this systematic, and follow Strauss in so doing. This issue, the distinction between the natural science and the civil science, can become slippery. This is due to the fact that Hobbes's physics extends into the domain of perception and desire: and desire, for Hobbes, is the basis of a hedonistic ethics. That is really the crux of the issue as it generates intense debate among more recent scholars. The civil science is not based upon a hedonistic ethics, many commentators insist. Hobbes expressly severs his ethics from his political science, they say; and so he does. These commentators do not tend to notice that Hobbes is following Francis Bacon in so doing, however. "Civil and moral philosophy do not so adhere to one another," Hobbes writes, "but that they may be severed. For the causes of the motions of the mind are known, not only by ratiocination, but also by the experience of every man that takes the pains to observe those motions within himself" (*English Works* I.73).

According to Sorell, "Hobbes's claim that the principles of civil philosophy are accessible from a starting point in acquaintance with the passions and not only from a starting point in the prior sciences, may be regarded as the claim that civil philosophy enjoys a certain autonomy in the scheme of the sciences. Civil philosophy may seem to depend on the other sciences, but that is an impression created by approaching civil philosophy by the method of synthesis" (55). "But he never shows in his political writings that it is important to know the motions of the mind—the passions—as motions of the mind in order to understand the principles of politics. It seems doubtful that any inkling of the supposed mechanical nature of the passions is in fact necessary for grasping the politics" (56). "Hobbes's Scheme of the Sciences," in Tom Sorell, editor. *The Cambridge Companion to Hobbes*. Cambridge: Cambridge University Press, 1996. In the same volume, Yves Charles Zarka argues the reverse of Sorell's position. "With regard to method, then, Hobbes sees two possibilities: either a deductive progress from first principles that can be grasped from experience and the observation of our internal motions. But the disassociation of moral philosophy from politics, which separates politics from other branches of science, does not imply a complete autonomy for politics. Even if immediate experience is able to reveal the principles of politics, these principles themselves can find an ultimate grounding only by being referred back to the concepts of first philosophy" (74). In "First Philosophy and the Foundation of Knowledge." The commentators fear that the fruits of Hobbes's theory of reason will be lost, if the natural science is not kept strictly separate from the civil science. The commentators tend to believe that Hobbes favors a rather robust theory of individual choice among individuals in the civil science, and perhaps Oakeshott helped to establish this trend of thinking. Bernard Gert, for one, wishes to make it very clear to us that the natural science of Hobbes and the civil or moral philosophy must be kept very distinct. Hobbes's moral philosophy, in Gert's view, cannot be reconciled with the egoism of causal pleasure and pain; there is an admixture of reason in Hobbes's conception of morals, Gert argues, and this means that the distinction between the moral philosophy and the natural science must be upheld. "Although Hobbes often equates emotions or passions with appetites and desires, when he is more careful, he is aware that the former are only a subclass of the latter. In addition to our emotional desires, we also have rational desires, our long-term desires for real goods. Hobbes is referring to these rationally required desires when he says 'the greatest of goods for each is his own preservation. For nature is so arranged that all desire good for themselves. Insofar as it is within their capacities, it is necessary to desire life, health, and further, insofar as it can be done, security of future time.' Hobbes is not claiming that everyone desires self-preservation more than anything else because he explicitly notes that 'most men would rather lose their lives (that I say not, their peace) than suffer slander'" (163). From "Hobbes's Psychology," in the *Cambridge Companion*.

5. According to Epicurus, the only two real things in nature are bodies and void. The only other things which can be said to exist, are characterized by Epicurus as 'properties or accidents' of the two true natures. "Moreover, the totality is made up of bodies and void; for in all cases sense-perception testifies that bodies exist, and it is by sense-perception that we must infer by reasoning what is non-evident, as I already said. And if there did not exist that which we call void and space and intangible nature, bodies would not have any place to be in or move through, as they obviously do move. Beyond these two things (viz. bodies and void) nothing can be conceived, either by a comprehensive grasp or analogously to things so grasped, at least not if we mean grasped as complete natures rather than as what are termed properties or accidents of these two things" (*Letter to Herodotus. In The Essential Epicurus.* Eugene O'Connor, editor, # 39–41). Lucretius embraces Epicurus's doctrine in this regard. "Therefore apart from void and matter no third substance can remain to be numbered in the sum of things, neither one that falls within the range of the senses nor one that mind can grasp by reasoning. For you will find that all things that can be named are either properties of these two things or else you can see that they are accidents of them" (*On the Nature of the Universe*, 1.443. Translated by Ronald Melville. Oxford University Press, 2008). Since perception cannot grasp the actual atoms, all of perception will fall into the category of 'accidents.' This is indeed the terminology that Hobbes relies upon; and in Hobbes's formulation, that which we perceive is not allowed to qualify as knowledge of any actual objects external to us.

6. I have discussed Epicurus's theory of the prolepsis at length in *Machiavelli and Epicureanism: An Investigation into the Origins of Early Modern Political Thought*. Lanham: Lexington Books, 2012, 74–81.

7. *The Elements of Law Natural and Politic*. Edited by J.C.A. Gaskin. Oxford: Oxford University Press, 1994, 26.

8. A reason for this could be that our philosophers and scientists have simply accepted this interpretation of perception. Cees Leijenhorst, in "Sense and Nonsense on Sense" (Patricia Springborg, editor. *The Cambridge Companion to Hobbes's Leviathan*. Cambridge University Press, 2007) and one of the more respected recent commentators on Hobbes's science, has the following to say. "One of the most important philosophical problems of the first half of the seventeenth century was the question to what extent mechanical principles applying to the natural world would apply to the human world of cognition and volition. Hobbes had a simple but radical answer: they do apply. As we shall see in fuller detail later, his philosophical program was designed to reduce all human psychology to mechanics" (86). "The subjectivity of sensible qualities was an integral part of seventeenth century mechanical thinking in general. If the natural world is reduced to nothing more than matter in motion, the other side of the coin is simply that all the qualities we perceive in the world are subjective mental phenomena. In reality they are nothing but some form of corpuscular motion that our brain translates into a corresponding sensible quality" (90). "In Hobbes's philosophy of science, sense perception and imagination belong to physics, the science that studies natural phenomena, finding its basis in the most wondrous of all phenomena, namely the fact that things do appear to us by means of sense perception" (98). Leijenhorst denies that Hobbes is an atomist, as I will explore shortly. What should be pointed out at this time is that Hobbes's physics of perception depends first of all on his theory of *body*. It is the conviction in Hobbes's natural philosophy, and in its predecessors, that the real bodies in nature are too small to be perceived. This is not simply an arbitrary postulate. There is a theory behind the belief that the true bodies must be inaccessible to perception: that theory is one Hobbes partly articulates. It is also a theory that Plato addressed in some detail in his *Parmenides*. If Plato disproves the theory of body upon which Hobbes's science of body rests, then it will likewise lift the embargo upon ordinary human perception. Clearly, nothing could be more fundamental to the study of civil science than the theory which claims to determine what human perception is *capable* of knowing. In Hobbes's civil science, the entire political world effectively disappears, leaving us only with abstract sketches of bundles of pain and pleasure which Hobbes refers to as men.

9. *Leviathan*. Chapter 1, #1–2, 9. Oxford: Oxford University Press, 1998.

10. Michael Oakeshott. "Introduction to Leviathan," 22–23. In *Hobbes on Civil Association*. Indianapolis: Liberty Press, 1975. Oakeshott, in this quote, refers to a passage from the *Elements of Law* in *EW* IV. Hobbes there gives us his definition of sense, but he nowhere says that it is 'absolute knowledge.' Granted, Oakeshott qualifies his statement. 'Within its limits," Oakeshott says, Hobbes insists that sense perception constitutes "absolute knowledge." What though are these limits? They are severe. "Originally all conceptions proceed from the action of the thing itself, whereof it is the conception: now when the action is present, the conception it produceth is also called sense; and the thing by whose action the same is produced, is called the object or sense" (3). "By our several organs we have several conceptions of several qualities in the objects; for by sight we have a conception or image composed of color and figure, which is all the notice and knowledge the object imparteth to us of its nature by the eye. By hearing we have a conception called sound, which is all the knowledge we have of the quality of the object from the ear. And so the rest of the senses are also conceptions of several qualities, or natures of their objects" (4). "Because the image in vision consisting of color and shape is the knowledge we have of the qualities of the object of that sense; it is no hard matter for a man to fall into this opinion, that the same color and shape are the very qualities themselves; and for the same cause, that sound and noise are the qualities of the bell, or of the air. And this opinion hath been so long received, that the contrary must needs appear a great paradox . . . I shall therefore endeavor to make plain these points: that the subject wherein color and image are inherent, is not the object or thing seen. That there is nothing without us (really) which we call an image or color. That the said image or color is but an apparition, or alteration, which the object worketh in the brain, or spirits, or some internal substance of the head. That as in vision, so also in

conceptions that arise from the other senses, the subject of their inherence is not the object, but the sentient." The final measure of the severity of Hobbes's definition of sense perception needs to be measured against his definition of the origin of *language*. People believe that names indicate objects and their natures; Hobbes dismisses this with contempt, and claims that his philosophy entitles him to found his own definitions for names.

11. Oakeshott, along with Macpherson, reacted strongly to Strauss's characterization of Hobbes's teaching. Macpherson however doesn't indicate much familiarity with Hobbes's natural science to begin with, and Oakeshott's significant reaction to Strauss concerns the latter's views on Hobbes's moral and political philosophy. The academic division of labor here takes a toll on political philosophy. One problem is that the politicized physics of Epicureanism, which Machiavelli was the first modern to revolutionize, tend to fall into a no-man's land where no part of the scholarly division of labor contemplates the intellectual history adequately, if at all. Cf. my *Machiavelli and Epicureanism: An Investigation into the Origins of Early Modern Political Thought*. Lanham: Lexington Books, 2012. Oakeshott ends up making an argument that Hobbes is indeed a systematic thinker, but this analysis is essentially bereft of serious investigation into Hobbes's scientific philosophy. Strauss, who argues that Hobbes's natural science will cause us to overlook Hobbes's actual moral and political philosophy if we view him as a truly systematic thinker, is only disputed by Oakeshott on a superficial level. Oakeshott has proclaimed it the philosopher's purpose in life to create a 'system' of ideas that embraces all varieties of experience. Macpherson for his part equates the new mathematical physics with the rise of bourgeois man; but Oakeshott against this theory adduces that the philosopher is a distinct kind of man which arises out of the natural temperament of the individual rather than any evolving out of socio-economic circumstances. Again, those philosophers who make the argument that Hobbes's natural science is distinct from his political and moral philosophy, are concerned to indicate that reason has some role to play in and against 'mechanism.' Since mechanism, that is, materialistic physics, tends to reduce the human being to an automaton, and since the scholars in question wish to emphasize that Hobbes has some theory of significant rationality that he wants to identify and protect for human beings—for these reasons (among others) there is considerable resistance to the argument that Hobbes's science is all of a piece. Oakeshott can be seen to argue that Hobbes entertains a systematic perspective, but he only does this by ignoring the better part of the science that is operative in Hobbes's system. Oakeshott does not seem to be conversant with the atomistic tradition of physics, or with the Eleatic-Epicurean-Lucretian tradition of philosophy generally speaking. This is a great weakness. Oakeshott seeks to trace Hobbes back to certain developments in Medieval nominalism, a train of thought that continues to attract admirers. It does seem rather an extraordinary characterization of Hobbes's philosophy, to indicate that it comes to balance in some sort of defense of individuality. Yet this is what Oakeshott gives us to work with. One can say that Oakeshott defends the claim that Hobbes is a systematic thinker, but only by ignoring, or failing to recognize, the fundamental bases of Hobbes's scientific system. "From beginning to end there is no suggestion in Hobbes that philosophy is anything other than conditional knowledge, knowledge of hypothetical generations and conclusions about the names of things, not about the nature of things. With these philosophy must be satisfied, though they are but fictions. Indeed, philosophy may be defined as the establishment by reasoning of true fictions. And the ground of this limitation is, that the world being what it is, reasoning can go no further" (xxvi–xxvii). "My contention is, then, that the system of Hobbes's philosophy lies in his conception of the nature of philosophical knowledge, and not in any doctrine about the world. And the inspiration of his philosophy is the intention to be guided by reason and to reject all other guides: this is the thread, the hidden thought, that gives it coherence, distinguishing it from faith, 'science' and experience" (xxvii). One must say that for Oakeshott, the thread which is consistent in Hobbes's diverse writings is this one characteristic of philosophy, which appears to originate in something that a man may know about the nature of his mind, but not about the nature of the world.

12. *EW* I. 17.

13. "Introduction to Leviathan," in *Hobbes on Civil Association*, op. cit., 23.

14. *Thomas Hobbes' Mechanical Conception of Nature*. Copenhagen: Levin & Munksgaard, 1927.

15. In Kathleen Freeman. *Ancilla to the Presocratic Philosophers*. Cambridge: Harvard University Press, 1996, 43.

16. *The Essential Epicurus*. Translated by Eugene O'Connor. New York: Prometheus, 1993, paragraph 39, p. 21.

17. *Man and Citizen*. Edited by Bernard Gert. Indianapolis: Hackett, 1998, 91.

18. Frithiof Brandt. *Thomas Hobbes's Mechanical Conception of Nature* (1927). Brandt believes that Ferdinand Tonnies, who discovered the manuscript among Hobbes's papers (along with one other) at Chatsworth, gave it a title by himself ("the Short Tract"), when he attached it as an appendix to his 1889 edition of Hobbes's *Elements of Law*. Brandt makes the argument here, as regards the issue of substance, accident, and sense perception as 'touch,' that Hobbes builds on the Aristotelian tradition. "Thus Hobbes takes the Aristotelian conceptions, accident and substance, and consistently maintains them. . . . Those species that originate from the objects of sense (in casu, the object of sight), are thus substantial species of corporeal nature. In other words, Hobbes imagines the act of illuminating as an act of emanation consisting of particles. It is such an emanatory point of view he desires to maintain as against one of medium" (17). Brandt is regarded as the gold standard on Hobbes's natural science among modern commentators, but these statements make it clear that Brandt's comprehension of Aristotle is questionable. There is nothing inherent in an Aristotelian substance which might make it incapable of self-movement. Needless to say, an Aristotelian substance must have all four of the Aristotelian conceptions of cause: matter, form, efficient cause, and final cause. There is no possibility of Hobbes's substances having forms or final causes. Final causes indicate a principle of motion internal to the natural object; whereas substances in Hobbes's theory, beginning with the *Little Treatise*, cannot so much as move themselves, but must receive their disposition to motion from other bodies said to touch it.

19. Brandt would like to characterize Hobbes's philosophy, once and for all, as the philosophy of 'motion.' A philosophy of 'motion' is like a philosophy based on shape, or 'extension,' which are in themselves but *parts* of bodies. No human being has ever had experience merely of 'dimensions,' and deigned to refer to that as 'body.' The 'motion' that Hobbes ascribes to body, even in his *Little Treatise*, is not like any motion that human faculties are acquainted with. Hobbes's conception of motion is anchored, not in observation or factual perception, but in theory; and the theory is anchored in convictions of *metaphysics*, not common sense. Brandt reflects the modern prejudice: he is only too ready to surrender the physics of 'final causes,' but he does not possess adequate familiarity with Hobbes's theory of body in order to be able to evaluate the validity of this indictment of the much derided 'final causes.' The fact of the matter is that Hobbes's conception of body makes a radical departure from our experience. Brandt is willing to acknowledge that Hobbes could have been influenced by the atomistic currents present in Bacon's philosophy, but he ultimately decides that this issue cannot be resolved, that the evidence is insufficient. What seems clear to this writer is that Brandt lacks the familiarity with the atomistic doctrines necessary to make this judgment persuasively. According to Brandt, it is much more likely that Hobbes reached his theories of physics in an independent refutation of Aristotle, rather than that he did so by studying the theories of body in Bacon, Descartes, Lucretius, Epicurus, Democritus, or even Machiavelli. "In the first place, we are of the opinion that in Hobbes's *Little Treatise* as such there is much to suggest that he, by criticizing the prevailing Scholastico-Aristotelian conception, independently arrived at his theories. We have in fact observed how Hobbes continually adapts the Aristotelian possibilities but demolishes them by criticism to end on his own ground, motion pure and simple. Having established this by the previous analysis we shall not repeat ourselves. In our opinion no justifiable criticism can be directed against our characterization. Moreover it is worth noting that it is absolutely in agreement with Hobbes's autobiographies to presume that he independently and originally arrived at his theory of motion" (73). Brandt's fervor however rings hollow. Brandt himself does not spend one page evaluating Aristotle's devastating criticisms of Democritus. If he had, he would have been in a position to notice that Hobbes does not spend one page addressing Aristotle's indictments of Democritus. Brandt would have at least learned, for one thing, that the characterization of 'prime matter' that Hobbes ascribes equally to Aristotle and Democritus, is one that is in no way compatible with Aristotle's theory. Moreover, Hobbes's entire theory of 'motion' depends on his characterization of body. The fact of

the matter is that the atomistic theory of body itself verges on the incorporeal: for this is body which cannot be perceived, which is alleged to be indestructible; and as we will prove in due course, Hobbes's characterization of 'motion' is fully dependent on a theory of body which is incompatible with such body being a true whole of parts. At the same time, it seems entirely dubious for Brandt to profess such helplessness as regards evaluating Hobbes's exposure to the atomistic teaching of Francis Bacon, for instance. It is established that Hobbes helped to translate several of Bacon's later works into Latin (Croom Robertson, *Hobbes*. London: William Blackwood, 1910, 19); that he served as his listening companion, helping the master to work out his thoughts on various walks in his gardens (Richard Peters. *Hobbes*. New York: Penguin, 1956, 15–19). All Brandt can say is that Hobbes's theory of sense perception bears some resemblance to the atomic theories. One hazards to surmise that Brandt himself simply is not familiar enough with the atomistic conception of body, and the critiques thereof, to make that theory of body the object of his commentary. "Is there any historical source which Hobbes may have known which emphasizes the points of view applied by Hobbes in the *Little Treatise*? Yes, indeed there is. Hobbes's main points as to the act of sense are to be found in the atomic philosophy that teaches both the subjectivity of sensible qualities and conceives of the external act of sense as an emanation of particles. This point will ultimately be mentioned" (73), "To establish this, however, we do not feel competent. Thus the ultimate result is undecided. That Hobbes was acquainted with the atomic currents that put in an appearance just at the very time when he must have written the *Little Treatise* is possible, but impossible for us to prove" (77). At this point Brandt gets a bit daffy. "We here encounter a reasoning which we would like to lay stress on and to which we shall subsequently return, viz. that Hobbes's conception of matter tends towards dissolution into a conception of motion. Here again it is his unique insistence on and carrying out of the idea of motion which characterize him. The hardness and softness of the bodies which—seen from a primitive point of view—are accounted the primary properties of matter, are derived from motion. This point in our opinion is one of the sublimest in Hobbes's natural philosophy, showing as it does, how absorbed he is in his fundamental point of view" (113). That motion produces body, that body is a product or attribute of motion, is rather like arguing that horses are the product of galloping. Hobbes relies upon a theory of body that is said to be unavailable to the senses for detection; but this does not mean that Hobbes believes that motion could ever be the cause of body. Hobbes's theory, we reiterate, is first of all a philosophy of *body*. Hobbes's conception of motion is dependent on his theory of body; and as we shall prove, both the theory of body, and the concomitant theory of motion, are untenable.

20. Dr. Herbert dismisses the notion that Hobbes is a 'naïve' atomist. He denies, that is, that Hobbes is a metaphysical atomist. Hobbes's bodies are infinitely divisible. Moreover his bodies (like Bacon's atoms) have impulses, not traditional Democritean or Epicurean parts. "Hobbes is through and through a materialist, but not (as we shall see) in the unreflective modes provided by ancient atomism or even modern mechanical materialism. Hobbes held firm to the thought that to be means to be something, that is to be determinate. There is nothing that is neither here nor there, neither this nor that. In short, he rejected the logic of summum genus. On the other hand, his claim that physical reality is reducible to motion, not to simple bodies, and that motion is not simply the spatio-temporal change of position attributed to otherwise inert bodies but, rather, is conatus, makes it impossible to interpret him as a naïve atomist. The function of conatus in his account of determinate reality makes his physics a dynamics" (14). *Thomas Hobbes*: *The Unity of Scientific and Moral Wisdom*. Vancouver: University of British Columbia Press, 1989. Hebert is determined to argue two contradictory points. On the one hand, Herbert characterizes the ancient atomists as 'naïve,' that is, as taking something on faith that argument has not prepared the grounds for. Bacon, Descartes and Hobbes, in Herbert's characterization, developed a science which "tended to abstract itself from the metaphysical issues of the ancients and their commitment to the existence of absolutely solid bodies which, somehow, can be known. The 'points' or 'particles' of seventeenth century mathematical physics are more the conceptual foundations of modern physics than its metaphysical foundations" (28). In the first place, the capacity of philosophy, both ancient and modern atomism, to abstract from ordinary perception and the bodies commonly known, is attributable to one thing and to one thing only: theory. That is, *metaphysics*. The arguments which persuaded Democritus and Leucippus of the atomist hypothesis were derived from Parmenides' towering influence, which,

while certainly disputable, can hardly be characterized as thoughtless or as taken on faith. Cf. my *Machiavelli and Epicureanism: An Investigation into the Origins of Early Modern Political Thought*, chapter 1. Lanham: Lexington Books, 2012, for a discussion of the origin of the eleatic atomism, and for Plato's refutation of the theory. It will be proved in this chapter that Hobbes fully depends on the same metaphysical postulates of eleatic atomism that Herbert attempts to separate from Hobbes's argument. Hobbes is fully dependent on these arguments. The only difference is that Hobbes is not so forthcoming with these metaphysical postulates. Nevertheless, they are present, and we will indicate them. The abstract theory of body that Herbert retails is in fact incoherent. Body, whatever else it is, is not "the outward endeavor of any motion of sense, that is, the phenomenal presence of sense that an object is outside the mind or sense" (51). Hobbes does not believe this formulation that Herbert would ascribe to him, as will be proved. Cees Leijenhorst, *The Mechanisation of Aristotlelianism* (Leiden: Brill, 2001, 66–67) argues that the 'species' in the Hobbesian *Little Treatise* are not corpuscles, and that Hobbes firmly rejects atomism. This is because, Leijenhorst argues, infinite divisibility of times and continuums is insisted upon by Hobbes. "First, Hobbes's species are not corpuscles. It is true that a certain atomist echo is present in the *Short Tract*. Hobbes takes up a problem that already occurs in Epicurus. If light is produced by the continuous emission of species, how can light sources be materially inexhaustible? Hobbes could not solve the problem, but instead suggests that light sources may be 'supplied with fuel, by converting other bodies or species adjacent into themselves.' Despite this particular echo from the atomist tradition, Hobbes emphatically rejects atomism in *the Short Tract*. Hobbes extensively discusses a very complicated version of the old argument concerning Achilles and the 'snayle.' He concludes that 'the fault of this argument lies in this, that it supposeth a least part in line and time, where there is no minimum in either.' Thus, Hobbes maintains the old Aristotelian, anti-atomist tenet that the continuum (at least potentially) is infinitely divisible. So whatever the nature of Hobbes's species may be, they are certainly not corpuscles" (66). Leijenhorst's argument however is unpersuasive. The 'species' are 'emanations' from the 'substance.' The first question is, what is the nature of Hobbes's substance? The nature of Hobbes's substance is that it is not capable of undergoing any movement by itself. It cannot move itself. Included in the movements which Hobbes has struck from his definition of substances, are coming-into-being and passing away. Any body which lacks the capacity to come into being or pass away is already, by definition, *atomos*: uncuttable. The reason why Hobbes allows that there are no least parts in bodies, is because the division that Hobbes contemplates, is said to occur *only in the mind* of he who undertakes the contemplation. The fact that a man may imagine infinite parts in any body whatsoever, for Hobbes, does not in the least indicate that this is *true* of any body in nature. In fact, Hobbes does not allow that we can have knowledge of any body in nature such that we can do more or other than 'imagine' what its properties are. Except that Hobbes wields the postulate, which this chapter will examine in considerable detail, as to the indestructible nature of matter. Only God can actually destroy matter, Hobbes insists. Only a God can bring matter into being. Excepting for these extraordinary events, which no human being has ever witnessed, none of the objects in nature, all of which Hobbes characterizes as 'matter,' are conceded to possess the possibility of coming into being or passing away. In fact this possibility is expressly rejected by Hobbes in both *Thomas White's De Mundo Examined* and *De Corpore*; and the same is implicitly argued for in the *Little Treatise* or *Short Tract*, by Hobbes's declaration that the only possible source of movement in a body is a possible source of local movement; and that this source cannot come from the body itself, not if the body is at rest at any time. "But that magnitude for which we give to any thing the name of body is neither generated nor destroyed. For though we may feign in our minds that a point may swell to a huge bulk, and that this may again contract itself to a point; that is, though we may imagine something to arise where before was nothing, and nothing to be there when before was something, yet we cannot comprehend in our mind how this may possibly be done in nature. And therefore philosophers who tie themselves to natural reason, suppose that a body can neither be generated nor destroyed, but only that it may appear otherwise than it did to us, that is, under a different species, and consequently be called by other and other names; so that that which is now called a man, may at another time have the name of not-man; but that which is once called body, can never be called not-body." (*De Corpore*, in *English Works* I.116). "That, whereunto nothing is added, and from which nothing is taken, remaines in the same state it was." (*Little Treatise*, section 1, principle 1). "A body, and the magnitude, and the place thereof, are divided by one and the same act of the mind; for to divide an extended body, and the extension thereof, and the idea of that extension, which is place, is the same with dividing any one of them; because they are coincident, and it cannot be done but by the mind, that is by the division of space." (*EW* I.108) "That space, by which word I here understand imaginary space, which is coincident with the magnitude of any body, is called the place of that body; and the body itself is that which we call the thing placed, differ. First in this, that a body keeps always the same magnitude, both when it is at rest and when it is moved; but when it is moved, it does not keep the same place. Secondly in this, that place is a phantasm of any body of such and such quantity and figure; but magnitude is the peculiar accident of every body; for one body may at several times have several places, but has always one and the same magnitude. Thirdly in this, that place is nothing out of the mind, nor magnitude anything within it" (*EW* I.105). These remarks by Hobbes can be explained rather easily. Hobbes is distinguishing, in the above comments, between bodies that we have experience of, and bodies as they actually exist. Bodies as they actually exist, we do not have experience of, in Hobbes's view. That would be directly in line with the atomist view. And also in line with the atomist's view is Hobbes's resolute claim that the magnitudes of bodies in their real existence, which are beyond human perception and experience, cannot and do not change. Unless I mistake, this means that they have no parts at all. Because they cannot be really divided, otherwise they would really have to gain and lose magnitudes. Of those bodies that we have experience of, Hobbes denominates these as mere creatures of our mind, such as may, or may not actually exist of themselves, independent of our minds. Time and place, as pertains to those bodies we perceive, are merely actions of our minds, which indicate not the true properties of any object in nature. Leijenhorst does come to the understanding that in Hobbes's actual definition of body, these bodies can not be either generated nor destroyed; that there is no motion, for the real bodies in nature, pace Hobbes, except local motion; but Leijenhorst has not been able to see why these premises enforce the atomist hypothesis. In fact, they do.

21. Liejenhorst is only the most recent commentator to attempt to make Aristotle the focus of Hobbes's scientific speculations. Thomas Spragens did so before him, and even Brandt seems to think that Aristotle is the philosopher against whom Hobbes is working. To be sure, the metaphysics of body that Hobbes retails, as early as the *Little Treatise*, is anti-Aristotelian, but it possesses a long heritage of positive doctrines that scholars tend to ignore. Thomas Spragens (*The Politics of Motion: the World of Thomas Hobbes*. Lexington: University of Kentucky Press, 1973) is attempting to argue, and not well, that Hobbes is enacting a transformation, rather than a replacement, of the Aristotelian system. Leijenhorst really does not make arguments terribly different from Spragens. The operative concept in Spragen's eyes is 'motion.' Aristotle has a theory of what the moderns call 'teleological' motion, which makes motion sound as if it involved a First Communion, a Bah Mitzvah, or some such religious celebration. In Spragen's view, rest is the course of motion in Aristotle's physics, since fulfillment means 'repose' or achieved, thus concluded motion. Spragens has very little to say about the formal cause. The focus on the final cause by itself enables him to stay abstract. Bizarrely, Spragens thinks that the final cause in Aristotle requires an external source of movement for a natural body. This is the opposite of correct. The whole point of Aristotle's physics is bodies that have a source of movement internal to themselves, given by the formal cause, although for the sake of a final cause. "Hobbes accepted the underlying framework of the Aristotelian paradigm, and this component was perpetuated in his thought as the tacit matrix of his own paradigm. He rejected, however, the focal model of the Aristotelian paradigm and replaced it with a quite different model which he found more satisfactory. This overall pattern may be characterized, then, as a paradigm transformation—a change in which the paradigm is systematically altered, but altered within the channels established by the original paradigm" (45). In Spragen's analysis, the key to Hobbes's system of physics is his theory of 'motion.' The novelty of Hobbes's system, Spragens insists, is that the motion characteristic of bodies in Hobbes's system is directionless. It has no true beginning point, and no destination. Spragens would do better to simply observe that Hobbes has chosen to demote perceived motions of actual bodies to mere 'imaginations' and 'phantasms' undergone by human beings. That is,

Hobbes rules them out of account, with his criteria of evidence, as pseudo-knowledge, as in fact information with no probative bearing. This doesn't bother Spragens. "Any account of Hobbes's thought, therefore, must begin by examining the substance of, and the reasons behind, this infatuation with motion. Only after such an examination can the implications of Hobbes's natural philosophy for his political philosophy be assessed knowledgeably" (55). "Together, these two basic assumptions make up the proposition that Hobbes tries to maintain throughout his consideration of everything from physics to politics: all change is really nothing but change of place" (63). Spragens, like Brandt and like Leijenhorst, puts the cart before the horse. For Hobbes's conception of motion cannot tell us anything at all, until we have first inferred and analyzed Hobbes' s conception of what a body is; for it is Hobbes's conception of what a body is, and must be, that predetermines his arguments in terms of what motion shall be defined to be. Is there any difference really between the 'local motion' undergone by Hobbesian bodies, and the random motion undergone by the atoms, along with their serendipitous 'curves'? "Where Aristotelian movement was finite, Hobbesian movement is infinite. It has no order, no structure, no end or limitation. It is endless, aimless motion. It is not 'from . . . to,' as Aristotle assumed, but rather an endless chain without a goal. Hobbes therefore deletes from change the connotation of fulfillment which formed a part of Aristotle's definition of motion as the actualization of potential" (63). It is false to suggest that Aristotle 'assumed' that natural objects have a directed path of natural motion. Aristotle *perceives* that this is so.

22. *EW* I.36.

23. Plato, *Parmenides* 143–145.

24. Brandt (*Thomas Hobbes's Mechanical Conception of Nature*, op. cit., 54) makes here the important claim that not only is the *Little Treatise* the occasion of Hobbes's reawakened interest in philosophy; nor only that the *LT* has as its principal focus the new doctrine about 'motion' as the principal thing; but very significantly, that the *LT* is primarily taken with the question of the cause of sense perception. Hobbes's physics of motion, in this claim, originates with his interest in the cause of sense perception; which one must say is certainly not Galileo's focus. Galileo is preoccupied with the issues of acceleration and impetus, such as can be seen to concern the movement of projectiles. This revelation about Hobbes, if accurate—and I have no reason to doubt it—illuminates the political origins of Hobbes's physics, that is, its Epicurean-Lucretian nature. Now Machiavelli had already brought the will-based physics of Lucretius to politics. That Hobbes, after his brief 'Humanist period,' arrives at his principal of motion as the cause of everything, in the *LT*, with a focus on perception, reveals that the subsequent plan for the trilogy *De Corpore* (1655), *De Homine* (1657), and *De Cive* (1642) was itself developed out of the principle concern with perception, that is, in the states of fact determination, the very marrow of Epicurean physics. It should furthermore be observed that, in proportion as Hobbes is revising Epicurus, his protestations notwithstanding—*that* Hobbes is not the 'philosopher of motion.' He is one of a succession of philosophers (political) who argue that real body cannot be perceived, because real body must be indestructible. 'Motion' is motion of these imperceptible bodies in Hobbian perception. But by simply omitting the subject from his doctrine, and focusing on its predicate of motion, Hobbes adds a layer of mystery to things, that is, aggressively seeks to deprive our minds of access to the nature of the body that is his true premise. The fact that the *LT* is based on particle effluence—orthodox Epicureanism (the eidola) only clinches the case. It is also clear that Hobbes is worried about challenges to the eidola theory, that he is already contemplating a medium based theory of perception. By the *EL*, Hobbes has not only adapted a mediumistic theory of perception. He has also incorporated a dose of Cartesian skepticism. Brandt (55–62) next proceeds to question the philosophical influences at work on Hobbes. Brandt prefers Aristotle. It is at this point that Brandt's ignorance of the neo eleatic tradition of philosophy becomes apparent. Francis Bacon, urged quite plausibly as the principal influence by one of Brandt's sources, possessed a full fledged doctrine of atomism, as demonstrated in the first chapter. Bacon did more than any Early Modern philosopher before Gassendi to reinvigorate the Democritean tradition of atomism, and furthermore, Bacon is full of animus against Aristotle. The doctrine of particle effluence that Hobbes presents is classically atomistic. Brandt however chooses to focus, as evidence, upon Bacon's late *Sylvara Sylvarum*. The theory of sense conveyed in *SS*, Brandt says, advanced a mediumistic theory of perception that focuses on incorporeal things rather than Hobbes's particles. Brandt does not

mention *The New Organon* and its radical indictment of ordinary perception, which is the issue, as Brandt has proved, that the *LT* is preoccupied with.

25. Tom Sorell, introduction, 3. *The Cambridge Companion to Hobbes*. Cambridge: Cambridge University Press, 1996.

26. *Machiavelli and Epicureanism: An Investigation into the Origins of Early Modern Political Thought.* Lanham, MD: Lexington Books, 2012, chapter 1.

27. I have discussed these early natural philosophers in two previous works. I will not undertake any needless repetition here. The discussion of Parmenides, the Milesians and the pluralists is contained in chapters 1, 2, 3 and 5 of *Plato versus Parmenides: The Debate over coming-into-being in Greek Philosophy*. Lanham: Lexington Books, 2010. Plato's evaluation of the eleatic atomistic philosophies, along with the doctrines of Epicurus and Lucretius, are treated in *Machiavelli and Epicureanism: An Investigation into the Origins of Early Modern Political Thought*. Lanham: Lexington Books, 2012, chapters 1–3.

28. *EW* I.105.

29. Brandt concludes his book (*Thomas Hobbes' Mechanical Conception of Nature*, op. cit., 372) saying that Hobbes almost fails to be a significant figure in the history of mechanistic science. Hobbes's significance, Brandt concludes, is in Hobbes's demolition of the authority of sense perception. Exactly. But Brandt himself does not appreciate the extent. In between 'phantasm' (from which nothing can be learned) and naming (sciences), there is a chasm. How does naming reach external objects if it has no other bearing except phantasm? Hobbes releases the embargo on perception after he has imposed his new authority to name. Not that this theory entitles him to. Brandt at the end of the day is not conversant with the nuts and bolts of Epicurean atomism, He is basically concerned to establish a place for Hobbes in the history of modern science. Which means that Brandt's unprecedented work on Hobbes's science has to be overcome. It can't be the gold standard if it is made out of bronze. It is to be noted that Hobbes seeks to instruct us as to where and when natural philosophy begins. Galileo, Hobbes says. Hobbes, as Brandt points out (378–380) did not even understand Galileo's experimental method. Galileo was an unsophisticated atomist, theoretically speaking. Hobbes won't even admit his substantial debt to Bacon, because in Bacon the deeper philosophical history of natural philosophy is made evident. Brandt himself discounts Hobbes's dependence on Bacon, but Brandt, in this matter, seems to underestimate the evidence.

30. I must return again briefly to Oakeshott's view. Hobbes, Oakeshott insists, is a philosopher who speculates principally upon the nature of the mind, *not* about the world. "My contention is, then, that the system of Hobbes's philosophy lies in his conception of the nature of philosophical knowledge, and not in any doctrine about the world" (27). "Introduction to Leviathan," in *Hobbes on Civil Association*, op cit. This is simply incorrect. Hobbes's statements about the indestructibility of body are not statements concerning the limits of human knowledge. It is a statement that overthrows all of the limits to human knowledge which truly *are* founded in perception. The genealogy of the theory has been indicated.

31. *EW I*, Author's epistle to the reader.

32. "From the same fountain spring those insignificant words, abstract substance, separated essence, and the like; as also that confusion of words derived from the Latin verb est, as essence, essentiality, entity, entitative; besides reality, aliquiddity, quiddity, which could never have been heard of among such nations as do not copulate their names by the verb is, but by adjective verbs, as runneth, readeth, or by the mere placing of one name after another; and yet seeing such nations compute and reason, it is evident that philosophy has no need of those words essence, entity and other the like barbarous terms" (*EW* I.34).

33. *EW* I.20.

34. Scholars have intermittently commented on Hobbes's audacious claim to be able to found a language that did not arise out of the common use of names, which must already be extant. Croom Robertson (*Hobbes*. London: William Blackwood and Sons, 1910) makes the following observations. "What Hobbes fails to see is that if names are both marks and signs—marks to the thinker himself, signs of his thoughts to others—they are signs before they are marks, since it is only as men have a life in common and form by intercommunication a social conception of things that they become swift to mark what goes on within them" (83). "He errs also, as we now understand, not seeing that speech has in the first instance a purely natural

origin among men as they are constituted, however it may later receive all manner of voluntary, and in a sense, arbitrary developments" (84). Tom Sorell (*Hobbes*. London: Routledge and Kegan Paul, 1986) indicates the importance of Hobbes's theory of the application of names as it pertains to science. "Although the imposition of names is the first step in the preferred method of science, Hobbes has very little to say about how the names are imposed or how they can be ordered into affirmations. Such comments as he does make suggest that the imposition of a name is accomplished at will and privately. A single speaker simply takes a sensible mark and, in affixing it to an object, makes it into a reminder for himself of a conception raised by the object. So taken, a name is a mark that means something only to the one who imposes it on a thing. For a name to convey the imposer's conception of a thing to someone else, it must occur within the context of a speech. It is a notably psychologistic sketch of name-imposition and the merest gesture toward the way in which naming contributes to communication, but it is important that Hobbes intends no more than a sketch or gesture. There is no evidence that he was concerned to elaborate a free-standing theory of reference, let alone a full-blown philosophy of language" (38). Hobbes sets up his doctrine of names in the cradle of his natural science, this much is clear. Yet it is not intended to exert its force, this theory of naming, in the field of natural science alone. It is intended to exert its force in the definition of names which apply to moral and political matters: to the definitions of manners, virtues and vices, of what it means to deliberate, of what it means to be just or sociable. In the domain of natural science proper, once again, scholars observe that Hobbes left very little mark on the development of continental science, or even in the English Royal Society. Yet it is as the cannon for a new rhetoric that Hobbes's doctrine of 'marks' and 'signs' needs to be understood. Douglas Jesseph ("Hobbes and the Method of Natural Science." *Cambridge Companion to Hobbes*, op. cit., 102) writes that "Hobbes's methodological doctrines had little influence, except insofar as they were the objects of attack by members of the Royal Society. Continental scientists working in the tradition of Descartes shared both Hobbes's commitment to plenism and his reliance upon hypotheses in explaining phenomena, but Hobbes had little or no direct influence on the development of Continental science." Jesseph is skeptical about Hobbes's doctrine of language, precisely because it seems to forswear any obligation of the language to indicate truths about the objects external to us, that is, to be about anything more than language itself. "But even taking this into account, it seems that Hobbes is tempted toward an improbable conception of science that secures the truth and certainty of scientific knowledge at the cost of restricting it to the analysis of language. Furthermore, Hobbes's stress on the arbitrariness with which names are 'imposed' on the world seems to commit him to a strongly conventionalist conception of scientific truth in which the truth of a scientific statement will amount to nothing more than speaker's agreement upon the definition of the terms it contains" (100). In the field of moral and political philosophy, the defects of Hobbes's nominalism would not be so glaring, nor would many of Jesseph's complaints about Hobbes's method seem to apply. For in the domain of political and moral philosophy, names evoke passions; and it is through the evocation of passions that Hobbes's civil science is designed to function.

35. Leo Strauss says that "What he learned from Plato's natural philosophy was that mathematics is 'the mother of all natural science.' By being both mathematical and materialistic-mechanistic, Hobbes's natural philosophy is a combination of Platonic physics and Epicurean physics" (4). From "The Spirit of Hobbes's Political Philosophy," in K.C. Brown, editor. *Hobbes*. Cambridge: Harvard University Press, 1965. Strauss's contention is misleading. Plato's natural philosophy arises from perception first of all, for it is in perception where the forms are initially learnt. Plato's theory of physics, his theory of causation, is well known to be bound up with the forms. Numbers themselves for Plato are forms, whereas the science of mathematics is *not*. Cf. *Phaedo* 100b; *Theaetetus* 198–200.

36. I have discussed this at length in *Plato versus Parmenides: the Debate over Coming-into-Being in Greek Philosophy* (Lanham: Lexington Books, 2010), chapter 4 passim.

37. *Man and Citizen*, op. cit., 39.

38. *Thomas White's De Mundo Examined*, op. cit., 34.

Chapter Four

Hobbes's 'Right of Nature' and the Politics of Agony

Human nature is very much the issue that Hobbes raises. Hobbes's civil science has a theory of human nature, and this theory is designed to reshape public opinion. The trouble is that Hobbes's conception of human nature contradicts reality. This brings up the inconvenient issue of truth, which Early Modern Philosophy is determined to render an irrelevant question. Hobbes wants to make human beings organize themselves around a moral principle which is at radical variance with human nature.

Human beings are mortal. On this much, at least Hobbes in his civil science and the general opinions of the human race are in accord. This indicates that self-preservation is a limited good. Human beings have natural limits, and passing away is one of them. Thus an overriding determination to survive at all costs is a doomed strategy, impossible of realization. Hobbesian morality would rivet individuals to the project of self-defense as their ultimate goal.

Human life is guided by dimensions of psychological experience that are informed by the limited amount of time we have to live. It is Hobbes who is unrealistic, when he argues that human beings have no summum bonum, or greatest good for which to live.[1] Every individual is fated to choose a greatest good, based upon the particular circumstances of each life and the knowledge that it ends. By nature, then, Aristotle is correct. Not only do human beings need to have a greatest good which they choose in freedom; this greatest good must also beckon towards *fulfillment*. Contemplation, no matter how many amusements a society may possess, is by nature a part of the greatest good that human beings live for. This means that people must not only choose greatest goods, but that these greatest goods must prove satisfying upon reflection, and in memorable experience. No matter how large the population of the planet may grow, no matter how adept science may get at prolonging life, these facts will remain transcendent ones.

It is really not strange that the major Early Modern philosophers wage a war against the distinction between 'necessity' and free will. None of the philosophers contemplated in this study is really comfortable with the notion of a free rational choice that isn't somehow forced. This is a psychological affliction endemic to the atomists: they have chosen, for their greatest good, to avoid thoughts that cause anxiety, in the Epicurean fashion. This is an empty dream, to say the least, not to mention a lifeless one.

The neo-Epicureans would have a hard time justifying their thesis in argument. Therefore, they insist that such a mental disposition is fated or 'necessary,' that is, unavoidable. We are under no obligation to indulge the fantasies of philosophers, no matter how ferocious their wills may be. The rest of the human race is under no obligation to default on its own conceptions of the greatest goods, merely in order to subsidize the delusions of the neo-Epicurean worldview.

By nature, man is in fact a social and political animal. Hobbes despises this proposition. He is committed to reversing human nature through what he conceives of as 'education.' In the current state of society, it is no wonder that education is constantly at the center of public complaint and debate. The Early Modern project of utility in mind has seeped deep into the recesses of the modern intellect, enervating its spirit and blunting its purpose. It is thought that the only productive education is one that issues in technological inventions or managerial schemes. Our educational mission is therefore twisted, in being too committed to transient ends, and negligent of guiding purposes.

Ever since Lucretius, the Epicurean sort has been looking for ways to bridge the gap between the goals of their master and the values of the rest of the human race. This is the cause for the great shift in modernity that Machiavelli brought about: that is, the proclamation that conflict in society is 'good,' and that human beings are better governed by their passions because this is more 'realistic.'[2] To govern human beings by their passions is only a realistic mode for philosophers whose opinions could never endure the test of open public scrutiny. 'Indirect' government, which Bacon so celebrates, is the order of the day by 'necessity.' Indirection in government is necessary to those whose goals can't win the day in the court of human deliberation and rational persuasion.

The geometry of Hobbes's civil science falls with great harshness on the human soul. It is not really difficult to manufacture a chamber of anxiety when one has acquired the powers of the ring of Gyges; that is, when the new philosophy has allocated to itself the authority to redefine the meanings of names, confounding the ordinary significations. It really is a case of *mis*direction rather than indirection. In youth, the imperative quality of greatest goods in terms of life goals is experienced as obvious and paramount. Yet the vocabulary of the new philosophy paints these impulses as immature, or impractical; as vain, and in need of remediation.

For the reader, some rough indications of the Hobbesian tenets should prove useful. In the Hobbesian universe of significations, conflict is among the very best of things. It helps to isolate people. Pride and honor as ordinarily understood are among the very worst things that Hobbes can imagine, precisely because they savor of mental independence, courage, and therefore of truth. Fear is the diadem in the crown of Hobbesian motivations, because it is the passion most consistent with the absence of self-awareness. The specter of futility is at last 'very heaven,' because it reflects the misanthropy which has always been at the root of sincere Epicurean perspectives.

HOBBES'S CIVIL SCIENCE

Hobbes's political science is very much about emotions. Based on the last chapter, I take it as proved that Hobbes rejects the veracity of the perceptual evidence that ordinary human beings possess as regards the actual objects in the external world; and that he similarly dismisses any insights into human nature which the ordinary opinions are capable of obtaining. This does not mean that Hobbes is disinterested in the feelings of the people. To the contrary: Hobbes's entire political and moral philosophy is designed to *make* human beings feel, very intensely.

The essence of the new political science that Machiavelli developed and that Hobbes is attempting to perfect is a kind of soul-craft. Machiavelli's political science depends upon the reality of the soul: that is, it depends upon the reality of the agony that is suffered by he who is driven to surrender honor. Machiavelli's political science, like Hobbes's political science, consists in the setting forth of the ways and means necessary in order to reverse the authentic equation of human nature: individuals must somehow be driven to choose self-preservation *instead* of their honor. If this can be accomplished, the individual will be thereafter a servant of this regime. If this can be accomplished, the individual has consented to be set up as a sentinel against the possibility of her own freedom and happiness. Machiavelli believes that his political science can bring this about, and Hobbes too believes that these goals are within reach.[3]

Early Modern philosophers love to ridicule Plato, when he makes remarks such as 'justice is just.' What can Plato possibly mean by such foolishness, they wonder. Allegedly, this logic of identity does not constitute knowledge. It does not produce any new effects, or bring into existence any power that did not previously exist. There is certainly nothing experimental about the proposition. Yet as regards the equation of human nature that we are now discussing, Plato's argument that 'justice is just' makes a very great deal of sense, and a tremendous amount of difference in people's lives.

The political science of Machiavelli and the political science of Hobbes cannot succeed in reversing the deepest psychological and spiritual principles of the human soul, unless they can conceal realities with rhetoric. Both Machiavelli and Hobbes undertake to make the claim to justice, *the* human attribute according to Aristotle, *appear* as a sinister thing. Hobbes's undertaking, especially, is determined to make the individual's assertion of honor appear to be synonymous with aggression, and to threaten the entire commonwealth with ruinous war and strife. The claim to justice must be converted, through rhetoric, into the great sin and curse of the human race: Hobbes intends for this human impulse to be viewed as the quantum that must be beaten back and overcome, in order for peaceable social life to prevail. Plato would point out that 'justice is just.' In other words, justice cannot produce injustice. Justice, Plato would insist, cannot be the cause of disorder and war.

The meaning of names is the essence of science, Hobbes argues; and this knowledge cannot be acquired by experience. "By this it appears that reason is not as sense, and memory, born with us; nor gotten by experience only, as prudence is; but attained by industry; first in the apt imposing of names."[4] As for the common run of men, they are fated to trust and dependence, not equality, as regards this most devastating of powers, "for neither error, nor nonsense, can without a perfect understanding of words, be detected."[5]

Despite the fact that Hobbes has foresworn the ability to know the truth about any perishable objects per se, his science of definition claims to possess absolute certitude. This is the legacy of the 'skepticism' of Early Modern Philosophy, insofar as it traces back to atomistic, Epicurean roots. In the Hobbesian formulation, we cannot know the truth about any external object whatsoever; yet it is alleged that the philosopher can know the absolute truth as to what he *creates*, when he ordains a name to represent some mental entity. The Early Modern philosopher, through this ruse, is able to fabricate the appearance of infallibility in his skillful deployment of names. In the meanwhile, this modern philosophy undertakes to call into question every last judgment made by human perceptual faculties, the only faculties we have by which to navigate the world about us.

Those who would contest Hobbes's definitions will never succeed in an effort to engage him in discussion. Hobbes insists that his will is a sufficient foundation for the definitions of names. "So that in the right definition of names, lies the first use of speech; which is the acquisition of science: and in wrong, or no definitions, lies the first abuse; from which proceed all false and senseless tenets; which makes those men that take their definitions from the authority of books, and not from their own meditation, to be as much below the condition of ignorant men, as men endued with true science are above it."[6] Those who would dispute Hobbes's definitions don't phase him in the least. "On the other side, if reasoning aright I win not consent (which may very easily happen) from them that being confident of their own knowledge weigh not what is said, the fault is not mine but theirs. For as it is my part to show my reasons, so it is theirs to bring attention."[7]

What Machiavelli taught the Early Modern philosophers is the wholesale damage that the mere abuse of names can do to public order and to individual lives. Philosophy possesses the ability to make arguments which have cast adrift from the ordinary opinions. These arguments, the ordinary opinion can neither follow nor refute. Philosophers can even make calumny appear to be disinterested and objective. Machiavelli instructed the Early Modern philosophers in the arts of 'speaking ill.'[8] The definitions and science of names that Hobbes strings together, from his definition of the human passions to his enumeration of the human virtues, are all of them exercises in this sophistic practice. Hobbes is a true experimentalist in the realm of domestic politics. The polity is his formless matter; and his rhetoric, anchored in equations of logic, conveys the form that he would impose upon it. Hobbes's definitions themselves seek to inflict injury upon man, an injury *upon* honor. Hobbes's use of speech is both surgical and savage. It is the great source of molestation that the Hobbesian commonwealth brings into being.

For Plato and Aristotle, it is injustice that is the cause of war. For Hobbes, there is no such thing as 'injustice' until it has become a name minted in the manufactories of the new science. In the case of Hobbes's political science, he undertakes to make it seem that justice is a luxury too far to reach for. The conditions of life that Hobbes's civil science seeks to bring about are such that the individuals would not be in a position to be able to care about their honor. If one is truly 'moral' in the Hobbesian sense, one is wedded to selfishness as to a creed: and since there is nothing of generosity or courage in selfishness, there cannot be in it anything of justice or honor either.

It shall be the project of this chapter to set forth the ways and means of Hobbes's civil science, and to unfold more fully the relationship between the new science of nature and the same. For this latter enterprise, and for certain other insights as well, the writer is highly indebted to the work of Leo Strauss. Strauss, in my view, in his early scholarship on Hobbes, achieved insights into that philosopher which have not been equaled before or since. I must also admit to the reader that in my considered view, Strauss feels himself to be kin to Hobbes; that Strauss shares the shibboleths of the Hobbesian faith. Such an observation will not go down easily with the majority of readers, for Strauss carefully cultivated an image which suggested that he was ill-disposed towards modernity, and rather a champion of 'classical rationalism.' It is also widely known that Strauss professed that he had the right to lie about his true opinions, in order to guard himself against persecution. The subject of the relationship between Strauss and Hobbesian philosophy could well take up a good chapter of its own. Yet this book is not the place for such. Suffice it to say that I have absolutely no doubts that Strauss is devoted to the Hobbesian political science, and that he is in truth quite allergic to the actual teachings of Plato and Aristotle.

This does not interfere with the fact that we must indeed often learn from those whose points of view we do not at all share. Strauss is a philosophic warrior cast in the mold of the Early Modern thinkers. He is a late addition to the line of Machiavelli, Hobbes, Spinoza, and Hume. Strauss brings alive, like no other twentieth-century writer, the true agenda of modernity. The fact that Strauss happens to be wholly committed to the Modern agenda does not subtract any value from the caliber of his insights.

HOBBES ON EQUALITY

Hobbes certainly makes pretension to an equality among all human beings in the state of nature which he represents to us. The discussion of Hobbes which has taken place thus far in this book should, at the very least, make such a claim appear suspect. Even when Hobbes speaks directly to the issue of equality, he takes pains to except certain domains of power. Men are not equal in their possession of science, Hobbes allows. Science is the greatest power of all, in his view, even though its strength often fails to attain to *eminence*.

Hobbes, famously, in that part of his teaching which all human beings are thought to be capable of understanding by introspection, asserts that it must be the general rule of political culture that all shall be regarded as equals. "Nature hath made men so equal, in the faculties of the body, and mind; as that though there be found one man sometimes manifestly stronger in body, or of quicker mind than another; yet when all is reckoned together, the difference between man and man is not so considerable, as that one man can thereupon claim to himself any benefit, to which another may not pretend, as well as he."[9] Hobbes is nothing, however, if he is not a careful writer. "And as to the faculties of the mind (setting aside the arts grounded upon words, and especially that skill of proceeding upon general, and infallible rules, called science; which very few have, and but in few things; as being not a native faculty, born with us; nor attained, as prudence, while we look after somewhat else) I find yet a greater equality amongst men, than that of strength."[10] We are concerned with the exception. Since Hobbes's entire political science is crafted out of names and definitions, it suffices to observe that he does not espouse people to be near so equal as he suggests.

Hobbes makes the observation that the sciences are not very powerful, because not 'eminent' or understood. "The sciences are small power, because not eminent; and therefore, not acknowledged in any man . . . For science is of that nature, as none can understand it to be, but such as in good measure have attained it."[11] While it is true that the rank and file of the human race will not be able to understand Hobbes's science, or appreciate it in its detailed glory, this by no means indicates that the science lacks power. For Hobbes has demonstrated the *trust* which is the fate of the non-philosophers. "Want of science, that is, ignorance of causes, disposeth, or rather constraineth a man to rely on the advice, and authority of others. For all men whom the truth concerns, if they rely not on their own, must rely on the opinions of some other, whom they think wiser than themselves, and see not why he should deceive them."[12]

Hobbes, especially in moral matters, is committed to producing the largest possible distance between the new definitions and the ordinary understanding. The scientific definition of names will calmly emanate from learned tribunals, destined to penetrate and circulate among the ordinary opinions that are articulated. In this mixture, whereby public opinion is formed, the ordinary opinions will become suspended and therefore thwarted. Since society must have order, a stalemated situation is not possible. The ordinary opinions will bend in the direction of the scientific definitions, and the public opinions will reflect this.

It is the ordinary understanding of right and wrong which Hobbes sets out to suspend and confuse. Hobbes makes this dramatically apparent as he embraces the skeptical postulates of Descartes. The would-be philosopher must renounce all of the evidence of his senses. In fact, Hobbes counsels his tutee to feign that the entire world has been annihilated, leaving in existence only the philosopher himself and the images and ideas that persist in his head. From this position, Hobbes's philosopher becomes free to assign new names and definitions to the several objects. From Hobbes's testimony, when he names, he simply affixes these to the images and ideas in his own brain. Technically speaking, therefore, it would not be possible for him to be wrong, in Hobbes's view.

Hobbes begins the *Elements of Law*, *Leviathan*, and also *De Homine* with studies in physics or sense perception. The *Elements of Law* is the first work Hobbes wrote for publication (1640), two years prior to *De Cive*. In fact, all of the components of the mature Hobbesian system are present in the *Elements of Law*. Hobbes takes his earnest student by the hand, as it were, in both *the Elements of Law* and *Leviathan,* and together they approach an archway. Under this archway the two of them walk, uttering a profession of faith as they pass beneath it. In this profession of faith, the Hobbesian philosopher professes estrangement from the world of the objects of ordinary perception; in this profession of faith, the Hobbesian philosopher enacts a foundation for the use of names which has denied its moorings in the perceived objects common to all individuals. Hobbes's political philosophy therefore begins with this method.

> For the understanding of what I mean by the power cognitive, we must remember and acknowledge that there be in our minds continually certain images or conceptions of the things without us, insomuch that if a man could be alive, and all the rest of the world annihilated, he should nevertheless retain the image thereof, and of all those things which he had before seen and perceived in it; every man by his own experience knowing the absence or destruction of things once imagined, doth not cause the absence or destruction of the imagination itself.[13]

The power cognitive to which Hobbes refers fully encompasses the civil science. Those scholars who insist that it is Hobbes's view that the civil science can be located through introspection by the meanest of minds is therefore far from accurate. Hobbes feels it necessary to banish from his acknowledgment the entire mental culture of the ordinary speakers. The names to which Hobbes fastens the several objects of human experience do not equate with the ordinary assignment of name to object. This is not a deliberative rhetoric that Hobbes prepares. Furthermore, based on the thesis of this chapter, the sort of introspection that Hobbes intends for the individual human being to undergo is meant to issue in submission and anguish. These are clearly not the outcomes that individuals would ordinarily arrive at, when conducting an independent reflection upon their own experiences, memories, and opinions.

Strauss,[14] Tom Sorell,[15] and others have made the claim that Hobbes maintains that everyman can understand this political philosophy simply by introspection. Hobbes does indeed urge us in this direction. As Hobbes writes in the Epistle Dedicatory to *De Homine*: "Having completed this section, *On Man*, I have finally fulfilled my promise. For you now possess the prime Elements of my philosophy in all its divisions and subdivisions. Moreover, it happens that the two parts whereof this section consists are very dissimilar. One is very difficult, the other very easy; one consists of demonstrations, the other of experience; one can be understood by few, the other by all."[16] It should be pointed out that the first nine chapters of *De Homine* are taken up with discussions of optics, Hobbes's science *of perception*. Hobbes, who by his doctrine of perception would take away the power of men to speak effectively for themselves, through his doctrine of philosophic definition provides himself with the power to speak for them.

STRAUSS, HOBBES, AND THE ESTRANGEMENT FROM NATURE

How does the human being end up allied with that doctrine which Hobbes denominates the 'right of nature'? "Nature hath given to every one a right to all; that is, it was lawful for every man, in the bare state of nature or before such time as men had engaged themselves by any covenants or bonds, to do what he would, and against whom he thought fit, and to possess, use, and enjoy all what he would, or could get."[17] What does the teaching of the Early Modern philosophers and their new vision of nature have to do with the cultivation of an attitude, or disposition in the human being, such as could issue in the 'discovery' of the 'right of nature' so advertised?

Strauss helps to supply that analysis. It makes a great deal of sense, once a reader has sifted through Strauss's study of Hobbes. The proto-atomistic philosophy cultivated by Bacon, Descartes, and now Hobbes is virulently opposed to the Aristotelian teachings about formal and final causes. Formal and final causes, in truth, are the very means by which human beings know their world. The form is the intelligible part of any object; and the final cause is the good condition, excellent development, or healthy unfolding of that type of object. The Early Modern teachers denounce and repudiate these doctrines. Nature has no purpose in the teaching of this physics. There is only matter in mechanistic motion, nothing more. This version of nature does not intend for generation to occur; it does not have any intended destiny for the different forms of objects to fulfill; and worse than that, it is indifferent to man. Man, in the eyes of the new physics, is an orphan. Strauss captures this psychological dynamic of the new physics with great strength.

Strauss, evaluating the impact of Hobbes's natural science on his political science, is the first to put the accent on the *helplessness* that the individual is meant to feel. Nature, Hobbes's nature, Strauss tells us, not only has no concern for man; this very same nature is contemptuously negligent of man and his needs, his hopes. So much so that the man properly disposed by the teaching of modern science will find himself increasingly *driven* to a condition of desperation. If nature is hostile and indifferent to man, then man must (and should) accordingly become aggressive toward nature, including human nature: his own, and that of the others. If nature has played a dirty trick on man by situating him in a universe that is deaf to his needs, ruthlessly exposing him to pain and baffling suffering, then man truly has nothing to lose by becoming a 'free agent.' By 'recognizing' Nature's indifference to him, Strauss argues, Hobbes's individual becomes 'free' to assert himself as against every other authority that ever was or could be. In fact, man is driven to this, Strauss argues, although he does not illustrate the manner in which this pressure comes about. The point is that Strauss himself has discovered the link between Hobbes's natural science and the natural right teaching: *If* nature has betrayed man, or has nothing to do with man, then man has nothing to lose by marshalling his will against nature, and against every conceivable form of authority which might interfere with his looking out for himself.[18]

> Since man is at the mercy of a fate utterly unconcerned as to his weal or woe, a fate which one may call God's irresistible power, because man experiences only force, and not kindness from the overwhelming power of the universe, he has no choice but to help himself. He has to live, not in gratitude, but in the serious and oppressive consciousness of his freedom, of himself as a free being, of his capacity to free himself. Constantly aware of the desperate seriousness of his situation, it will not occur to him to be proud of his freedom, and therefore, he will, above all, be on his guard against taking that freedom as the object of his speculations, against contemplating himself in his freedom and taking pleasure in it. It is better and more becoming to the situation of man to deny that freedom theoretically by mechanistic physical science, and to assert it practically by the conquest of nature, and particularly of human nature, with the help of that science. [19]

The new physics undertakes to disorient human beings in the world. It is in the shadow of this physics that Hobbes proposes to lead the human race into darkness. For both Plato and Aristotle, knowledge begins from the ordinary opinions, from obvious facts, in order to seek for more perfect and difficult kinds of knowledge. In Hobbes, this dependence of science on the ordinary opinions is smashed. "For we may not, as in a circle, begin the handling of a science from what point we please. There is a certain clue of reason, whose beginning is in the dark; but by the benefit of whose conduct we are led as it were by the hand into the clearest light."[20] Strauss explains that "Man can guarantee the actualization of wisdom, not in spite of, but because there is no cosmic support for his humanity. He can be sovereign only because he is forced to be sovereign. Since the universe is unintelligible and since control of nature does not require understanding of nature, there are no knowable limits to his conquest of nature. He has nothing to lose but his chains, and for all he knows, he may have everything to gain."[21]

Hobbes's right of nature cannot be so much as understood without a comprehension of the doctrines of physics that he builds upon. In political terms, the representation of nature as something mysterious and enigmatic is designed to evoke *feelings* in the populace. Early Modern Philosophy's attack on the reality of the formal and final causes in nature are omnipresent in its hypotheses; and these hypotheses are among the most powerful elements in the constitution of the new public opinion. Which is to say that it is part of Hobbes's purpose to make man feel *helpless and alone*. These emotions, as Strauss has powerfully argued, as it were propel the individual towards the gospel of natural right. In order for the individual to feel his 'freedom' in *this* sense, he must first of all find himself desperate. It is only in the condition of desperation where man is available to the postulate that his every moral impulse is to be measured upon his ability to extend and measure his power in the world. Or, at least, this is the case for the non-philosophers. Those who set up the experimental model are obviously free to extend their power in their very hypotheses and teachings.

It is not pleasant for man to feel desperate. Furthermore, it is not without consequences for a man to 'choose' to make himself a moral free agent in the Hobbesian sense. When Hobbes defines liberty as the power to move unobstructed, he is writing for the desperate man. Here, in the definition of the name that is above all others the watchword of freedom in the Modern philosophy of life in society, one beholds the denuding of language by the banishment of the forms and final causes from human understanding. "Liberty, or FREEDOM, signifieth (properly) the absence of opposition; (by Opposition I mean external impediments of motion;) and may be applied no less to irrational, and inanimate creatures, than to rational."[22] The applicability of the concept to 'inanimate' objects helps to ensure the meaninglessness and uselessness of the sort of liberty that Hobbes's system is designed to produce.

No matter how many inducements human beings experience to get them to undergo this radical and unnatural transformation that Hobbes proposes, it cannot ever occur absent some moment of true choice. This is to say, that it is the individual's will that Hobbes is concerned with. For it is only in the fastening of the will to the right of nature and its corollary moral impulses that the possibility of resistance can be eliminated. Thus Hobbes's society is full of harsh reprimand for those who cleave to affection, honor, trust, and good faith. It is also full of seductions for those who are willing to strike the proffered deal. Desperation is intended to engulf the isolated individual. It is for individuals whom Hobbes writes. In order to become an individual in *this* sense, that is, where the moral imperatives have pulled loose from the common feelings of praise and blame, the individual must in fact reject the better part of *himself.* It is a politics of agony that Hobbes undertakes to formulate; and a science which is obligated to a common set of facts with the people over whom it would exert influence could never accomplish such an enterprise.

The 'recognition' of nature as indifferent to man, as savoring of accident merely, gives man the license to do whatever he feels he must. This 'God' construed purely as power reduces to the command to feel desperate, desperate in a way which excuses man from the customary decencies.[23] In this chain of mental events man becomes free, in Strauss's view; man is the world's 'proletarian,' who has nothing to lose but his chains.

> Now since our basic principle is that every man is not only right, but naturally compelled, to make every effort to win what he needs for his own preservation, anyone who tries to thwart him for the sake of luxuries will be to blame for the war which breaks out, because he was the only one who had no need to fight; and is therefore acting against the fundamental law of nature.[24]

From the vantage point of desperation, then, honor can be made to appear a ridiculous luxury or trifle. In the point of view that Hobbes's science seeks to anchor the moral world, where caution and urgency are forever flooding in upon the senses, it is possible to *portray* goodness as a moral defect. Yet, justice is just. Goodness is good. True honor never can be dishonorable. In order to swallow the lie that honor is dishonorable, turbulent passions must impede access to calm judgment. In order for this to occur, the people must live forever near to chaos. This is the direction of the Hobbesian institutions, to be examined more carefully below.

The vast inequality between Hobbes the philosopher and the generality of human beings is conspicuous on precisely this issue of passion. Hobbes belongs to the Epicurean tradition. As such, he seizes for himself the prize that Epicurus vouchsafes to his adherents: that prize is to *overcome* the fear of death. In fact, for the Hobbesian neo-Epicurean philosopher, the entire new teaching about nature is serving important psychological purposes. The neo-Epicurean philosopher wants, above all else, to be shielded against the customary authority of ordinary praise and blame. The Epicurean sage cannot live strictly in accordance with his own pleasures and pains if there are duties and obligations which would make him obey a different master. Since the rest of the human race can never be reconciled to Hobbesian postulates through persuasion, they must be led into situations which result in the kind of manners and mores which are roughly hospitable to the Epicurean imperatives.

Lucretius led the way in this Epicurean teaching. In the famous book VI of his poem, Lucretius observes how the plague of Athens led the people to discard all of their holy rites and ceremonies. Under the duress of desperation, people struggle to survive; and in this they exhibit a rough kinship with the gospel of selfishness according to Epicurus.

> For reverence now and worship of the gods counted for little,
> Present grief was all. No longer too the ancient customs stood
> Of burial, which the city was wont to use. Confusion and fear
> Were now everywhere, and in sorrow each buried his own as
> Circumstances allowed. And sudden need and poverty inspired
> Them to many actions horrible and shameful.[25]

When the people fears for its life, when the terror of anonymous death is rampant, the appetites are let loose and reason is subverted, along with custom and piety. One need only compare Hobbes's own attitude toward death with the attitude of death that he prescribes for the denizens of his 'commonwealth.' For Hobbes, as for Spinoza, the free man doesn't think of death. The student of Hobbes's natural science may become perplexed, when he begins to study the political science. In the physics, as discussed above, Hobbes would drill it into the reader that there is truly no such thing as perishing of bodies in rerum natura. This conviction, which hurls itself against the entirety of our waking thoughts and most sober reasoning, is clung to by Hobbes with fanatic insistence. In our examinations of the book on Thomas White and in *De Corpore*, we have seen Hobbes insist that it is an improper use of speech to discuss passing away and coming into being, birth and death. However, when we turn to Hobbes's *political* science, all we hear about is 'preservation' of the self. So great is the fear for preservation, Hobbes indicates that there is absolutely no impermissible thing that can interfere with this impulse and its goals.

These attitudes only appear contradictory. They make sense when it is understood that one attitude belongs to the neo-Epicurean philosopher, and the other to the generality of the human race that is under the governance of neo-Epicurean philosophers. These two antithetical attitudes towards death are addressed in the quotes just below:

> Birth and death are sometimes taken in the respective senses of 'creation out of nothing' and 'destruction' or 'reduction to nothingness.' These processes, however, can be carried out by divine omnipotence alone, for we cannot in any way comprehend in our imagination how something can be created from nothing or how nothing can result from something. Sometimes, indeed frequently, the terms are taken to mean 'generation' and 'corruption,' that is, change. This is not, however, 'any change you like,' but only 'a change which makes us assign to, or remove from a thing the name that answers the question, 'what is the thing?' . . . Certainly, in no way can a body perish, but it can be changed and to the degree that it must then be called by another name. Again, the premise 'what can perish is mortal' is not a correct use of terms, for the death of any substance is not perishing but a certain prescribed change of percipient bodies into non-percipient.[26]

The neo-Epicurean philosopher therefore proposes to live in a vastly different mental landscape than the one he prepares for the generality of men. Those who understand the new physics attack the reality of death. Yet in the configurations of the new civil science, the people are to be brought well nigh unto a state of terror. "Amid so many dangers," Hobbes writes, "therefore from men's natural cupidity, that threaten men every day, we cannot be blamed for looking out for ourselves; we cannot will to do otherwise. For each man is drawn to desire that which is good for him and to avoid what is bad for him, and most of all the greatest of natural evils, which is death; this happens by a real necessity of nature as powerful as that by which a stone falls downward."[27]

What we can see, in Hobbes's description of the human beings in his fear-fraught world, is that the *formal and final causes have been stripped away from man*. Man is not allowed a purpose, in Hobbes's political science. He must always pursue desire, but no desire is allowed to constitute happiness, or *satisfaction*. Men will no longer have time for contemplation in Hobbesian society. They won't have the *leisure*, for even leisure in Hobbesian society must take place under the canopy of the permanent state of anxiety. The people will always be on the road, as it were; where the road is endless, and harbors no destinations. "So that in the first place, I put for a general inclination of all mankind, a perpetual and restless desire of power after power, that ceaseth only in death."[28]

For Aristotle, against whom Hobbes is arguing, happiness is precisely that state of soul which is *not* a means to further ends. "Now such a thing happiness, above all else, is held to be; for this we always choose for the sake of itself and never for the sake of something else, but honor, pleasure, reason, and every virtue we choose indeed for themselves (for if nothing resulted from them we should still choose each of them), but we choose them also for the sake of happiness, judging that through them we shall be happy. Happiness, on the other hand, no one chooses for the sake of these, nor, in general, for any thing other than itself."[29]

In Hobbes, we behold the reverse values. The individual must be absolutely driven into a morality that begins and ends with the *body*. Man must regard himself as unto a machine of pleasure, a relentless whirring of gears and springs that is incapable of distinguishing between the high and the low, the truly good as distinct from the merely pleasant. "Therefore the first foundation of natural right is that each man protects his life and limbs as much as he can."

Commentators typically regard Hobbes as upbraiding the prideful, assertive impulses among human beings, in order to pursue peace. Yet how distant is Hobbes's very prose from such a confession of vulnerability. The very essence of Hobbes's literary style is bombastic. It is confrontational and proudly so. Hobbes's political and moral philosophy is as aggressive and piercing as his physics is remote, abstract, and to most readers, unintelligible. Hobbes's attitude towards his detractors is telling. "I found my book very sharply criticized: on the ground that I have immoderately enhanced the civil power, but by Churchmen; on the ground that I have taken away liberty of conscience, but by Sectarians; on the ground that I have exempted sovereigns from the civil laws, but by lawyers."[30]

It is probably fairest if we examine Hobbes's personal *conduct* as it pertains to his chosen profession, that of philosopher. Above we have noted Hobbes's 'borrowings' from Descartes' attack on the authority of sense perceptions, and they are not few. Yet Hobbes engaged in a relentless and pointed quarrel with Descartes as to which one deserved to be called the founder of the new science. In short, the quarrel between Hobbes and Descartes was over what truly deserves to be called 'empty self-esteem' and 'vanity.' People in Hobbes's state of nature are condemned for their pride, which is there dismissed as a juvenile desire for pre-eminence and glorying. Yet Hobbes engages in an escalating battle with Descartes for the accolade of founder of the new science, never for a moment regarding such impulses as anti-social or irrational. Where human beings truly can outgo the others in some power, by a sufficient margin, then Hobbes endorses the rivalry and prepares to offer the garland of victory and honor to the winner. From Hobbes's point of view, power is honorable regardless of the means it employs to obtain its ends. It is certain that Descartes cannot compete with Hobbes in the field of rhetoric. Hobbes takes himself therefore to be so far superior to Descartes. Hobbes is indicating a moral order that is entirely at variance with the ordinary sense of right and wrong, one which he exemplifies in his personal conduct and attitudes.[31]

MORAL CULTURE IN HOBBESIAN SOCIETY

In order to gather insight into the nature of the moral regime that Hobbes's philosophy prepares, it makes sense to examine some of the nuts-and-bolts aspects of his laws of nature. The laws of nature in Hobbes emerge *out of* the right of nature. The concern for the isolated self which is the cornerstone of the right of nature therefore lives and breathes in the laws of nature; and it is important for the reader to have an opportunity to weigh the changes in manners and norms that Hobbes intends to establish. For Aristotle, of course, ordinary praise and blame establish the moral norms of the community and indicate the signification of virtue. In Hobbes, philosophical postulates overrule ordinary praise and blame. It could hardly be otherwise, if the task of the new regime is to demonize honor and pride as ordinarily understood.

Readers may well object that Hobbes only intends to root out noxious forms of pride, such as the individual who is a braggart, and who would incite his fellows to violence. Scholars have often held that Hobbes is simply appalled by the bad manners of human beings, and that it is his earnest attempt to tame these wicked humors. By contrast, I here argue that Hobbes's city goes out of its way to *mortify* decency; to inflame and even to torment integrity wherever it shows its face. I furthermore argue that Hobbes's new prototypes of moral culture solicit and cultivate obnoxious dispositions, attempting to *ignite* resentments and offensive feelings.

Hobbes can certainly be referred to as the anti-Aristotle. For Aristotle, man is a political animal, for whom society is the natural seedbed for growth, development and fulfillment. In Aristotle, human emotions are part of human nature, and they bear intelligence. It is Hobbes's point of view which views man as an animal who is apolitical by nature; and from whom the social and political predispositions (which to Aristotle are still just potentialities) must be rooted out, over and over again.

For Hobbes, society must be created by convention, by the artificial, because nature has not seen to man's need for society. This is an awfully convenient position for Hobbes to embrace. It points up however the somewhat Draconian nature of the moral education that institutions are designed to instill in individuals in Hobbes's city. It would not be going too far to describe the education Hobbes bases his culture on as a form of punishment: man must be made to feel violent resistance *from* society at the very core of his being. In Plato and Aristotle the emotions are full-fledged faculties in the human personality. Upon their proper cultivation resides the human opportunities for happiness. In Hobbes, the emotions are dismissed as mere 'perturbations' of the mind.[32] In Hobbes, ordinary emotions constitute obstacles to reason, unless these emotions have been elicited by a managerial scheme directed by the new civil science itself.

Hobbes proceeds with his polemic. In Hobbes's definitions, man must be driven to fear, total fear as the proper moral attitude, because it is the only thing that can counteract his pride, which is defined simply as the cause of war. "However, no large or lasting society can be based upon the passion for glory. The reason is that glorying, like honor, is nothing if everybody has it, since it consists in comparison and preeminence."[33] "But man, whose joy consisteth in comparing himself with other men, can relish nothing but what is eminent."[34]

By 'glory,' Hobbes is indicating all other imperatives which fall outside of the ethos of self-preservation. Unless and until the individual is brought before the altar of the right of nature, where desperation is *embraced* in some fashion, that individual will remain attached and anchored in the relationships, identities and affections natural to human life. It is by the authority vested in Hobbes to set forth authoritative definitions that the entire category of ordinary virtue is dissolved into the epithet of 'glorying.' This is not to say, either, that Hobbes himself lacks for his own theory of counter-virtue. For Hobbes, power rooted in the attitude of self-preservation, if and when it sufficiently eclipses the power of the others, can properly be called honor. How such power is obtained is of no matter to Hobbes. For power, not happiness, is the final end in the Hobbesian moral metaphysics.

> Honorable is whatsoever possession, action, or quality is an argument and sign of power.[35] Nor does it alter the case of honor, whether an action (so it be great and difficult, and consequently a sign of much power), be great or unjust: for honor consisteth only in the opinion of power.[36] The passions that most of all cause the difference of wit, are principally, the more or less desire for power, of riches, of knowledge and of honor. All which may be reduced to the first, that is, desire of power. For riches, knowledge and honor are but several sorts of power. And therefore a man who has no great passion for any of these things; but is as men term it indifferent; though he may be so far a good man, as to be free from giving offence; yet he cannot possibly have either a great fancy or much judgment . . . For as to have no desire is to be dead: so to have weak passions, is dullness.[37]

Nor is it any customary doctrine that Hobbes adopts when he declares that *power* is the ultimate and the only true currency in the affairs of human beings. It is no exaggeration to observe that Hobbes regards the theorist of artificial body, or commonwealth, as the greatest power known to man. Hobbes regards himself in these assertions as the most powerful, and therefore as the most honorable. "The greatest of human powers, is that which is compounded of the powers of most men, united by consent, in one person, natural or civil, that has the use of all their powers depending on his will, such as is the power of a commonwealth."[38]

The institution of exchange is in Aristotle's view the very foundation of justice in society. As the human need for one another is the original cause for society, and as interdependence is the human characteristic par excellence, so the governing of exchange is as fundamental a matter as there is in the struggle for human happiness. In ordinary opinion, 'equity' signifies more than something accidental or purely circumstantial. Aristotle's definition, which Hobbes has condemned, insists that equity involves a more exact application of the laws of justice than governmental laws are able to indicate. The judge, in other words, is the party whose business it is to restore equality where transactions have cast it to the hazard. "Of particular justice and that which is just in the corresponding sense, (A) one kind is that which is manifested in distributions of honor or money or the other things that fall to be divided among those who have a share in the constitution . . . and (B) one is that which plays a rectifying part in transactions between man and man."[39] "If the unjust is unequal, the just is equal, as all men suppose it to be, even apart from argument."

The right of nature as Hobbes teaches it begins and ends with the self. Justice as Aristotle defines it puts the relationship between human beings first. The right of nature has no place for justice, and this is again why Hobbes tends to view justice through a Calliclean lens, as somebody else's good. Hobbes's man retains some rights of nature even inside the commonwealth. In fact, Hobbes's individual retains rights of nature even *against* the commonwealth. Yet it is the attitude of the Hobbesian man, his curt dismissal of the ordinary decencies, which commands attention. For the Hobbesian man, any agreement he can get another to go along with is to that extent valid and binding. It does not matter how equal the exchange has been. "Whatsoever is done to a man, conformable to his own will signified to the doer, is no injury to him."[40]

Obligations in Hobbes's commonwealth do not follow facts. They follow agreements, formal arrangements. A man entering into Hobbes's commonwealth makes an agreement with the sovereign to submit to him, and to obey his laws. Against the sovereign, a man may therefore do wrong. How is it then between two individuals inside the commonwealth, who have not made some explicit agreement between themselves? Hobbes tells us that a man in such a condition may cause loss to such and such an individual, by cheating him or injuring him; yet that he cannot be said to *wrong* that other individual, since no agreement is in place. There's a distinction destined to fester in the soul of many a wronged man, who is herein barred from employing that word to characterize his experience.

Hobbes illustrates his argument with barely concealed glee. "It follows that a wrong can only be done to someone with whom an agreement has been made, or to whom something has been given as a gift, or promised by agreement. And hence, a distinction is very often made between loss . . . and wrong. . . . So too in a commonwealth, if one harms anyone with whom he has no agreement, he causes loss to the person he maltreats, but does a wrong only to the holder of authority over the whole commonwealth. For if the victim of harm should claim to have been wronged, the person who did the action would say, 'What are you to me? Why should I act at your pleasure rather than my own, since I am not preventing you from acting at your discretion, not mine?' I do not see how one could fault that response, since there has been no agreement.'"[41]

One can think about this individual, empowered by the regime of natural right to be able to treat so dismissively the emotions of his fellows. Hobbes has given him voice. Is this the voice of peace? When we come to the matter of more serious injuries, such as murder, we have a second opportunity to evaluate how Hobbes's new culture is going to *feel.*

When Hobbes sketches out the alleged misanthropy of the human race, he does not speak in tones of awe and trembling. He speaks with a cutting edge, drenched in hyperbole, aiming to injure, to ridicule, to mock.

> How, by what advice men do meet, will be best known by observing those things which they do when they are met. For if they meet for traffic, it is plain every man regards not his fellow, but his business; if to discharge some office, a certain market-friendship is begotten, which hath more of jealousy in it than true love, and whence factions may sometimes arise, but good will never; if for pleasure and recreation of mind, every man is wont to please himself most with those things which stir up laughter, whence he may, according to the nature of that which is ridiculous, by comparison with another man's defects and infirmities, pass the more current in his own opinion. And although this be sometimes innocent and without offence, yet it is manifest they are not so much delighted with the society, as their own vainglory. But for the most part, in these kinds of meetings, we wound the absent; their whole life, sayings, actions are examined, judged, condemned.[42]

It is not to be denied that human beings are subject to quarrel. According to Aristotle, the first fact of the human race is interdependence. Justice, in its most elementary form, involves exchange: *giving and taking.*[43] According to Aristotle, justice instructs that equal things be exchanged for equal things.

Nobody should come away from a transaction with more or less value than he had at the outset of it. Quarrels over exchange, over men and women, over political opinions, over religious faiths: such conflicts are indeed endemic to the human race. Yet they may be ameliorated, and governed. Most conflicts are not truly intractable; and even those conflicts which appear to be intractable can be modulated, restrained: once again—they can be governed, at least potentially. If human happiness is confined to the pleasure and pain which are distinct for each individual, then the connection between the quality of human relationships and happiness is easily lost sight of.

Hobbes certainly advances the Epicurean teaching. It is true of course that the original Epicurean community cultivated close relationships with one another; but they were joined together by a *creed.* The Epicurean creed consists in the proposition that human beings can be useful to one another's solitary ends, not that they can potentially fulfill one another, and themselves, in their relationships. Ordinary opinion will never be separated from the latter belief; and it is for this reason why Hobbes seems adamant that society must be kept in a permanent state of agitation.

These definitions and skillfully doctored names are the 'unforseen injury' which alone can bring a man before the altar of Hobbesian natural right.[44] The new mighty moral teaching is waiting to embrace him. He need only consent to be desperate, or cease fighting against his own desperation, in order to turn himself over to the new authority. In a man's baptism into the new moral attitude, he need think of nothing but what advantage his own power can bring him. It is the bonds of love, duty, honor, obligation which must be laid aside. It is the virtues of justice, courage, moderation and wisdom which must be betrayed. This transformation is not inconsiderable; and there is certainly a point of no return. The new moral culture that Hobbes is attempting to drill into social members tears at the moorings of the natural virtues and dispositions enumerated above. The new moral valuations are so cutting, and betoken such insidious injuries, that they stand a very good chance of forcing the individual into the chamber of transformation.

> The value, or worth of a man, is as of all other things, his price; that is to say, so much as would be given for the use of his power: and therefore it is not absolute, but a thing dependent on the need and judgment of another. . . . And as in other things, so in men, not the seller, but the buyer determines the price. For let a man (as most men do) rate themselves at the highest value they can; yet their true value is no more, than it is esteemed by others.[45] The manifestation of the value we set on one another, is that which is commonly called honoring, and dishonoring. To value a man at a high rate, is to honor him; at a low rate, is to dishonor him.[46] To be conspicuous, that is to say, to be known, for wealth, office, great actions, or any eminent good, is honorable; as a sign of the power for which he is conspicuous. On the contrary, obscurity is dishonorable.[47] And therefore to be honored, or feared of many, is honorable, as

> arguments of power. To be honored of few or none, dishonorable. Dominion and victory is honorable, because acquired by power; and servitude, for need, or fear, is dishonorable.[48]

In *actual* human experience, the achieved virtue of individual human beings does not elicit jealousy, or resentment, or spite. It elicits admiration. Courage, justice, moderation, wisdom are hard and difficult things. To possess them is to benefit many others besides oneself. Hobbes, again resorting to his authority of definition, specifies a conception of virtue which is itself distinctly vicious. In Hobbes's definition, virtue is a crass competition of one human being with another. Yet justice, courage, wisdom and moderation are not vicious attributes. None of these attributes seeks to make another human being appear the less, or is f*undamentally* concerned with making the virtuous individual seem to be the more. Justice does not mean the same thing as self-glorification; courage does not mean that one seeks to make others appear as cowards. Yet these are interpretations of the ordinary significations of names, and these are not valid currency in Hobbesian vocabularies.

To suffer a near relative or friend to a murder is a shattering, nay a devastating experience. If we have not experienced this trauma by ourselves, many of us have known someone well enough who has so suffered. For an individual so afflicted, the wound is irreparable. The punishment of the assailant *for* that deed, *because* he did it, would seem to be the most modest concession to the injured feelings of the afflicted that a human society could manage to bestow. Aristotle certainly thinks as much. "Therefore, this kind of injustice being an inequality, the judge tries to equalize it; for in the case also in which one has received and the other has inflicted a wound, or one has slain and the other been slain, the suffering and action have been unequally distributed; but the judge tries to equalize things by means of the penalty, taking away from the gain of the assailant."[49]

In Hobbes's commonwealth, however, this most rudimentary aspect of justice, that the assailant be punished for what he did, *because* he did it, is waved away. No sense crying over spilled milk, Hobbes mumbles. More to the point: in the suffering of such a grievous wound, the honor and dignity of the afflicted, and of the community as well, will be inflamed and conspicuously on display. Accordingly, Hobbes moves to arrest *it.* Not only does Hobbes forbid, as a vainly glorious trifle, that the punishment focus upon the act that has been done, the evil deed; Hobbes actually goes so far as to slander the anguish of the victims' survivors, to represent their grief and helpless rage as self-indulgent cruelty.

To demand justice, as ordinarily understood, as against an assailant who has slain someone dear to us—this sentiment, which can hardly be more indispensable to civilization than any other, does not square with the right of nature. The right of nature, even inside Hobbes's commonwealth, militates against punishment *for any crime committed.* This is held to be the case by Hobbes for several reasons, but principally because the impulse to revenge is held to be frivolous, gratuitous, and vainglorious. Revenge concerns what *is past*, the done deed. Punishment in Hobbes's state is committed to look forward, and not to pay attention to the deed that has been done in its own right.

> Now since our basic principle is that every man is not only right, but naturally compelled to make every effort to win what he needs for his own preservation, anyone who tries to thwart him for the sake of luxuries will be to blame for the war which breaks out, because he was the only one who had no need of fight; and is therefore acting against the fundamental law of nature.[50] The fifth precept of natural law is, that one should pardon another for the past, if he is sorry and asks pardon, provided one gets a guarantee for the future.[51] The sixth precept of natural law: in revenge, or punishment, consider future good, and not past evil. That is, it is only permitted to inflict a penalty in order to correct the wrongdoer or so that others may be reformed by taking warning from his punishment.[52]

What is at issue is something that Hobbes is trying to factor out of the culture of the new type of society that he envisions. The individual must not be allowed to feel that he or she counts for him or herself in *any* circumstances. To punish a murderer for the murder done would be to recognize the absolute value of the individual slain. This would be at cross-purposes with Hobbes's entire moral reformation effort. Thus punishment must look beyond the slain; beyond the kin of the slain; beyond those who have suffered the grievous wound. This could hardly be possible if Hobbes did not truly intend to mortify pride as a fundamental principle of his commonwealth.

Consider as well Hobbes's discussion of the *perpetrators* of crimes. In typical interpretations of the Hobbesian canon, it is thought that Hobbes is a humane individual; that he is truly concerned for peace. He is willing, as most commentators allow, to be severe with individuals who are the cause of war. At least, this is the appearance that Hobbesian rhetoric provides us with. Yet does it accord with the facts? One would expect that Hobbes would be absolutely ruthless with wrongdoers, if the above interpretations were accurate. Yet that is not the case.

Hobbes enters deeply into the psyche of those who have committed heinous crimes. He knows of their furious unwillingness to concede their guilt, to repent, or confess. These experiences would be felt to be most bitter by one who has contemptuously trampled upon the life of another. Yet Hobbes shows himself in his attitude towards such individuals. The right of nature itself, since it makes preservation of the self the *sole* origin of moral virtue, actually has no place for these qualities of repentance, confession, or admission of guilt, unless there is some *profit* that is attached to the arrangement (such as pardon, or a reduced sentence).

Hobbes has written of the assailants, and how they feel about "their own facts." To be made to admit one's wrongdoing is a very hateful thing, Hobbes allows; and many are those assailants who had rather die than do it. Is it not a trifle, in the Hobbesian sense, for the killer, the thief, to refuse to own up to what he has done, for the sake of peace? But no: Hobbes extends the umbrella of his new culture's protections to this tender soul's sensibilities. "Revengefulness is that passion which ariseth from our expectation or imagination of making him that hath hurt us to find his own action hurtful to himself, and to acknowledge the same; and this is the height of revenge. For though it be not hard, by returning evil for evil, to make one's adversary displeased with his own fact; yet to make him acknowledge the same, is so difficult, that many a man had rather die than do it."[53] "A covenant to accuse oneself, without assurance of pardon, is likewise invalid. For in the condition of nature, where every man is judge, there is no place for accusation: and in the civil state, the accusation is followed with punishment; which being force, a man is not obliged not to resist."[54]

It is true that Hobbes bans and prohibits the public expression of contempt, ridicule, and other forms of provocation which grate upon men's nerves. However, I think we are proving here that Hobbes's *civil science* is exempted from such a law. Hobbes's every sentence pours forth vitriol and defamation upon the human race. Hobbes suffers nary a feeling or hidden emotion to evade his calculated scorn. This is the curse of the new physics that the generality of the human race can only marvel at. For it is the mystery of the new science which affords the deference to its appointed spokespersons, in almost any category of human experience and life.

> And because all signs of hatred, or contempt, provoke to fight; insomuch as most men choose rather to hazard their life than not to be revenged; we may in the eighth place, for a law of nature, set down this precept, that no man by deed, word, countenance, or gesture, declare hatred, or contempt of another. The breach of which law is commonly called contumely.[55]

Shall we hold the man who has been robbed, the man who has been assaulted, the man who has been slandered, to be subject to this policy of zero tolerance? One suspects that Hobbes has just that in mind. The evidence cited above seems ample warrant for this conclusion. As noted previously, men certainly aspire to honor, to good standing among others. To respect; but certainly not to dominion. The vast majority of men and women seek neither to command nor obey; and in their desire to be well thought of, men betray rather their *political* nature, that is, their natural claim to justice as the measure of their personhood; and this is indeed fused to men's desire for, not commodious life, but of a good and free life. Courage is the diadem in this crown of freedom: the only virtue that can guarantee justice, and the other virtues as well.

What Hobbes is doing, and what Strauss is not at all critical of, is to represent man's political nature as a belligerent one. Man shares in emotions with the other human beings, in Aristotle's view. Plato remarks as much to the depraved Callicles. "Callicles, if mankind did not share one common emotion which was the same though varying in its different manifestations, but some of us experienced peculiar feelings unshared by the rest, it would not be very easy for one of us to reveal his feelings to another" (*Gorgias* 481c). Hobbes's political science is based upon the opposite premise: that what one feels to be good, another feels to be evil.

For Hobbes, each man calls that 'good' which is pleasant to him, and calls that 'evil' which annoys him. In Aristotle, however, the individual must be tutored by ordinary praise and blame in order to properly match feeling and action to matters of fact. Hobbes's portrait of man is of the anti-political animal. He is hammering out a new artificial persona for man, one which will rebuke him for seeking communion with ordinary praise and blame, that is, with the bonds of a *shared* morality. Let us face the fact: Hobbes's morality of self-preservation cannot tolerate any such shared morality, since that which every man is pledged to defend with all his might, is *different* for each one.

The major movement in Hobbes's thoughts is in his reworking of the moral vocabulary of culture. In ordinary conversation, selfishness is not appreciated; ridicule is usually frowned upon; and gross competition is regarded as anti-social, even as repugnant. Hobbes sets down a new range of platitudes. He is not characterizing the ordinary opinions; he is attempting to supplant them. The result can only be puzzlement, anxiety, or worse.

> To see another fall, disposition to laugh.[56] To hold fast by another is to love. To carry him on that so holdeth is charity (Ibid.). The vain glory which consisteth in the feigning or supposing of abilities in ourselves, which we know are not, is most incident to young men, and nourished by the histories, or fictions of gallant persons.[57] Grief, for the calamity of another, is pity; and ariseth from the imagination that the like calamity may befall himself.[58] Desire, by doing hurt to another, to make him condemn some fact of his own, revengefulness.[59] Virtue generally, in all sorts of subjects, is somewhat that is valued for eminence; and consisteth in comparison. For if all things were equal in all men, nothing would be prized.[60]

Ordinary opinion easily distinguishes between petty, toxic rivalries on the one hand, and ennobling competitions after excellence on the other. Virtue is virtuous. True virtue does not depress anybody's status, much less their relative value in society. On the other hand, Hobbes's new teaching that the value of a man coincides with the price that he may fetch on the market, condemns to indignity the vast majority of the population. To make wealth and power the measure of virtue in a society is to condemn the generality of human beings to a futile and desperate effort to escape the infamy of obscurity.

The competition that Hobbes's regime of natural right seeks to advance is the effort to get ahead of the others purely for the sake of getting ahead of them. No means are too crude, so long as the state has not expressly prohibited them. The focus turns away from the tasks to be accomplished, and devolves almost entirely upon the people who are competing. To come out on top will be akin to honor, though the contestants be competing to jump over a dunghill. "Continually to out-go the next is felicity. And to forsake the course is to die."[61]

VANITY VERSUS FEAR

Strauss's earlier study, *The Political Philosophy of Hobbes: Its Basis and Genesis*, supplies the vanity-fear hypothesis as the proper way to gain insight into Hobbes's moral argument. According to Strauss, Hobbes defines honor, pride, and the claim to justice itself as 'vanity': as the grasping at meaningless straws. While men assert their right to pride, or vanity, the city descends into civil war, the war of all against all. By contrast, the moral teaching of Hobbes consists in the directive of fear: the fearful man realizes that he is all alone in nature, that no one is looking out for him but himself. Nature, having betrayed man, has a manifest indifference to his fate. This experience of nature, which Early Modern Philosophy undertakes to furnish to man, is thought to result in the 'just' attitude: fear. From the vantage point of fear, honor is a luxury that cannot be taken seriously.

> As we have seen, the very title of the book *Leviathan*, in which the two passages just quoted occur expresses the opinion that pride, far from being the origin of the just attitude, is rather the only origin of the unjust attitude. . . . Not pride, and still less obedience, but fear of violent death, is according to him the origin of the just intention. What man does from fear of death, in the consciousness of his weakness at the hands of other men, when he honestly confesses to himself and to others his weakness and his fear of death, unconcerned about his honor, this alone is fundamentally just.[62]

Strauss's view of the antinomy on display in Hobbes's civil science appears to be correct. For Hobbes defines the desires for pride and dignity to be the actual *causes* of war. Accordingly, the embrace of the right of nature and the law of nature to which it leads must *reverse* pride and dignity. Men are not like bees and ants, Hobbes observes. "First, that men are continually in competition for honor and dignity, which these creatures are not; and consequently amongst men there ariseth on that ground, envy and hatred, and finally war; but amongst these not so."[63]

Michael Oakeshott raises some useful questions about the Straussian interpretation. Strauss urges that the dichotomy that Hobbes sets up in his political and moral philosophy is one between vanity and fear, between the desire for honor and the desire for preservation. In Strauss's explication, Hobbes's moral law causes us, "like a stone moving downward," to embrace fear at the expense of pride *and* dignity. Oakeshott thinks that in Hobbes's view, man can avoid both death and dishonor. In Oakeshott's view, man is guided by Hobbes's laws of nature to safeguard both his pride and his life. This does not seem to be a compelling argument.[64]

It is the Hobbesian project to make the price of a human being's assimilation to society coincide with the forfeiting of honor. In truth, honor is not something human beings can bear to live without. The pain and anguish which are implicit in having to surrender and part ways with one's honor are experiences which can never leave a human being. To undergo such an experience is a trauma of vast proportions. It indicates the moment when happiness ceases to be possible; where life ceases to hold out the potential of sweetness and gaiety; where self-contempt is fated to become the alloy of any alleged good, possession, attribute, or relationship.

To so arrange a political culture such that the assertion of a claim to justice, dignity, and respect are automatically received as aggressive, belligerent, disreputable acts is to envision the creation of an anti-society. Such a society is compatible with any degree of luxury consumption, sexual liberty, or economic opportunity. Yet at the base of this cultural pyramid is the surrender of happiness; and a lethal despair as to the impossibility of pursuing truly good things of lasting value.

In the Hobbesian culture human beings are confronted with the most piercing punishments except and until they are willing to forsake as unrealistic the prospects of true joy and fulfillment in life. Beauty and justice fall into the ranks of the tabooed quantities. Man, in short, is fated in this landscape to waste his life. Thus the burden of the amusements, the decadent underside of the entertainments: they are available only as substitutes and palliatives. In truth, happiness is forfeit, and this is finally not something that man can come to terms with.

Hobbes's tendency to polemic is characteristic of the anti-deliberative character of his philosophy. "By nature, then, we are not looking for friends but for honor or advantage from them. This is what we are primarily after; friends are secondary. . . . All society, therefore, exists for the sake either of advantage or glory, that is, it is a product of love of self, not of love of friends."[65] The very rhetoric of Hobbesian political science itself is intended to be woven into the warp and woof of public life. Hobbes's definitions identify 'types.' It is not necessary to specify a man by name in order to compare him to something scurrilous. One need only compare scurrility to certain characteristics which a man happens to share. The logic of association of ideas will do the rest of the work.

The moral choice that Strauss identifies in Hobbesian moral philosophy, that between honor or preservation, is correct but also misleading. As indicated above, human beings will not choose preservation over honor or pride as a general rule. This 'choice' of Hobbes's devising only comes into operation when pride has been or appears to be *irretrievably* lost already. In truth, it is the *indictment* of man's honor which brings about the ultimate situation of desperation. Without pride, there is nothing that man can do except preserve himself, because in this condition he is actually dying.

The absence of peace is the cause of disorder, Hobbes tells us. This is not exactly helpful. Yet it is clear how the quality of pride, which is often linked to anger, falls afoul of Hobbes's new equation. It is possible to injure someone without being passionate; to commit an injustice without displaying qualities of anger or contumacy. Yet an injured party is far more likely to bristle; and it is upon those who bristle that Hobbes's civil science falls with the greatest harshness. "Philosophers have heretofore found no remedy for this situation. For as they have not observed that the goodness of actions lies in their tendency to peace, their evil in their tendency to discord, they have constructed a moral philosophy which is alien to the moral law, and inconsistent with itself."[66]

'War,' in Hobbes's view, is due to the fact that men and women define the just and unjust differently, based on their passing moods, irregularities of body, perturbations of the appetite or the like. "One must recognize that good and evil are names imposed on things to signify desire for or aversion from the things so named. Men's desires differ, as their temperaments, habits and opinions differ; one may see this in the case of things perceived by the senses, by the taste, for instance, or by touch or smell, but it is much more so in everything to do with the ordinary actions of life, where what one man praises, that is, calls good, the other abuses as bad; indeed the same man at different times praises or blames the same thing."[67]

Hobbes ultimately blames the state of war, the war of all against all, as originating in each human being's desire for pride. Aristotle has fastened the pursuit of claims to justice to the very possibility of a good life: Moreover, Aristotle identifies the capacity of the individual to make a claim to justice as the very threshold of the free personality. Hobbes's antipathy to Aristotle's conception of man as a political animal is conspicuous in this light. For it is the assertion of a claim to justice that Hobbes regards as belligerent.

Hobbes is attempting to reverse that equation; to make the claim to justice, which as ordinarily understood encompasses pride, issue in the consequence of mayhem. "If to the natural tendency of men to exasperate each other, the source of which is the passions and especially an empty self-esteem, you now add the right of all men to all things, by which one man rightly attacks and the other rightly resists (an unfailing spring of suspicion and mutual resentment); if you add also how difficult it is, with few men and little equipment, to take precautions against enemies who attack with the intention to overwhelm and subdue, it cannot be denied that men's natural state, before they came together into society, was war; and not simply war, but a war of every man against every man."

"An empty self-esteem"; a "natural tendency of men to exasperate each other." This is Hobbes's revision of Aristotle's political animal. Aristotle's political animal must learn first of all that he is part of a whole; that pleasure is at times a great deceiver; and that he must habituate himself to face fearful things, with courage, if he is to become capable of standing against the wicked, and in favor of the good. "So too is it with the virtues; by abstaining from pleasures we become temperate, and it is when we have become so that we are most able to abstain from them; and similarly too in the case of courage; for by being habituated to despise things that are fearful and to stand our ground against them, we become brave, and it is when we have become so that we shall be most able to stand our ground against them."[68]

The people whom Aristotle observed, however, were not under any commandment to abstain from honor. Even for people who do not consciously seek to cultivate their honor, there is a penumbra of dignity from which they will never willingly retreat. For this reason, Hobbes must oppose the ordinary opinions as well as the philosophies or Plato and Aristotle. "For commonly truth is on the side of the few, rather than of the multitude."[69] Philosophers who take the ordinary opinions, anchored in perception as they are, for true, are indicted by Hobbes. This includes Plato and Aristotle most prominently. "And yet every man thinks that in this subject he knows as much as any other; supposing that there needeth thereunto no study but that it accrueth to them by natural wit: though they play, or employ their mind otherwise in the purchase of wealth or place. The reason whereof is no other, than that in their writings and discourses they take for principles those opinions which are already vulgarly received, whether true or false; being for the most part false."[70]

Consider Strauss on the issue of popular morality in Hobbes. "Thus Hobbes's political philosophy is directed not only against the political science of tradition, but against all norms and values which are based on opinion, against any and every system of morals which is popular and prescientific. The ideal of exact political philosophy means, therefore, that only science discloses to man the obligatory aims of his volition and action."[71] Hobbes's science of definition undertakes this moral revolution. The new definitions are not meant to instruct. They are meant to subvert, to incite, to redirect the captive passions into a brew that reason can no longer handle. Only in this way can the founding of natural right be approached. Only in the condition of extreme crisis can a man be led to this crossroads, where he must, under agonizing pressure, either resign from his absolute values or bear the epithet of outlaw.

Hobbes characterizes man in his state of nature as a boaster, and he even makes a distinction in the state of nature between morally better and worse types of individual. The better type, Hobbes argues, is 'temperate.' He follows the law of nature: which is to say he allows to everyone else the same thing that he arrogates to himself, that is, the right to *all*. In a competition between Aristotle's argument that human society originates in the need human beings have for one another, and the Hobbesian formulation of autonomous human beings at liberty to seize what power they can, Aristotle's is the voice of sobriety and dispassionate reason.

Hobbes is attacking those who *display* or insist upon 'honor and respect,' the so-called vainly glorious. It is not the characteristic of the vainly glorious to actually engage in fights, Hobbes admits. These are men who puff up their chest, but who ultimately give way when challenged. Yet it is these, who regard themselves as possessing some value apart from that which others place upon them, that Hobbes indicts.[72] It is pride that Hobbes sets out to enervate in his state; beginning with the state of nature, and never to relent.

> In the state of nature there is in all men a will to do harm, but not for the same reason or with equal culpability. One man practices the equality of nature, and allows others everything which he allows himself; this is the mark of a modest man, one who has a true estimate of his own capacities. Another, supposing himself superior to others, wants to be allowed everything, and demands more honor for himself than others have; that is the sign of an aggressive character. In his case, the will to do harm derives from vainglory and overvaluation of his strength.[73] Vain-glorious men, such as without being conscious to themselves of great sufficiency, delight in supposing themselves gallant men, are inclined only to ostentation; but not to attempt: because when danger or difficulty appears, they look for nothing but to have their insufficiency discovered.[74]

The vainglorious man is therefore someone Hobbes holds in contempt: not because he is a bully, but because he attempts to hold by his pride. Most men, as Hobbes correctly perceives, will back down if the adversary is overwhelming. Hobbes's political science means to be that adversary. The 'vainly glorious,' those whom Hobbes most demonizes, are observed to be pathetic because they lack the power to back up their claim to respect. In Hobbes's land, honor consists in power and nothing more. This is certainly at loggerheads with the ordinary conception of honor, and thus it is no surprise that Hobbes condemns ordinary praise and blame. Derision is Hobbes's attitude toward the majority of men. He does not fear them as hoodlums. He disdains them as all talk, as individuals who do not know that only that respect can be asserted which is backed by real power. "Vainglorious men such as estimate their sufficiency by the flattery of other men, or the fortune of some precedent action, without assured ground of hope from the true knowledge of themselves, are inclined to rash engaging; and in the approach of danger, or difficulty, to retire if they can: because not seeing the way of safety, they will rather hazard their honor, which may be salved with an excuse; than their lives, for which no salve is sufficient."[75]

As Strauss calls our attention to repeatedly, Hobbes characterizes his Leviathan state as a dominion "over the proud." "Hitherto I have set forth the nature of man, whose pride and other passions have compelled him to submit himself to government; together with the great power of his governor, whom I compared to Leviathan, taking that comparison out of the two last verses of the one-and-fortieth of Job; where God having set forth the great power of Leviathan, calleth him king of the Proud."[76]

According to Strauss's account of Hobbes, the individual's very desire to think well of himself constitutes a 'dreamworld.' Man's aspirations towards honor, towards excellence, towards fulfillment, utterly threaten the Epicurean point of view. The Epicurean cannot afford to regard life as too much of a potential good; for this would in turn require that he expose himself to the insecurities which attend to hope. Hope is not the characteristic of the Epicurean persuasion. Calculation is.

Strauss's evaluation of the human predicament as seen through Hobbes's eyes is partly accurate. First, we have the individual in his alleged 'dreamworld.' In this dreamworld, the individual *matters*, not only to himself but to others. At least it is fair to say that in a Hobbesian regime, with its peculiar institution of distributive justice, and its warped view of penal law, this individual is going to experience the world as a shock. The Hobbesian institutions at least are inhospitable to the self-respect which the individual seeks to cultivate. Hobbesian institutions teach the individual that he must rely upon his own force, rather than agreed upon standards of right and wrong.

Strauss proceeds to tell us that when the individual leaves his dreamworld, and discovers the opposing will of other individuals, that a dialectic of war results. First the two individuals are at loggerheads, supposedly because they both seek after an eminence which neither of them can possibly possess. Resistance, that is, withholding of recognition, leads to contempt. This contempt leads into violence, in which men want not to kill, but to triumph over one another; this process must go on ad infinitum for Strauss's Hobbes, at which point of realization man 'remembers the body' and the need to protect it, and dismisses the feelings of the mind as 'powers invisible' and meaningless.[77] This, Strauss argues, is the dialectic between vanity and reason in Hobbes's morals and politics.

Strauss gets the dialectic partly right. It is the signification of 'vanity' that is really crucial. Nowhere else is it so conspicuous how Hobbes's dominion over language takes effect. In the ordinary conversation and experience of mankind, the human desire to earn the respect and affection from others bespeaks the very social and political nature of human beings. It concerns all that is beautiful and natural among individuals; and it is accordingly most manifest in the young. Is there some magical threshold that individuals cross, where their desire to interact with others suddenly becomes a toxin, and a tocsin, of war and treachery?

There is something almost poignant in Strauss's description of Hobbes's argument on this issue of vanity and imagination, the 'dreamworld.' In this strictly rhetorical dimension of Hobbes's devising, 'imagination' denotes a self-deception, a pretension to self-importance that is not borne out by the facts of power. According to Strauss, in the imaginative state, man dreams of "glittering triumph," of having the whole world at his beck and call: a dreamworld where he is victorious as if in some great battle, and that this dreamworld is obnoxious to the like dreams of others. For so long as each man lingers in his dreamworld, it is no problem; but when men interact, dreams mix with expectations. Man wants deference, and his fellow wants the same from him. This predicament advances from contempt to hate, and of a sudden man's life is exposed to mortal danger.

At this point men feel the terror of impending violent death. They hear all of a sudden the wisdom, not of the 'useless' feelings in the mind, but of the body. At this point men come under the law of nature, whereby they surrender their vanity, (since it is the quantity allegedly responsible for false hopes) and pledge to live only in accordance with the morality of self-preservation so construed: as the forfeit of the feelings of the mind, that is, the surrender of pride in exchange for survival. It is a probing, majestic exegesis of the depth psychology of Hobbian liberalism: but it is a fairy tale.

> Absorbed in the race after the happiness of triumph, man cannot be aware of his dependence on the insignificant primary good, the preservation of life and limb; failing to recognize his bodily needs, man experiences only the joys and sorrows of the mind, that is, imaginary joys and sorrows . . . He can awake from this dreamworld and come to himself only when he feels in his own person—by bodily hurt, the resistance of the real world. By damnorum experientia man becomes reasonable. But if experience of injury is to show him not only the limit of his physical powers, about which he does not greatly care, but the limit of his insight and intelligence—the injury must be unforeseen . . . Because man by nature lives in the dream of the happiness of triumph, of a glittering, imposing, apparent good, he requires a no less imposing power to awaken him from his dream: this imposing power is the imperious majesty of death. . . . Why and how does the natural vain man gain this experience? Only by the answer to this question does one grasp the real reason which causes Hobbes to recognize the origin of morality not in the fear of death itself but in the fear of violent death.[78]

What additional education does the vainly glorious man need, if we accept Hobbes's definition? Honor and pride, spirit, are the only qualities available to human beings to summon in a struggle against oppression. Hobbes is much determined to anticipate and squelch this last domain of possible resistance. It is not that the vainly glorious man poses any kind of threat to public peace; it is that he poses some threat, to Hobbes's aspirations of dominion. To seek to dominate men is not great power, because men who know themselves to be oppressed will eventually rebel. Yet to attack a man's capacity to *resist* oppression, is a much more formidable political science. Hobbes, as Machiavelli, is seeking for the sort of control that has eliminated the possibility of resistance.

It is, in the end, the power of classifications that drive human beings towards bedlam in the political sciences of Machiavelli and Hobbes. It is not seriously disputable that Machiavelli depends upon accusation, in several forms, as the basis for political authority in his political science. It does not seem reasonable to suspect that Hobbes has arrived at a similar conclusion by accident. Hobbes does not mention Machiavelli to my knowledge; yet Hobbes's very definition of knowledge or sapience consists in the ability to offend other people from a posture of absolute superiority.

> As much experience, is prudence, so is much science, sapience. For though we usually have one name of wisdom for them both; yet the Latins did always distinguish between prudential and sapientia; ascribing the former to experience, the latter to science. But to make the difference appear more clearly, let us suppose one man endued with an excellent natural use, and dexterity in handling his arms; and another to have added to that dexterity, an acquired science, of where he can offend, or be offended by his adversary, in every possible posture, or guard: the ability of the former, would be to the ability of the latter, as prudence to sapience; both useful, but the latter infallible. But they that trusting only to the authority of books, follow the blind blindly, are like him that, trusting to the false rules of a master fence, ventures presumptuously upon an adversary, that either kills or disgraces him.[79]

NOTES

1. *Leviathan.* Edited by J.C.A. Gaskin. Oxford: Oxford University Press, 1996. Chapter 11, article 1: "To which end we are to consider, that the felicity of this life, consisteth not in the repose of a man satisfied. For there is no such finis ultimus, (utmost aim), nor summum bonum (greatest good) as is spoken of in the books of the old moral philosophers."
2. *Discourses on Livy.* Translated by Nathan Tarcov and Harvey Mansfield. Chicago: University of Chicago Press, 1996, book I, chapter 4.
3. I have discussed Machiavelli's political science at some length in *Machiavelli and Epicureanism: An Investigation into the Origins of Early Modern Political Thought.* Lanham, MD: Lexington Books, 2012, chapters 4 and 5 passim.
4. *Leviathan*, chapter 5, article 17.
5. *Leviathan*, chapter 11, article 18.
6. *Leviathan*, chapter 4, article 13.

7. *Elements of Law.* Edited by J.C. A. Gaskin. Oxford University Press, 1994. Chapter 1, article 3.

8. *Mandragola*, Prologue. In *Machiavelli: The Chief Works and Others.* Translated by Allan Gilbert. Volume Two. Durham: Duke University Press, 1989, 778. "Yet if anyone supposes that by finding fault he can get the author by the hair and scare him or make him draw back a bit, I give any such man warning and tell him that the author, too, knows how to find fault, and that it was his earliest art."

9. *Leviathan*, chapter 13, article 1.

10. *Leviathan*, chapter 13, article 2.

11. *Leviathan*, chapter 10, article 14.

12. *Leviathan*, chapter 11, article 17.

13. *Elements of Law Natural and Politic*, chapter 1, article 8, p. 22. In *Leviathan*, the first chapter on sense perception accomplishes this profession of faith: "And though at some certain distance, the real and very object seem invested with the fancy it begets in us, yet still the object is one thing, the image or fancy is another. So that sense in all cases, is nothing else but original fancy, caused (as I have said) by the pressure, that is, by the motion, of external things upon our eyes, ears, and other organs thereunto ordained" (*Leviathan*, chapter 1, article 4).

14. Leo Strauss. *The Political Philosophy of Hobbes: Its Basis and Genesis.* Translated by Elsa Sinclair. Chicago: University of Chicago Press, 1984, 7.

15. Tom Sorell, *Hobbes*. London: Routledge and Kegan Paul, 1986, 11.

16. *Man and Citizen.* Edited by Bernard Gert. Indianapolis: Hackett, 1998, 35.

17. *The Citizen*, chapter 1, article 10. In *Man and Citizen.* Edited by Bernard Gert. Indianapolis: Hackett Publishing, 1998.

18. *The Political Philosophy of Hobbes: Its Basis and Genesis.* Chicago: University of Chicago Press, 1937. Strauss, in this passage, is clearly counseling scholars to ignore the scientific teaching of Hobbes as the proper preparation for studying the political science. This doesn't signify much, insofar as the modern philosophers, among whom Strauss numbers, prefer to operate 'in between the lines.' It is Hobbes's moral and political philosophy which moves Strauss, which he regards as the apex of modern teaching. That teaching can't be effective if scholars understand the doctrines of nature upon which they rest; and that is the reason why Strauss does not encourage the investigation into Hobbes's natural science as a means to understanding the political thought. The proof is still in the pudding. Strauss himself makes that link understandable, although not obvious, to his careful readers.

19. Strauss, *Political Philosophy of Hobbes*, 125.

20. *Man and Citizen*, Epistle Dedicatory, 92.

21. Leo Strauss. *Natural Right and History.* Chicago: University of Chicago Press, 1965, 175.

22. *Leviathan*. Chapter 21, article 1.

23. Richard Tuck's surprising essay on Hobbes's utopianism ("The Utopianism of *Leviathan*," in Tom Sorell and Luc Foisneau, editors. *Leviathan After 350 Years*. Oxford: Clarendon Press, 2004) makes the claim that Hobbes approximately wishes to deplete the passions of the people in his ideal city. They are, Tuck astonishingly says, to be protected especially against the passion of fear. "If Hobbesian man were to live according to the laws of nature, he would not only renounce his individual judgment to his sovereign, but would live a strikingly passionless life; indeed, Hobbes sometimes implied that the elimination of passion might be necessary before one could effectively follow the law" (134). "Luxury and conspicuous consumption are a constant target for Hobbes, as one would expect given what I have been saying, for they arise from emulation, envy, ambition, and pride, all passions which are to be eliminated" (136). "By incorporating themselves into this artificial man who feels no fear, the citizens are themselves able to live a life freed from it. But they are also going to live a life without pride, their sense of their own individual importance reduced to nothing in the face of their commonwealth. We have to recognize the strangeness and the radical character of this vision, if we are properly to understand the role which Hobbes's theory played in the construction of modern politics" (138). Tuck's argument can be responded to in the following manner. In the first place, desire must be accounted a passion; and in Hobbes's view, that individual who lacks great desires is to that degree a poor specimen of a human being. Not only is high desire stipulated by Hobbes as

the norm in his 'utopia,' but it is furthermore indicated that the individual may as well be dead if he ever is aloof from the pursuit of new desire. "The passions that most of all cause the differences of wit, are principally, the more or less desire of power, of riches, of knowledge and of honor. All which may be reduced to the first, that is, desire of power. For riches, knowledge and honor are but several sorts of power. And therefore a man who has no great passion for any of these things, but is as men term it indifferent; though he may be so far a good man, as to be free from giving offence; yet he cannot possibly have either a great fancy or much judgment. For the thoughts are to the desires, as scouts and spies, to range abroad, and find the way to the things desired: all steadiness of the mind's motion, and all quickness of the same, proceeding from thence. For as to have no desire is to be dead; so to have weak passions, is dullness" (*Leviathan* chapter 8, articles 15–16). That man will have to give up courage as ordinarily understood; justice as ordinarily understood; and certainly temperance as ordinarily understood, all go without saying. Yet Hobbes never indicates, in his discussion of the virtues in chapter 8, that these are to be set aside in order to observe the laws of nature. Only those 'virtues' which interfere with the preservation of the self are to be squelched in Hobbes's commonwealth; indeed, it is precisely the purpose of Hobbes's political science to *unleash* great desires, great selfishness, hitherto suppressed passions into social circulation. Enjoyment is limited by Hobbes *to the hunt*. Repose is no part of Hobbes's theory. "Nor can a man live, whose desires are at an end, than he, whose senses and imagination are at a stand. Felicity is a continual progress of the desire, from one object to another; the attaining of the former being still but the way to the latter. The cause whereof is, that the object of man's desire, is not to enjoy once only, and for one instant of time; but to assure forever, the way of his future desire" (*Leviathan* chapter 11, article 1). Thirdly, Hobbes's discussion of the sort of freedom of the will that is to be acknowledged in his state is strictly compatible with fear. The Leviathan state itself is brought about by fear, and Hobbes obviously regards that deployment of human will to be binding. Nor does Hobbes ever make it appear that all pride must be crushed or suffocated in his state. Pseudo-pride, that is, valuing oneself in accordance with what the market will pay for one's services, is obviously going to result in a status hierarchy. Luc Foisneau ably traces the new contract-based conception of justice in Hobbes, and how this ridicules the virtue of justice (an equality defined by Aristotle). It may appear to Tuck that Hobbes expects individuals to ignore their accidental advantages in the market place, and to forswear seeking for advantage in exchange. Nothing could be further from the truth. Hobbes is sometimes thought to be the bourgeois philosopher precisely because his doctrines seek to overthrow the careful regulated exchanges governed by the Common Law. Foisneau argues that "The loss of confidence in the theoretical relevance of justice went hand in hand with the emergence of the free market as a new means for determining value in seventeenth century England. Although one must beware of the kind of oversimplification that may arise when interpreting a political work within the context of a complex economic and social situation, it does help to see Hobbes's theory of justice in the context of a new social and economic order based on the free market. The rejection of the theory of justice in the *Elements of Law* and *De Cive* hinges largely on a new principle of valuation, which replaces the principle of fair exchange with the law of supply and demand as a price-fixing mechanism. No need any longer to ask whether an exchange is fair: the matter is settled if there is a willing buyer" (105). From "Leviathan's Theory of Justice," in Foisneau and Tom Sorell, editors. *Leviathan After 350 Years*. Oxford: Clarendon, 2004. In fact, if we study the list of passions and manners in Hobbes, we find, not the neutralization, but the humiliation of the ordinary decencies. Hobbes never expects the majority of men to truly be fit for society, and thenceforth what passes in Hobbes for 'education' is certain to revolve around fear. The prohibition on 'revenge,' in terms of Hobbes's conceptualization of this emotion, involves almost the banishment of substantive equality from the principles of the penal code. This is certain to raise the passionate temperature of the population, just as unsupervised economic contracts will. Besides, if fear ever dissipated, men would resume trusting their own faculties and perceptions, using their own judgment on right and wrong, which Hobbes's philosophy can never permit to occur.

24. *On the Citizen*, chapter 3, article 9.

25. Lucretius. *On the Nature of the Universe*. Translated by Ronald Melville. Oxford: Oxford University Press, 2008, book VI, lines 1276–1283.

26. *Thomas White's De Mundo Examined*. Translated from the Latin by Harold Whitmore Jones. London: Bradford University Press, 1976, 63.

27. *De Cive* I, 7. Translated by Richard Tuck and Michael Silverthorne. Cambridge University Press: 1998.

28. *Leviathan*, chapter 11, article 2.

29. *Nichomachean Ethics*, book 1, article 7.

30. *On the Citizen*, preface, 15. Translated by Richard Tuck and Michael Silverthorne. Cambridge University Press, 1998.

31. The 'doubt' with which Hobbes begins his philosophical investigations is borrowed from Descartes. "For the understanding of what I mean by the power cognitive, we must remember and acknowledge that there be in our minds continually certain images or conceptions of the things that are without us, insomuch that if a man could be alive, and all the rest of the world annihilated, he should nevertheless retain the image thereof, and of all those things which he had before seen and perceived in it; every man by his own experience knowing that the absence or destruction of things once imagined, doth not cause the absence or destruction of the imagination itself" (22). *Elements of Law*, part 1, chapter 1, article 8. Hobbes does not mention Descartes here or elsewhere when he makes this formulation, but it was certainly Descartes' intellectual invention. This original position of skepticism whereby one imagines the world annihilated but one's own ideas still in existence more than parallels Descartes' extreme skepticism at the outset of the *Meditation*s. True, Hobbes deduces differently from the position (Descartes proves only the existence of the incorporeal soul, whereas Hobbes, who does not believe in an incorporeal spirit, deduces the knowledge of the body). Yet Hobbes does not acknowledge Descartes anymore than he acknowledges Bacon for his atomistic meditations. All of which is to say that Hobbes regards himself to be in a sort of state of nature with this rival for very limited eminence. As 'force and fraud' are the 'cardinal virtues of war' to Hobbes, so the claiming of authorship for other's ideas is fair game. We never see Hobbes relinquish his pursuit of extreme glory. This tells us something about this Leviathan: if his state is to enable governments to 'tame the proud,' this taming does not extend to the ruthlessly acquisitive spirit of Hobbes's own attitude. Inside the Leviathan state this newly ordained pride, a residue of the right of nature, will be the complement to Hobbes's own lupine disposition towards rival philosophers. Hobbes demonstrates, in his interactions with Descartes, as he illustrates in his feud with Wallis, that power is the honor that he respects. Whatever equality Hobbes may 'profess' before the congregation of men assembled in his imaginary commonwealth, his actions speak much louder than his words. Hubris, belligerency, contumacious discourse: these are the characteristic marks of the theoretician of natural right. This indicates well enough that for those who do not need to be brought to natural right through coercion, that the morality of natural right is their long suppressed dream. Both of the above mentioned postulates are, of course, anathema to ordinary opinion. There is scarcely a clergyman, or university professor, or non-philosopher, who would fail to be up in arms at such propositions. The community would even be more apoplectic if it contemplated the following statement made by Hobbes: "Nor does it alter the case of honor, whether an action (so it be great and difficult, and consequently a sign of much power) be just or unjust: for honor consisteth only in the opinion of power" (10.48, *Lev.*). Evidently the two men met in 1641 through the offices of Mersenne, who passed on some of Descartes' letters to Hobbes. At this time Hobbes would have published his *Elements of Law* with his mediumistic theory of optics. Descartes' *Dioptrique* had been published in 1637, and evidently in the 1641 meeting, Descartes suggested that Hobbes had plagiarized from him. This competition for honor and glory is obviously relevant to Hobbes's moral theory: it shows what he *practices*. Frithiof Brandt discusses this episode in *Thomas Hobbes's Mechanical Conception of Nature*. Levin and Munksgaard, 1928, 184–185. In 1645, as Brandt gives us to know, Hobbes appears to have finished a draft of at least the first twenty-eight chapters of *De Corpore*. However, Hobbes now lays this work aside, in order to produce a new work on optics. Brandt wonders why. His conclusion is that the publication of Descartes' *Principia* provoked Hobbes to compose a new treatise on optics, for Hobbes had already claimed to be the founder of two sciences: optics and civil science. The rivalry here between Descartes and Hobbes is instructive. Hobbes, hungry for glory, proclaims himself not the co-founder of optical science, but the first founder. Towards this end, he refuses to meet

with Descartes in 1645. He is thus pursuing, three years after the publication of *De Cive* and six years before the publication of *Leviathan*, glory at the expense of equality. Is this not the opposite of the moral teaching that Hobbes has set forth for the majority of the human race? Except that for Hobbes, glory is possible to obtain, by the great, as being a deed few or none other can do. The disposition towards war with Descartes, not the surrender of pride, is Hobbes's clear attitude. Indeed, this attitude is displayed in *De Cive's* introduction. Hobbes calmly goes to war with the English establishment across the board: churchmen, universities, politicians.

32. *On Man*. In *Man and Citizen*. Edited by Bernard Gert. Indianapolis: Hackett Publishing, 1998, chapter 12. "Emotions or perturbations of the mind are species of appetite and aversion, their differences having been taken from the diversity and circumstances of the objects that we desire or shun. They are called perturbations because they frequently obstruct right reasoning" (55). For Aristotle, the emotions are vulnerable to distortion, *yet they retain a reasonable core*. It is possible for people to feel too much or too little of an emotion in a certain circumstance, but in general the passions indicate facts or apparent facts. For Aristotle, practical reason depends upon the emotions for its very instruction on the issue of right and wrong.

33. Ibid.

34. *Leviathan*, chapter 17, article 8.

35. *Leviathan*, chapter 10, article 37.

36. *Leviathan*, chapter 10, article 48.

37. *Leviathan*, chapter 8, articles 15–16.

38. *Leviathan*, chapter 10, article 3.

39. *Nichomachean Ethics*, book V, article 2.

40. *Leviathan*, chapter 15, article 13.

41. *On the Citizen*, chapter 1, article 4.

42. *On the Citizen*, chapter 1, article 2.

43. Aristotle, *Nichomachean Eth*ics. Oxford University Press, 1998, v. 5: "These names, both loss and gain, have come from voluntary exchange; for to have more than one's own is called gaining, and to have less than one's original share is called losing, for example, in buying and selling and in all other matters in which the law has left people free to make their own terms; but when they get neither more nor less but just what belongs to themselves, they say that they have their own and that they neither lose nor gain. Therefore the just is intermediate between a sort of gain and a sort of loss, viz. those which are involuntary; it consists in having an equal amount before and after the transaction."

44. *Leviathan*, chapter 44.2: "And from hence it comes to pass, that men have no other means to acknowledge their own darkness, but only by reasoning from unforeseen mischances, that befall them in their ways."

45. *Leviathan*, chapter 10, article 16.

46. *Leviathan*, chapter 10, article 17.

47. *Leviathan*, chapter 10, article 44.

48. *Leviathan*, chapter 10, articles 38–39.

49. *Nichomachean Ethics* book V, article 4.

50. *On the Citizen*, chapter 3, article 9.

51. *On the Citizen*, chapter 3, article 10.

52. *On the Citizen*, chapter 3, article 11.

53. *Elements of Law*, part 1, chapter 9, article 6.

54. *Leviathan*, chapter 14, article 30.

55. *Leviathan*, chapter 15, article 21.

56. *Elements of Law*, part 1, chapter 9, article 21.

57. *Leviathan*, chapter 6, article 41.

58. *Leviathan*, chapter 6, article 46.

59. *Leviathan,* chapter 6, article 34.

60. *Leviathan*, chapter 8, article 1.

61. *Elements of Law*, part 1, chapter 9, article 21.

62. *The Political Philosophy of Hobbes: Its Basis and Genesis*. Translated by Elsa M. Sinclair. Chicago: University of Chicago Press, 1952.

63. *Leviathan*, chapter 17, article 7.

64. Michael Oakeshott. "The Moral life in the Writings of Thomas Hobbes." In *Hobbes on Civil Association*. Indianapolis: Liberty Fund, 1975. Oakeshott has certainly been influenced by Strauss's interpretation of Hobbes, but he gets lost. Pride is the natural condition of man, Oakeshott maintains. This is to say that, for Oakeshott, Hobbes believes that the supreme joy in human life, the supreme pleasure, is to be honored by others. "The existence of others of his kind, and the impossibility of escaping their company, is the first real impediment in the pursuit of felicity; for another man is necessarily a competitor. This is no mere observation, though its effects may be seen by any candid observer; it is a deduction from the nature of felicity. For, whatever appears to a man to belong to his felicity he must strive for with all his powers, and men who strive for the possession of the same object are enemies of one another" (36–37). "And further, competition does not arise merely when two or more happen to want the same thing, for when man is among others of his kind his felicity is not absolute but comparative; and since a large part of it comes from a feeling of superiority, of having more than his fellow, the competition is essential, not accidental. There is, at best, a permanent potential enmity between men" (37). Oakeshott uses names of common signification, but he does not grasp that Hobbes has altered that signification. To be 'honored' by men in common speech is to be admired and loved by men. To be courageous is indeed to be great. To be just is almost to have a tincture of the divinity. To be temperate, in one's desires, is to radiate humility. To be wise is to have one's understanding respected and sought after. The ordinary signification of pride and honor is equivalent, for the most part, to the classical philosophical understanding. This is to say, that the virtues are emblematic of excellence in political being: these qualities deepen, strengthen, and perfect human relationships. Yet this is not what Hobbes, not what Oakeshott mean by pride and honor. Hobbes's conception of pride is anti-social: it is the braggart, the boaster, the conceited, the weasel, the petty and small minded. Hobbes says that men live for the chance to lord it over other men: to be worshipped due to excess, not of quality, but of power. Hobbes's definitions of the virtues makes them identical with what ordinary people call vices. Nobody wants to be found out to be a poser. Nobody wants to be found out to be greedy or vicious or double-dealing or dishonest. But Oakeshott, submitting to Hobbes's reinvention of the signification of names, says both that men seek to diminish one another by their achievements, and that they desire to be honored. These two cannot really go together. Men are not, as Hobbes says, poisonously anti-social in their deepest desires: rather, their deepest fear is to be outcast, unaccepted. The other appetite is fear, Hobbes says, per Oakeshott. But fear of what? Violent death, says Oakeshott. And what specifically does the fear of violent death amount to, in Oakeshott's view? Shameful death. "And whereas animals may fear anything which provokes aversion, with men the chief fear (before which all others are of little account) is fear of the other competitors in the race. And whereas with animals the ultimate dread is death in any manner, the ultimate fear in man is the dread of violent (or untimely) death at the hand of another man; for this is dishonor, the emblem of all human failure. This is the fear which Hobbes said is the human passion 'to be reckoned with': its spring is not a mere desire to remain alive in adverse circumstances, not is it a mere aversion from death, least of all from the pain of death; its spring is aversion from shameful death" (87–88). No, this is quite incorrect. Pride is the fear of shame. Fear of violent death is Hobbes's *antithesis* to pride: desperate, like the denizens of Athens in Thucydides's portrait of the Peloponnesian War and Lucretius's book VI, men surrender pride and trample piety, along with their dignity. To be *fearful* of a shameful death would keep people attached to their self-respect, which Hobbes denounces as vainglory. Men are all naturally possessed with dread of shame, but in Hobbes's view, this is what unfits them for society. Civil science must shatter this pride, en toto. Terror can do it: terror by not knowing where one is, or how to safely proceed, in a situation governed by the new civil science. I think all readers of Oakeshott will agree that it is fair to observe the following things. That in Oakeshott's Hobbes, human beings are truly 'tamed' in the process of coming to the laws of nature. Oakeshott says that Hobbes equivocates about the laws of nature: that, for the vulgar, Hobbes says that God is the author of the natural laws, and that they are always binding. For the sophisticated, as Oakeshott counts himself, the laws of nature can only arise from that which we choose, or consent to truly (127). "Our study of Hobbes has reached some conclusions which most readers will find it difficult to avoid. It seems clear that he believed that a

rational disposition in human beings was to be identified as an endeavor for peace. And peace meant acknowledging all others as our equal, keeping our promises, not displaying contempt and hatred, and not endeavoring to outdo all others in order to have the elation of being recognized to occupy first place. This manner of living is suggested by reason, which also suggests the means by which it may be instituted and maintained: it is the civitas. The reward for its accomplishment is emancipation from the constant fear of violent and shameful death at the hands of other men. And, so far, the sufficient cause or motive for endeavoring peace is found in fear of shameful death: fear prompts reason and reason discloses what must be done to avoid the circumstances which generate fear" (127). Here we come to the real core of Oakeshott's Hobbes. Which is to say that men, "fearful of shameful death," give up their desire to outdo each other; truly recognize each other as equals; decline to "take more than their share." And live voluntarily without the desire of preeminence. Hobbes's laws of nature lead us to Aristotle as it were—which fits in with Oakeshott's claim that Hobbes is truly a 'late Scholastic voluntarist.' Where to begin? When Hobbes gives his variation on Christianity's golden rule—"Do unto others as you would have them do unto you," he says "do not do that to another, which thou would not have them do to you." Is there any reason why 'every man for himself' cannot be this postulate of equality that Hobbes endorses? For Hobbes makes 'right' a claim of self, appointing oneself judge, sole judge of the means to self-preservation, first. The laws of nature, which Oakeshott avoids talking about in specific terms, do not repudiate the 'right.' What is equal about the doctrine that those who have suffered a wrong must forgive and forego justice (as ordinarily meant)? Clearly, those inclined to commit vicious violence obtain an unequal advantage; for their malevolence is shielded, not by Christian charity, but by the phony accusation that the demand for justice is itself violent. "A seventh (law of nature) is, that in revenges, (that is, retribution of evil for evil), men look not at the greatness of evil past, but the greatness of the good to come. Whereby we are forbidden to inflict punishments with any other design, than for correction of the offender, or direction of others. For this law is consequent to the next before it, that commandeth pardon, upon security of future time. Besides, revenge, without respect to the example, and profit to come, is a triumph, or glorying in the hurt of another, tending to no end . . . ; and glorying to no end is vainglory, and contrary to reason; and to hurt without reason, tended to the introduction of war, which is against the law of nature, and is commonly styled by the name of cruelty" (*Leviathan*, chapter 15, article 19). Secondly, does not Oakeshott follow that Hobbes has erased Aristotle's distributive justice? In Aristotle, as in ordinary opinion, equal value (objectively determined) must be exchanged for equal value, so that each man 'has his own,' that is, neither loses nor gains. In Hobbes's model, barring express contract between two parties, the 'value of a man is,' not the value of all other men, but 'his price.' To buy cheap and sell dear is the natural right inside Hobbes's city, and Hobbes is offended by the suggestion that any body should deal justly in truth with us if our need is greater than the other fellow's in exchange. Lastly, let us just consider Hobbes's hypocrisy. Would Hobbes allow another man to deprive him of the significations he means to convey with his words? In other words, would Hobbes consent to observe the common vocabulary already in effect between individuals, and to conform his use of names thereto? Hardly.

65. *On the Citizen*, chapter 1, article 2, pages 23–24.

66. *De Cive* . Translated by Richard Tuck and Michael Silverthorne. Cambridge: Cambridge University Press, 1998, chapter 3, article 32.

67. *De Cive* chapter 3, article 31.

68. *Nichomachean Ethics*. Oxford: Oxford University Press, 1998. Book 2, article 2.

69. *Elements of Law*, part 1, chapter 13, article 3.

70. *Elements of Law*, part 1, chapter 13, article 3.

71. Strauss, *Political Philosophy of Hobbes*, 138.

72. Bernard Gert is very much in the spirit of Oakeshott. ("Hobbes's Psychology." In Tom Sorell, editor. *The Cambridge Companion To Hobbes*. Cambridge University Press, 1996). Hobbes, these men come to recognize, has been a major influence in modern civilization. This is to say that if we can't contest Hobbes successfully (and few have), that we must learn to make him seem more amiable. This Hobbes cannot be the egoist that he is thought to be, Gert insists. Hobbes cannot have meant to indicate that man is a beast, out only for himself, incapable of altruism. Indeed, it is not hard to find many passages in which Hobbes indirectly

testifies to the social nature of man. Gert zeroes in on indignation and shame as forces more powerful than the fear of death, to prove that Hobbes does not believe that man is an egoist. Nor should we take the state of nature for an actual portrait of man, Gert insists. Hobbes knows it is only a generalization. Above all, Gert argues, Hobbes makes a distinction between a rational desire and passionate desire, a distinction that makes it possible to discriminate good from bad objectively. Rational desire, Gert indicates, leads away from the state of war, while passionate desire or 'covetousness' leads into it. "In both works (Lev. Ch. 6 and D.H. ch. 13), once Hobbes has the concepts of appetite and aversion, pleasure and pain, his account of the individual passions completely ignores the relation between human behavior and his materialist philosophy. He simply proceeds by way of introspection and experience, along with liberal borrowings from Aristotle's account of the passions" (160). "Hobbes is not claiming that everyone desires self-preservation more than anything else because he explicitly notes that 'most men would rather lose their lives (that I say not, their peace) than suffer slander'"(163). "Hobbes never claims that actual persons, those who were raised up in families, are psychological egoists. Indeed, as the above quotes indicate, he denied such a view" (167). "Hobbes's disturbing statement about power is only a claim that all people tend to be concerned about their future; it explains pension funds and medical checkups more than it does anti-social power grabs. Indeed, for Hobbes, the former are not only more common than the latter, they are also more rational" (168). I think we must grant Gert what he says. Hobbes, after all, declares that man does not fit his natural reason, and the peace it leads to, by nature (*De Cive*, note 1, chap 1). The majority of men are said to live most or all of their lives 'unfit' for society, for Hobbes's conception of society. Hobbes's conception of society, via natural reason, is egoistic. Men build the rational society by relating to each other for purposes of 'advantage' or 'profit,' not 'pride.' Man however holds tightly to honor. Or, we could say, he holds tightly to dignity. He refuses to regard his worth as the amount of money that he can fetch on the market. He refuses to equate his worth with his power. He contests the more powerful as if he was their equal. And so he must be driven to surrender his pride, his 'glory,' his' honor,' by the contrivance of science. Hobbes does not say that this can ever be done once and for all; it must be done relentlessly and continually. Man regards himself as having an absolute value, which he will not subordinate to any merely relative ranking. This is the source of 'altruism' that Gert finds evidence of in Hobbes's testimony. But there can be no doubt but that Hobbes means to reduce man to passionate (not rational) desire, that is, to make him respect fear of death as the truth, to dislodge the sense of absolute worth and dignity he has in himself, as the source of the social convenant based on reason. Gert is quite mistaken to think that Hobbes intends to build on the natural social animal which exists in human beings. Hobbes expressly reverses Aristotle's definition of the zoon politikon at the outset of *De Cive*. Furthermore, even in the new list of passions that Hobbes produces in the *E. of L.* and in the *Leviathan*, it can clearly be seen that Hobbes is redefining social man in egoistic terms. The definition of the passions are all filtered through the prism of selfishness, and these are scientific postulates.

The fact that shame is more powerful than fear of death does not mean that the dread of shame cannot be brought into the service of the fear of death. It is precisely the object of Hobbes's rhetoric to accomplish this. His very definition of sapience is fused to the absolute ability to destroy one's opponent with disgrace (a gift of Hobbes's new science of definition). Witness those Grecian ladies, referred to by Hobbes, and then by Gert. These were suicides, they put fear of death second. But this was amended. By stripping naked the suicides, and disgracing the corpses, other young women were conditioned not to kill themselves, that is, to put fear of death first. Gert wants to conceive of Hobbes as a 'rational egoist,' as opposed to a 'psychological egoist.' This means, to Gert, that Hobbes only requires that our desires be in line with our long-term preservation; that they eschew immediate emotional threats to self-preservation; but that within these constraints, Hobbes allows us to value almost anything. This is incorrect. All desires are desires for power; and power only exists if one has considerably more of it than rivals. "What all men have equally, is nothing worth." Now Hobbes upbraids the man who is content with his modest desire. For he is lacking in vitality. He is inferior to the man of great desires, such as Hobbes himself, who regards himself as the power of creating the terms of civil order (clearly not an equal thing). Gert will say that power, to Hobbes, is simply a way to procure more desire. That is, that delectation is the god (hence diversity). But we say that

Hobbes denies that 'felicity' can ever be a state of enjoyment; enjoyment can only be in 'prospering,' accumulating more power, as only new power can make the old power secure. It is the lust for power that cannot rest anywhere, this is why no 'summum' can be put on happiness or its goods. In the end, most men do not want to live this way. They want enjoyment now. They want a limit to striving, not a perpetual submission to anxiety and restlessness. To this men will never be brought by reason. Hobbes is brought there by perverted reason. It is he who must deny 'what is.' Gert is sadly mistaken in thinking that natural right is an example of 'rational egoism.' Strauss is correct: the right of nature is only discerned by terror, by disorientation, by the violent resistance of the new order to man's rational expectations (i.e., that he matters in himself, rather than through his power). It is terror which drives man to abandon and neglect what he loves, and who loves him. It is only in this constellation that the distinction between liberty and compulsion (necessity) disappears.

73. *On the Citizen*, chapter 1, article 4.
74. *Leviathan* chapter 11, article 11. Cf. chapter 11, article 12.
75. *Leviathan* chapter 11, article 12.
76. *Leviathan* chapter 28, article 27. Cf. Strauss, *Political Philosophy of Hobbes*, 12–13.
77. Strauss, *Political Philosophy of Hobbes*, 20.
78. Strauss, *Political Philosophy of Hobbes*, 18–19.
79. *Leviathan* chapter 5, article 21.

Chapter Five

On Spinoza's 'Substance' or 'Nature'

The revolution in philosophy accomplished by Descartes has tended to limit scholars' investigations into the early modern thinkers. Descartes, famously, essayed a new separation between the life of the mind and sense perception. There is no disputing the fact that Descartes helped to unleash a wave of philosophical thinking that wholly disenfranchises the knowledge, and the knowers, who are limited by ordinary experience or perception. The trouble is that Descartes himself, as indicated in a previous chapter, stands on the shoulders of a much earlier school of philosophers: the Eleatics.

When scholars begin their investigations into the early moderns with Bacon and Descartes, they are cut off from the deep philosophical heritage which traces back to the philosopher Parmenides and the early atomism of Democritus.[1] Those teachings in turn bring alive for a reader the power and the counterarguments made by Plato and Aristotle. It is not possible to come to the bottom of the challenge made to the authority of sense perception by studying Descartes' *Meditations*. For Descartes relies upon a theory of *body*. And that theory of body is one of indivisibility, as proved in the second chapter.

Spinoza tends to be discussed through a Cartesian lens. Which is to say that Spinoza tends to be classified as a thinker who has discovered laws of 'pure' thought: thoughts which do not depend on sense perceptions, which are reached by the suppression of perceptual evidence.[2] This involves a debate about body, about physics; and for a philosophical movement which seeks to assert the equal rights of human beings under the doctrine of liberalism, there is no debate that is more important to understand.

The Enlightenment is famous for attacking superstition and dependence on mere authority. The rights of human beings can only be made good when they can trust themselves, rather than to 'hearsay' as Spinoza indicates. Yet if the ordinary man and woman are limited to the information which sense perception and correlative opinions make available to them, what sort of equality and freedom is it that they will gain in the new era?

It is not enough to regard clerics, and individuals born to noble status and office, as the only vanguards of potential oppression. Philosophy itself is fully capable of enacting this role. The vast tradition of philosophy which is rooted in the work of Parmenides fully cradles and prefigures the doctrine of atomism which Democritus, Epicurus, Lucretius, and the Early Modern philosophers work with. Spinoza himself seems to advance both a Parmenidean argument involving one all-inclusive being, *and* an atomistic doctrine of simple bodies.[3] Both arguments rest upon the premise that real being is indivisible; that real body is indivisible.[4] If the argument about indivisible body is false, then it should cause us to think anew about the wave of 'liberation' that early Enlightenment thinkers insist is their purpose. Untruth does not bring freedom.

The early modern thinkers, with Spinoza prominent among them, are often regarded as relying upon a mathematical variety of thinking.[5] Spinoza's employment at least of a geometric model of argument is on vivid display in his *Ethics*. However, as Plato's teachings can make evident to us, it is a suspect proposition to indicate that a mathematics or geometry is even possible without the evidence made available to us through sense perception. At least, the pretension of mathematical logic to effect a binding chain of reasoning upon the reader depends upon a mathematics and a geometry that is itself dependent on what can be originally known from the authority of the senses. In this chapter I will argue that the theory of indivisible body upon which Spinoza's arguments rest makes it impossible for us to reach the nature of 'extension' which a true geometry is concerned with. The existence of geometries which have forsworn dependence upon the evidence of the senses is acknowledged; yet these forms of geometry rest upon the arbitrary postulates of philosophers; and as such, they are all contestable, and bind nobody's mind to any conclusions whatsoever.

Spinoza's *Ethics* is divided into five sections. The first section begins with a doctrine that teaches about a great unity of being. This great unity of being, Spinoza argues, is 'indivisible,' and it is also eternal.[6] The second section of the *Ethics* begins with a discussion of a different sort of body: 'simple bodies,' which, though 'infinite' in number, are also eternal.[7] Neither of these sections ever makes it clear how the thinker is going to arrive at a contemplation of the 'finite modes,' that is, the perishable bodies: the bodies of the men and women surely whose enlightenment is at issue. Spinoza never does succeed in reaching the perishable bodies until the fifth section of the work. The theory of knowledge by which he proposes to plant firm foot on the ground of knowing perishable objects is designated by Spinoza as 'intuition': which is to say that this is not a kind of knowledge that Spinoza will endeavor to explain to us.[8] It seems to remain tied to the knowledge of eternal bodies as developed in parts I and II of the *Ethics*. Yet there can be no doubt that it has all along been Spinoza's purpose to deliver a teaching *about* the perishable bodies.

In the *Ethics*, Spinoza makes known to us that it is his purpose to establish a new model of a human being. Spinoza undertakes to construct an ideal template as it were, which strives after 'perfection' in the human experience. "For since we desire to form the idea of a man which we may look to as a model of human nature, we shall find it useful to keep these terms in the sense I have indicated," Spinoza writes. "So in what follows, I shall mean by 'good' that which we certainly know to be the means for our approaching nearer to the model of the human nature that we set before ourselves, and by 'bad' that which we certainly know prevents us from reproducing the same model."[9]

Spinoza's arguments in the *Ethics* and in his other works are all fiercely dedicated to establishing and promoting a new ideal of the human individual. This is why it is so important for us to make certain of the validity of the original evidence upon which the reasoning proceeds. How shall we be able to evaluate the legitimacy of the new model of the human being that Spinoza is determined to enact, if we cannot first come to the bottom of the validity of the theory of body upon which his moral and political theory themselves must rest?

The philosophies of Parmenides and the Epicurean wing of atomism, two doctrines on vivid display in the *Ethics*, were always concerned with power and authority. They are doctrines concerned with the power and authority of philosophers, within and against communities of non-philosophers. The thesis of indivisible body itself, from its earliest incarnations in Parmenides right up through Hobbes's teaching, is fundamentally about the authority of the philosopher to be the self-appointed determiner of what the first facts *are*. Plato makes what may well be conclusive arguments as to the falsity of the doctrine of indivisible body, and the implications of Plato's argument could not be more important. For, if it can be established that it is the nature of body and being to be composite, it is possible in turn from this foundation to prove the authority of the sense perceptions and the information derived from them. If the original facts are to be determined by sense perception, as opposed to the philosophical postulate of indivisible body, then those first facts will need to be mutual facts: not facts gestated by some strange new version of mathematics, but facts which are equally accessible to all, and which require the philosopher to persuade, rather than to define his way into a common reasoning process.

Scholars do not sufficiently wonder about Spinoza's emphasis on eternity. "It is in the nature of reason," Spinoza argues, "to perceive things in the light of eternity."[10] What, after all, does the perfection of the human way of life have to do with eternity? To be sure, the human being has a limited amount of time in which to live and excel. This raises issues of fulfillment and perspective as important helpmeets to the development of a satisfying life. Yet these are considerations appropriate to the contemplation of the *limits* of human existence. When Spinoza advances his theses to the effect that the most real things are eternal, he is not accommodating these limits of the human being. He is denying the relevance of limited experience for the purpose of establishing truth of fact. Spinoza's emphasis on eternal beings is promoted as a challenge to the knowledge that we are able to acquire by sense experience and perception. These doctrines of the eternal in Spinoza are intended to overrule the relevancy of ordinary experience, insofar as the contemplation of truths are concerned.[11] This focus on eternity, Spinoza's focus on eternity, is meant to silence the non-philosopher. It is an attempt to consign to the rubbish heap the awareness and information that ordinary speakers are capable of acquiring for themselves, in order to subordinate these opinions to the dictates of a philosophy which is not content to build its doctrines out of evidence commonly available.

The men and women who must be the object of Spinoza's social contract are flesh and blood. They don't exist eternally. They bleed *sub specie durationis*. What has Spinoza gained by dismissing the knowledge that individuals have of themselves from his metaphysics, which is designed to end in—not dreams or meditations—but new forms of social organization? What he obtains is the *power of definition*. What he obtains is the power to coin language anew. Spinoza could not win his way to this new battery of definitions if he began with the evidence supplied by the ordinary assignment of name to object.

The stipulation of the great 'being' as the cause of all; or the stipulation of the 'simple bodies' as the cause of all, makes this proof, this movement *from* ordinary opinion *to* science, something that can be ignored. This is to say that Spinoza relies upon the Eleatic arguments in order not to have to demonstrate his definitions. If the Eleatic reasoning is defeated, then Spinoza no longer has that luxury. The Eleatic reasoning can be defeated. Plato showed the way. Plato's proofs make it possible to disprove *Ethics* I, and to proceed to a more probing analysis of Spinoza's *Ethics* and his philosophy more generally speaking.

Spinoza titled *Ethics* I "On God." H. Barker, in his essay on Spinoza's *Ethics*, protested that the use of the name 'God' to indicate the being of Spinoza's first section is likely to produce only confusion. Barker elects to focus on the name 'nature,' which Spinoza offers as an equally valid name for his 'being.'[12] It is the third name that Spinoza likes to affix to his 'being,' that of 'substance,' which is the preferred denomination for this study. "By substance I mean that which is in itself and is conceived through itself,"

Spinoza writes, "that is, that the conception of which does not require the conception of another thing from which it has to be formed."[13]

'Substance' originally indicated the Aristotelian term for the most real beings that exists in nature. "We call substances the simple bodies, that is, earth and fire and water, and everything of the sort, and in general bodies and the things composed of them, both animals and divine beings, and the parts of these."[14] Spinoza's doctrine of 'God' and 'nature' is indeed offered as a stiff challenge, even as a rebuke, to Aristotle's definition of substance. To concentrate on Spinoza's revision of the idea of 'substance' makes it possible to compare Spinoza closely with the older philosophical tradition that he is attempting to displace.

It may also be pointed out that the very first definition that Spinoza offers in *Ethics* I is of 'substance.' 'God' is not mentioned until the sixth definition. As Steven Nadler has argued, God is not the first thing that Spinoza attempts to define.[15] In the *Theological-Political Treatise*, furthermore, Spinoza informs us that it is not possible to know of God in any a priori way.[16] Knowledge of God is something that must be adduced, in Spinoza's view, from reasoning about *nature*. Spinoza's doctrine of substance is at the core of this new teaching about nature, and in this light I will proceed.

ETHICS I

Scholars of the *Ethics* sometimes make the observation that there is no one definition in *Ethics* I from which all other definitions and axioms flow, deductively. It is suggested that one must absorb all of the definitions together, in order to appreciate the larger deductive framework of the *Ethics*. This may well be true. However, that does not mean that all of the definitions are of equal importance. The very first definition of *Ethics* I is the cornerstone of Spinoza's philosophical system. It is the definition of substance, which fuses together the concept of substance with the essence of existence.

It is the definition which stipulates that a substance can only be that which has existence as its essence: that is, that the only substances must be objects which cannot be separated from the category of being, for which being is identity.

Spinoza, in a second definition of substance, indicates that it must be 'infinite.'[17] This suggests that 'substance' can have no limits. As a preliminary observation, it may be indicated that the definition of body in *Ethics* II seem to make it possible for there to be infinite substances. The simple bodies which comprise 'individuals' in *Ethics* II do not appear to come into being and pass away.[18] Spinoza makes observations in his correspondence to the effect that matter is indestructible.[19] If the simple bodies are indeed atoms, a disputed point, I do not see why they too would not qualify as substances. The simple bodies are not individually infinite, but taken together they are infinite. This is a subject that will be discussed more thoroughly in its place.

Spinoza's fusion of the concept of substance, or true being, with eternal existence in *Ethics* I, is supplemented by his definition of the possible kinds of knowledge human beings may possess. Those forms of knowledge which concern perishable objects, and which therefore do not themselves indicate the eternal, are dismissed as 'inadequate' knowledge, as 'confused' or 'mutilated' ideas.[20] Two other forms of knowledge, 'adequate' knowledge and 'intuition,' are both cemented to the category of eternity.[21] Both the categories of 'adequate ideas' and 'intuition' appear to consist in a process, whereby perceived perishable objects are *reconceived* as eternal entities. In the case of 'adequate' ideas, perishable objects are viewed through the lens of a proto-atomic construct of 'simple bodies' that are said to exist in patterns of motion and rest, constituting 'individuals.' Thus in the case of 'adequate' ideas, the mind is still dealing with bodies, that is, atoms, that exist *sub specie aeternitatis*. It is only in the category of intuition, in *Ethics* V, where the mind seems finally to be dealing with perishable objects, finite modes, in themselves; but even there, the perishable object is being viewed under the category of eternity.

What does it mean to view an object under the category of eternity? It means, in the case of 'adequate ideas,' to view a perishable body as a part of the 'extension' of the eternally existing substance. All of the familiar characteristics of said body, including its form (in the Platonic sense, which happens to coincide with the ordinary use of universals), are stripped away. One is left with an abstraction that is no longer recognizable. The perishable object as it is ordinarily known is replaced with the language of the eternal substance and its 'attributes' and 'modes.' To undertake the translation of the inadequate idea into an 'adequate' idea is then to banish from the mind the object that one has perceived, and to replace it with an object that has been *theorized.*

Everything, including everything in man, is said to be 'in' substance, either as one of its attributes, or as one of the affections of those attributes (modes).[22] If we think about an apple 'adequately,' we can only sunder the name from the image of the perceptible apple that we eat. Substance is said to be the 'cause' of everything; the effect cannot have anything that the cause did not supply. Spinoza's substance is indivisible and infinite. To think of an apple adequately is to think of it as 'in' substance; as a 'mode' of substance; or as an affection of the (also infinite) attributes of substance. There is no past, present or future in substance, no time. The object existing 'in' substance is conceded to possess 'extension,' or dimensions, and to allow for 'thought'; but these are the only two attributes of substance that Spinoza ever indicates that human beings can or do know.[23] The 'extension' of the apple and 'thought' about the apple are said to be a unity: for substance is alleged to be a unity; and therefore its attributes cannot be said to exist independently of one another. Spinoza's substance, like Parmenides's 'being,' is indivisible. The indivisibility of Parmenides's 'being' did not prevent the Eleatic from recognizing a number of other forms in his 'unity.' The same can be said of Spinoza and his substance.

That which ordinary opinion experiences, obviously, begins and ends with perishable bodies. Spinoza refuses to accept the perishable bodies as given: he searches after causes of the perishable bodies. The perishable bodies are not perfect. Their essence, that is, limit, is not identical with existence. Spinoza's first principle is that the name 'substance' shall be reserved to that which exists *as its essence*: that is, to that which is eternal. Aristotle invented the term 'substance' to indicate an object that can exist by itself in nature. Single, perishable objects are substances to Aristotle; attributes inhere in these.[24]

For Aristotle a substance has both accidental and essential attributes. Socrates wearing a hat is an accidental attribute. Socrates can be separated from the hat while remaining Socrates. Mortality is an essential attribute of Socrates. Socrates cannot be separated from the attribute of mortality. It also needs to be pointed out that Aristotle's substance is a composite of matter and form.

Spinoza sweeps Aristotle's definition of substance away. Only that which exists *eternally* by its own power is 'substance' for Spinoza. This substance is the only substance, in Spinoza's view, and it is the cause of *everything*. In his *Ethics* Spinoza stops short of ever formally conceding the reality of the perishable bodies. He doesn't prove the emergence of the perishables out of the eternally existent. Rather, he defines the perishables as 'modes' of the eternal; as if the perishable bodies were parts of the eternal, thus casting all the perceptible qualities of the perishables into shadow.

Spinoza, in his *Ethics*, does not talk about the genesis of his original premise: that the one and only true substance must have, as its essential characteristic, existence. Which is to say that Spinoza's substance can't not-exist. This is indeed Parmenides's idea. Parmenides did not offer an analysis of how he reached his one idea either. It can be pointed out here that there is no such thing as deduction which does not depend on a prior *induction*. Spinoza may not provide an explanation for the derivation of his major premise in the *Ethics*; but a good case can be made to the effect that Spinoza reveals his reasoning as to the major premise of substance in his *Treatise on the Emendation of the Intellect*.

THE 'ONE TRUE IDEA' IN THE EMENDATION OF THE INTELLECT

In the *Treatise on the Emendation of the Intellect*, Spinoza argues, we are in need of a true idea, borne out of the intellect itself—and not in any way delivered by perception or memory—to guide us in the direction of true deduction.[25] A true idea, a clear and distinct idea, has a couple of characteristics for Spinoza. It cannot be broken down or analyzed into smaller ideas, for one thing. In this sense, the 'clear and distinct' idea must be 'simple,' or irreducible.[26] In the *Emendation*, Spinoza makes it clear that he is in search of one clear and distinct idea in particular: an idea which can serve as the foundation for a deductive system, 'one true idea' that once established, can guide the mind to all other true ideas.

The *Emendation of the Intellect* breaks off very suddenly. Yet it is at an interesting place that it breaks off. Spinoza is in search of his guiding true idea. He needs this idea to guide him towards deductions. He disclaims perception, as we have seen. Spinoza does not produce very good arguments for dismissing the evidence of perception through sense faculties. His 'arguments' amount to the deployment of nasty sounding names such as 'mutilated,' 'confused,' 'inadequate.' To locate his 'true ideas,' Spinoza decides, (without argument or proof), that they must somehow *originate* in the 'intellect.'[27]

That the intellect must have *ideas* to evaluate seems obvious. Where are these ideas to come from if not from experience? If the mind refuses to take from experience the objects that constitute reality, how will it ever be possible for the intellect to know that its ideas *comprehend* reality? Spinoza, however, is not interested in developing ideas that simply capture and comprehend or theorize what is *true*. Spinoza's ideas are extensions of the power of his mind: they are intended to be imposed upon reality, to convert that reality into a material that will also be subject to the power of Spinoza's clear and distinct ideas.

Spinoza begins to expostulate about 'power' as the *natur*e of the intellect. 'Clear and distinct ideas' seem to be something the intellect produces. The criteria for a true idea, Spinoza says, is that it enables us to make deductions.[28] A clear and distinct idea, Spinoza argues, cannot be false. The investigation into the nature of clear and distinct or 'adequate' ideas will be treated of more at length below. For the moment, the issue is the one guiding idea that can enable the mind to deduce all other knowledge. Spinoza is searching for this true idea in his *Emendation*. The intellect, in the *Emendation*, is said to be expressly about *conceiving* ideas, not perceiving. "I say 'conception' rather than 'perception' because the term perception seems to indicate that the mind is passive to its object whereas conception seems to express an activity of the mind."[29] The intellect, in other words, for Spinoza, is an inventor. Throughout the *Ethics*, Spinoza evinces a horror at the dependence of the mind on external objects as the sources of its ideas. The intellect as inventor should give us some solid insights into Spinoza's argument. Upon what grounds should we even be interested in the 'inventions' of the intellect? What does an 'invention' of the intellect, a so-called 'conception,' have to do with truth? The reader, for one thing, isn't interested in anything else. The reader is interested in *truth*.

Descartes is the only philosopher that Spinoza deigns to mention much in his writings. He mentions Bacon. Machiavelli. All three of these philosophers are conversant with the atomistic doctrines. All three of these philosophers are fluent in the language of the 'simple bodies,' which (like Spinoza's single great 'being') are said to be indivisible. The heritage of Parmenides and Epicurus would be the most instructive in terms of explaining Spinoza's fascination with the will as the source of ideas, for this is what it is to construe the intellect as 'inventor' rather than as *analyst* of previously given materials. Spinoza makes it clear in *Letter 56* that he prefers the Democritean Greek tradition of philosophy as opposed to the one founded by Socrates, and developed by Plato and Aristotle.[30] This indeed situates Spinoza on the Eleatic side of the great philosophical divide. Spinoza's definition of the intellect, of his intellect and of all intellects, is that it is power: a creator. The 'cause' of Spinoza's idea of substance is nothing external to Spinoza's mind. *Spinoza* is the cause of his idea of substance. It is his mind's creation, that which his mind controls, that which obtains its bearing from Spinoza's definition of being.

Spinoza's idea of the intellect must be disputed. Spinoza is free to delude himself if this is what he prefers to do; but his definitions of intellect and truth are what are referred to in ordinary speech as 'wish.' Spinoza commends his postulates to us as somehow 'necessary.'[31] This language of necessity that Spinoza employs is most importuning. It all but forces itself upon us, needy and full of desire. Yet Spinoza's ideas need to be examined as proposed *common* truths. Reality is something very frequently, and very much, independent of our will. To know reality is what is indispensable: to know reality is what makes comprehensible to us the degree to which our will is free and potent in act. It comes as little surprise that Spinoza, who subjects the entire category of reality to his own 'conception' or will, prefers to dismiss the category of free will entirely as a popular delusion. 'Will cannot be called a free cause,' Spinoza argues, 'but only a necessary cause.'[32] If Spinoza had to admit that his own formulations were the free creations of his will, then that would indicate that they might not be true, or could be otherwise. This would upset their deductive power, to say the least.

In the *Emendation or Correction of the Intellect*, Spinoza says more about the 'one true idea' that the mind needs in order to establish its 'deductive' system. This 'idea' itself is said to be the product of intellect as power, as conception rather than perception. The idea that he settles on is capable of enabling the philosopher to claim to know a number of true things about body. Spinoza already is becoming occupied with the 'conception' of his one true idea to the effect that body is indivisible. This will in fact form for Spinoza, in his later works, the basis for his theory of what a substance is.[33]

Aristotle defines substance as that thing of which qualities are predicated. One does not predicate a substance of still other things. Hence, very loosely speaking, we have substances in which qualities inhere, and qualities that are dependent on substance in order to exist. Spinoza's theory of substance, in a very abstract way, traces out a similar distinction. For Spinoza, substance is that which exists by itself; and everything else that exists must exist in a substance. In the *Emendation*, Spinoza has not yet sketched out this much of his eventual definition of substance. However, he has begun to indicate the nature of what a true body is or must be; an observation that illuminates just to what extent the 'one true idea' of the *Emendation* is the forerunner of Spinoza's later full-fledged theory of substance.

Substance is the really existing thing in Spinoza's *Ethics*. Substance is that, and only that, the essence of which involves existence or cannot be separated from existence. Thus it is not possible, Spinoza argues, for us to think of a true substance that at sometime does not exist. Hence substance is eternal. The effect cannot have in it anything that is not already formally present in the cause, he argues. Thus as cause of itself, 'substance' is already eternal. The effect can't contain anything non-eternal therefore. Hence the world of perishable objects is not substance. There is but one substance, the affections and modes of which are said to be 'infinite.'[34]

The substance which Spinoza alternately calls God or Nature has being. It is also noteworthy, and somewhat novel, that Spinoza defines substance as *infinite*. The 'infinite' indicates that which is not all present, which is incomplete. Spinoza has said that there is only one substance. One means unity, which cannot be more than one, or less than one. If the infinite is taken to indicate a close approximation to unity, but still not quite it, then it is not in that aspect a unity. What is clear is that whatever has being must be some *one* thing; not some *almost* one. Spinoza himself denies that his substance is many. He insists that there is only one. "No attribute of substance can be truly conceived from which it would follow that substance can be divided."[35] But this one he defines as other than, less than, unity; or as more than, and thus not, unity.

Spinoza gives us a reason as to why 'substance' must be infinite. If substance were to be finite, Spinoza reasons, then that would mean that there is a limit to it. If it has a limit, then another being must exist to limit it. In other words, if one thing is a distinct, limited unity, then it is not 'everything.' For one can't ever reach everything, that is, it can never all be accounted for in our thought. Spinoza has insisted that only *that* is 'substance' (true being) for which existence is its essence. Which means it cannot not-exist.

In *Ethics* I, Proposition 13, Spinoza is arguing that substance must be indivisible.

This is what causes a dilemma for Spinoza's theory that the substance is also infinite. How would it be possible for the incomplete to be indivisible? Spinoza suggests that it is only crude notions of the infinite which would upset his theory.[36] Yet Aristotle's notion of the infinite is not crude. For Aristotle, the infinite is that which is never 'all there,' where all the parts are accounted for.[37] Some scholars insist that Spinoza means, by 'infinite,' that it is indeed 'complete' or 'all together.'[38] In any event, Spinoza is going to have to at least oppose Aristotle's definition of what the infinite is, and supplant it with one of his own. Or, if Spinoza is going to accept Aristotle's definition of the infinite and redefine the name 'infinite' so that it means 'complete' or finite, it raises the question as to why Spinoza would go to all this trouble for no purpose.

Spinoza's conception of a substance cannot be both one and infinite. For the names indicate distinct things. That which is a unity has a limit. The limitless by definition is not limited, is not some one. Now we do not speak of parts here. Spinoza indeed denies that 'whole' and 'part' are real things in nature.[39] We may say that Spinoza can be forced to admit the following. Either his substance is finite, or it is not one. If it is not some one thing, or even the only one thing, then it is no-thing. If we are describing reality even by Cartesian principles, that much is already proved. For what Descartes proves is that he (not some infinite) thinks; that he, some one consciousness, therefore exists while thinking; and this origin of knowledge therefore concedes that what has being is some unity. Spinoza could argue that his substance is an incomprehensible unity, because too vast. So long as it is conceded that Spinoza's substance is a unity, no problem is presented by his eccentric use of the name 'infinite.'

It may be that Spinoza's insistence that his substance be 'infinite' is a direct attempt to deny that this substance is a 'whole' of parts. Spinoza indicates, in his *Short Treatise*, that he does not believe that 'whole' and 'part' are true natures in reality. Spinoza regards 'whole' and 'part' as 'ens rationis,' or creations of reason. What is true is that a whole must be a unity. What is also true, in reality and nature, and independent of any 'inventions' of the intellect, is that any object that can be said to exist must be a unity, some one object.

The indivisibility and the infinity of Spinoza's nature are the irreconcilable things. That which is indivisible must be unity: but unity is that which the infinite can never be, since the definition of the infinite is that whose parts are always more. Spinoza may claim to us that man cannot fathom nature's indivisibility or infinitude. That they may be reconciled beyond our ability to grasp. In either case, Spinoza must either ascribe unity to his being or multiplicity to his being. If none of the many (infinite) parts of infinity is 'one,' then there is 'no-one' that can have being.

Steven Nadler isn't exactly rigorous in his treatment of Spinoza. The first serious issue that Nadler takes up is Spinoza's conception of 'substance.' According to Nadler, Aristotle greatly 'developed' this concept. This is something of an understatement.[40] Aristotle is the founder of the concept. It makes a great deal of difference, since nineteen odd centuries of philosophy took their bearing from Aristotle. Substance, to Aristotle, is no mere name for a 'thing.' Aristotle's concept of substance is attempting to ascertain the *most real* things, those which have the *most* being. For Aristotle, the most real beings are the perishables ordinarily perceived. For Aristotle, existence is *not* the essence of a substance. This indicates that Aristotle and Spinoza offer competing definitions of substance, and we must judge the competition.

Aristotle distinguishes between substance, the object that has qualities; and the qualities. The qualities cannot exist by themselves. 'White' cannot exist by itself. It needs an object to exist in. Thus, the horse is white. In Aristotle's definition, substance indicates that of which qualities are predicated. Nadler attempts to indicate how Spinoza has modified the concept. For Spinoza, we cannot call a 'horse' a substance. Why? Because the horse cannot exist by itself. What does this mean? In Nadler's view, the horse cannot exist by itself because the horse needs water and hay in order to survive. In other words, the horse cannot exist by itself because it is perishable, because it is not eternal.

Nadler doesn't offer any criticism of Spinoza's theory. Why does an object's being perishable, that is, its being in need of water and hay in order to survive, disqualify it from being a substance? Is not the living horse more real than the idea of the horse? Nor is Spinoza a Platonist. Plato, it is true, recognizes the pattern of a horse as a self-subsisting thing. Yet Plato does not thereby argue that the living horse is not a real object. Plato does not pursue justice for the sake merely of the 'idea' of men who are in need of justice. The selection and development of Plato's guardian class does not take place among disembodied beings; nor does Spinoza have any respect for Plato's theory of universals, since these are things identified in ordinary speech.

No, Spinoza would replace Aristotle's substance, the perishable horse, with either his one great infinite 'being,' or with patterns of 'simple bodies.' The ordinary people are not familiar with either one of these objects. Even if they did exist, which is not conceded, why would they be more deserving of the name substance than the perishable horse or act of courage? Spinoza claims that his indivisible bodies are characterized by 'conatus,' or the striving for power; that they are existing by virtue of their own power and assertion of power.[41] Yet it is not at all clear why eternally existing objects would need to strive for power, or need power in order to exist, in the first place.

Upon what grounds shall we say that eternal matter, or eternal form, are more real, or *as* real, as the enmattered forms of the perishable objects? Aristotle thinks eternal matter and eternal forms are *inferior* in being to generated natural objects. Upon what grounds does Spinoza reverse this? Not upon candid grounds, surely. First, Aristotle argues that nature intends generation. Spinoza argues that there is only one 'substance'; that it neither comes into being nor passes away, nor can it produce other substances. Which means that the horse, in its mortality, form, limits, needs, saddle, and name must fail to be included in the reality of the new definition of substance, except as 'inadequate' or 'confused' ideas of it. 'Adequate' knowledge of the horse does away with its particular perishable nature. It dissolves the horse into the abstraction of 'extension,' as if extension is eternal, or part of the eternal, or an 'attribute,' or a 'mode' of Spinoza's 'substance.'

In *Ethics* Part I, Propositions 7–8, Spinoza urges that his substance is 'necessary,' that it exists by 'necessity.'[42] 'Necessary' is the term which interests us now. Plato and Aristotle do not speak of 'necessary' when they observe a unity that has being in nature. Indeed, all the natures of which we have experience do not *necessarily* exist: for they *can* not-exist, either not come into being (like an insight that is never brought into a student's mind by a teacher), or lose being. Spinoza's substance has necessary being: what is the cause of this necessity?

The cause of this necessity is a figment of Spinoza's imagination, that is, of the power of his inventive intellect, as the arguments in the *Emendation* demonstrate. Spinoza chooses an inventive role for the intellect, over and against 'passive' roles for the intellect (i.e., being presented with real objects to think about, which it has not produced). There is no 'necessity' that is inherent in Spinoza's choice of 'conception' as opposed to 'perception.' Spinoza's characterization of his argument as 'necessary' reveals the blistering quality of his willfulness: it does not however reveal the adequacy or justification of that willfulness in terms of truth.

PLATO'S CRITIQUE OF THE INDIVISIBLE BEING THESES

It is indicated above that Spinoza's argument closely follows the contours of the argument set forth by Parmenides. Plato provides a critique of Parmenides's argument, which can be briefly stated. This critique will then be applied to Spinoza's thesis. Plato initially approaches Parmenides's argument by reformulating it. It is important to pay attention to this reformulation, and to probe its validity.

Parmenides advanced the claim that only being exists; but he could not manage to avoid adding a variety of other natures to his 'being,' in order to give his argument a chance to succeed. Thus Parmenides's being is also said to be 'perfect.' It is said to be an indivisible unity. This being is said to possess a circular shape. It is also eternal, finite, and it is without genesis. "It never was, since it is now, all together, one."[43]

Parmenides made the argument that 'being' is the only thing that exists. Plato reinterprets this argument to indicate that 'the all is one.' In the *Parmenides*, Socrates remarks to Parmenides that according to his argument, all of the things that ordinary people refer to as diverse and distinct are regarded by Parmenides as just one thing.

> I see Parmenides, said Socrates, that Zeno's intention is to associate himself with you by means of his Treatise, no less intimately than by his personal attachment. In a way, his book states the same position as your own; only by varying the form he tries to delude us into thinking that his thesis is a different one. You assert, in your poem, that the All is one, and for this you advance admirable proofs. Zeno, for his part, asserts that it is not a plurality, and he too has many weighty proofs to bring forward. You assert unity, he asserts no plurality . . . This is why your exposition and his seem to be rather over the heads of outsiders like ourselves.[44]

Plato wishes to make it clear that not even Parmenides's account of being qualifies as a single thing. All of the other forms we have mentioned are said to be contained in Parmenides's 'being.' Yet it is impossible, in Plato's view, for these other natures to be *derived* from Parmenides's 'being.' Each nature, in Plato's theory of causation, can have only one cause: the absolute nature of that same form. To wit: any instance of the pattern of unity in nature must be caused by the absolute form of unity, and by no other cause. Every pattern of a horse that exists in nature can only be caused by one thing, the absolute nature of a horse. Thus unity, in Plato's argument, cannot be derived from 'being.' Nor can 'being' be derived from unity. These are two distinct natures, the *causes* of which are distinct.[45]

Plato decides to focus on unity as the basis of Parmenides's argument, because Parmenides makes 'all things' dissolve into 'one.' Unity, if we contemplate it under the Socratic theory of causation by itself, can be nothing other than unity. It must be strictly a unity. It *cannot* have parts, it *cannot* be derived from any other nature.[46] Thus it cannot in itself *possess* the part of being.[47] In order for unity to posses the part of being, Parmenides's argument must be undone: for in accordance with Plato's theory of causation, unity and being are distinct, coequal natures.

This means that any unity that has the part of being must be a whole of parts. A whole which has any parts must of course have *all* the parts. It must come into being, and pass away; it must exist in time; it must have the extremities of a container of parts; it must exist in a place; it must be capable of motion (at the very least of alteration and coming into being or passing away); and many objects will have the power of locomotion as well. Furthermore the extremities will be perceptible; and therefore they can be named and talked about; and this means that the perishable bodies that the ordinary people name and talk about are indeed the real bodies, the real beings. All of these arguments, if they can be deduced from Spinoza's claims, would bring *Ethics* I to ruin.

Spinoza, as we have seen above, located his one true idea by power of his own inventive intellect. Spinoza disdains the dependence of the mind upon fact and evidence external to its own *will.* There is something infantile in this attitude, despite the fact that it has been demonstrated to exist in minds of great power. Spinoza's one true idea is that the real substance must be inseparable from the reality of being. It must be incapable of not-existing. This unity is what Spinoza calls 'God' or 'Nature' or 'Substance.' Yet Spinoza also cannot do without many other names that are required for him in order to advance his argument and make it at least appear plausible.

Spinoza's 'being' has need of 'attributes.' In non-invented speech, that is, in the ordinary use of names that Plato takes as the basis for his philosophic investigation into forms, attributes suggests qualities, predicates, *parts*. Not only does Spinoza's 'being' have attributes, we are informed; but there are indeed 'infinite' attributes. For some reason, it is said to be possible for human beings to know only two of these attributes: thought and extension. Spinoza, we can see, is beginning to pile up names on his side of the ledger. Extension, an attribute, is also said to be characterized by motion or rest.[48] Thus body has been added into the equation; and the predicates of body. Yet we are instructed that these in no way impede the indivisible unity of Spinoza's 'being.' 'Attribute,' Nadler tells us, for Spinoza, is just another word for substance.[49]

Being. Unity. Infinity Attributes. Modes. Extension. Thought. Motion. Rest. Indivisibility. This is Spinoza's version of unity. Parmenides's version includes being, indivisibility, rest, eternity, circularity, perfection, completeness, sameness with itself. There are more than a half dozen natures in each case. Plato's argument in the *Parmenides* is that the proper construction of Parmenides's argument is 'the all is one.' This is also the proper construction of Spinoza's thesis. Unity therefore is the nature that Plato focuses on. In Plato's theory of causation, there is only *one* cause for unity: the absolute nature of unity. The nature of unity is to be one, not many. Unity itself, accordingly, cannot be a whole of parts, since parts are many. Spinoza agrees. I mean to say that Spinoza, adhering to the definition of his substance as indivisible, feels compelled to deny the reality of 'whole' and 'parts' in nature. In order to be consistent in his theory, this he must in fact do.

Perfect the one cannot be, for the nature of unity is not the nature of perfection. This will be proved, but not yet. Attributes are not something that unity, strictly speaking, can have. For attributes are not merely plural (more than one) in Spinoza, but they too are said to be infinitely many. To be a unity is not to be an attribute, for an attribute depends on something, that is, on another nature, which indicates *more* than one, or *many*. Strictly speaking, what is a true unity cannot *be* many. Which is to say that unity itself cannot be many. Spinoza's claim of indivisibility enforces this logic for his 'nature.' What is unity can only be caused by the nature of unity. 'Modes' are not part of the nature of unity. For modes indicate diverse ways of existing. Nor can thought be a part of the nature of unity, since unity does not indicate thought.

Nor can unity have any extension, if unity is all that there is. For extension has length, breadth and depth. These are clearly parts. We have said that what is purely one cannot have parts upon pain of being many. What is a true unity cannot be in motion. Motion must be one of three kinds: locomotion, alteration, or coming into being. Spinoza's unity cannot come into being, as he admits. Nor can it produce other substances, as he admits. Nor, however, can Spinoza's unity possess 'affections,' which indicate alteration. For to alter is to cease a certain mode of existence and to become a different mode of existence. None of these things—'affection,' 'mode,' 'motion' or alteration, can be of the nature of unity. Nor can this one of Spinoza's locomote. For to locomote, one must be in a place. Spinoza does not talk about place that I know of in regard to his one; but in any event what is strictly a unity cannot have place. Place is a separate nature. Nor can the one be at rest. For to be at rest still requires being in some location.

Nor can this unity be always, eternally. For eternity indicates duration, existence in time. Time is not of the nature of unity. What is unity has no diversity whatsoever. Time is other than unity. This means also that Spinoza's unity cannot be the same as itself. For the nature of sameness is not the nature of unity. Nor can Spinoza's unity be different from itself. Sameness and difference indicate aspects of reality that are unlike unity. Unity cannot be perfect or complete either. For perfection and completeness are distinct natures from unity. They do not indicate the same nature as unity. Which brings us at last to the nature of being. Spinoza's one cannot exist. It cannot have the part of being except if it has the nature of a *whole*, and the nature of *parts*. Unity, by itself cannot exist. Spinoza's unity, as he defines it, does not exist. It cannot exist. If it does not exist, it cannot be thought of or perceived. If it cannot be thought of or perceived, then it cannot be properly named or spoken of.

Spinoza's nature, therefore, is a unity bereft of name, existence, and reality. It can neither be coherently referred to nor thought of nor even conceived of, much less desired or loved. It cannot be 'necessary,' or 'determined.' It cannot exist, unless it is conceded that it is divisible into whole and parts. Unity and being must be its first parts. "Whole' and 'part' are proved to exist as *natures*, not as fictions of reason ('ens rationis'). What is a whole must have extremities, that is, limits. A whole, insofar as it is a container of parts, must have extremities. It must have the parts of time. It must have the parts of place. It must have the part of shape. It must have beginning and end. It must accordingly come into being and pass away. This object has parts in it; but this object also must be in a place. Spinoza's definition of 'substance,' therefore, is false. For what is a real object must exist in time, in place, and accordingly, possess parts of not-being. For to exist now is the same as not-to-exist any longer in a prior moment, and not yet in a future one.

Let us return to Spinoza's 'true idea,' the thing whose essence is existence. The being whose essence is being. The being that is. This is the 'true idea' from which Spinoza claims to be able to 'deduce,' mathematics style, every other truth, as if irresistibly. Yet we behold this being, and try though we might, we cannot say that it is something that truly exists. If it truly exists, it must be some one being. If it is not one being, it is not-one being, or no being, or nothing. Except unity exist, and unity cannot be derived from 'being' by any 'logic' or witchcraft—the name 'being' indicates at best confusion.

If Spinoza concedes that unity must be part of this being, as he does: then we merely need to correct Spinoza. There is no possible origin for 'unity' except the nature of unity itself. Thus, the one 'true idea' indicates either a unity that has the part of being, or nothing at all. If the former, then the natures of 'whole' and 'part' are proved. Spinoza denies the reality of whole and part in his *Short Treatise*. He calls them fictions of the mind. If we wrest from Spinoza the concession that the only true object is a whole of parts, then we have already achieved all the evidence we need to prove that this object comes into being and passes away. Spinoza's version of 'substance' is disproved. Aristotle's theory of substance is vindicated.

Spinoza's two named attributes of the great unity, extension and thought, are also talked about in a most unsatisfactory manner by scholars. This concoction of Descartes', that body is reducible to mere 'extension,' is flimsy to say the least. That extension is indivisible, Spinoza's contention, just makes things worse. 'Extension' is another name for dimensions, for shape. Dimensions and shape are *extremities* of an object. The extremities of an object cannot exist absent the *guts* of an object: its unity, its being. Every real body (not a fabrication of the mind) is a whole of parts. Its original parts are: unity and being. Dimensions and shape are parts of a unity that has the part of being, that exists. A whole of parts, qua whole, does not merely have one extremity: dimensions. Or shape. It also has extremities in time: a beginning, a middle, and an end—or else it is no whole! If a beginning, if an end, then not eternal, not indivisible. Thus not an attribute of anything eternal, such as Spinoza's 'substance,' 'God,' or 'nature.'

As for thought, to say that it is coequal with body is impossible. Thought understands the limits of body, that is, can think of when body was not, of when it will not be. Thought can think of the potential of body, real body, to actualize its form. It is no wonder that Spinoza's bodies can do no more than uselessly preserve themselves, as if it could be the purpose of a body to be what it can never be: eternal.

Substance, attribute (extension and thought), mode, or affections: If the theory of substance fails, as it must, then so too do the theories of attributes and 'modes.' Only when it is rediscovered that the 'modes' are the true substances, that the 'inadequate' ideas are the only verifiable evidence of truth, will political philosophy become capable of swinging *this* pendulum of so-called enlightenment back to the other side.

So we approach Spinoza's definition of the self-sufficient thing, as that being which cannot be separated from existence, skeptically. Being that must exist is Spinoza's definition of this first of all true ideas. Spinoza alleges that it must be 'simple,' that is, indivisible. However, this remains to be proved. We have noted at the outset that the attempt to define God is not the project Spinoza has initiated with us. He is 'searching' for a true idea, which he says is both a creation of his will and self-evident.

It makes a difference whether this idea is called 'God' or 'Nature.' Where nature is the subject, human beings may investigate. Where God is the subject, human beings may not. Certainly, human beings cannot investigate the divinity first in the order of time. For the divinity does not come first in the order of human knowledge. The divinity cannot be discovered through sense perception, the original inlet of all our ideas. Yet Spinoza, in his attempt to create a definition of the truest idea, equally proposes to deprive us of the knowledge we come by through the senses. Thus we must know, or relearn, our ABCs, rudimentary facts, before we can pretend to be in a position to think about divinity. And thus we come back to Spinoza's attempt to establish the true idea, from which all other 'true ideas' must be, as it were, deduced.

The argument offered here, employing Plato's critique of Parmenides as a way to approach Spinoza's theory of 'substance,' is complicated and involved. No foolish risks are run in restating it, perhaps from a slightly different angle. An argument as powerful and complicated as that of Spinoza demands no less from us. The point is not to seem to disprove his theory, but to come to the bottom of it: to plumb the depths of its truth or lack thereof. Hence to proceed for a second time.

'Being exists' is a puzzling phrase. Yet Spinoza is not the first one to have formulated it in the history of western philosophy. That dubious honor belongs to Parmenides. It is held by Parmenides, and Spinoza infers it from his major premise, that any true being must be eternal being; and that eternal being cannot be divided. Thus for Spinoza, as for Parmenides, there is but one substance; and Spinoza has put a fine point on it: that it is not possible for this one substance to produce any other substances. "One substance cannot be produced by another substance."[50] Anything that exists therefore exists as an attribute or 'mode' of the one substance, that is, exists 'in' it; and since the one substance is indivisible and eternal, so too must be the modes. Both Parmenides and Spinoza leave us in the dark as to how we are to arrive at the existence of the perishable objects; but both certainly do intend to do so.

At least, this being cannot exist, as Plato surmised, except it indicate some unity. Parmenides does not introduce his 'unity' of the eternal being immediately. Spinoza moves a bit quicker to do so. Spinoza introduces this 'unity' in *Ethics* I. Spinoza trots out the nature of unity for his 'being,' in a 'definition.' Spinoza argues that his being is 'indivisible.' This is the same language that Parmenides employs. Both Parmenides and Spinoza want to claim that 'being' is the only true substance; but neither can do without a host of other natures in order to make their argument appear plausible. The relationship between 'one' and 'many' is one of the most famous parts of Plato's analysis of Parmenides's argument.

Socrates, in discussion with Parmenides and his favorite, Zeno, comments upon the prospect that 'one' and 'many' can 'mix.' On a certain level, of course, it is not hard to prove that the same object can be both one and many. That is, it is not hard to prove that the same object is both one and many, if the object in question is a whole of parts. As one man among seven, Socrates can be said to be 'one.' As a body with several sides to him, Socrates can also be denominated 'many.' Does this mean that what is 'one' can be 'many'? Socrates does not think so.[51] Socrates, taken as one man among seven, is one, and not in that respect many. Socrates, taken as a body with several sides, is in that respect 'many.' What can truly never be proved, Socrates insists, is that one *itself* can be many. As natures, one and many are opposites. They can never mix. One object can never be *both* one and many, in the same respect, at the same time, and with the other necessary qualifications.

The unity of being advanced by both Parmenides and Spinoza, however, claims to be indivisible. This means that the opposites, the unlike natures, are somehow going to have to become 'one.' This is what Plato's Socrates does not think can be accomplished. Towards this end, Socrates's theory of causation is introduced into the argument. What is unity, as a distinct nature, can have only one cause: the absolute nature of unity. This is the same for all of the different forms: for being, for circularity, for perfection. The cause of the form of being is nothing except the absolute nature of being. 'Unity' and 'being' have different causes, due to the fact that they are different natures. They can never be one in an *indivisible* way. Being can belong to unity only as part of a composite whole.

Spinoza argues that his 'being' has 'infinite' attributes, but that only two can be known to man: extension and 'thought.' It has been proved that 'being' by itself has no attributes—since it cannot exist of itself. Only unity that has the part of being can exist. This object has the attribute of being a whole. It has the attributes of parts. Spinoza has said that 'whole' and 'parts' are mere 'ens rationis,' creations of the mind. It has been proved that it is Spinoza's unity that is a mere figment of the imagination. Any real object in nature is a whole of parts.

The parts in any true object, Plato has proved, are infinite.[52] By infinite, one wishes to indicate that there is no limit, no limit that the mind can reach, and no limit inaccessible to the mind, of these parts. Yet the *whole*, while it serves as a container for infinite parts, also exists as that which *is* contained. If it is a whole, it must have, not a location in 'space,' but a *place*. In order to exist in a place, in order to be a whole at all, it must also have the other necessary extremities. It must have length, breadth, depth—that is, dimensions or 'extension.' These divisions however are not 'attributes' of any *eternal* thing. They belong to a *perishable* whole: for among the extremities of a whole are beginning, middle, and end; to which correspond time past, time present, and time future. The mind proves the existence of all these things. It does not create them. No, it comprehends them, identifies them, understands them. So there are indeed infinite parts in any object in nature, but none of these attributes depend upon the eternal as their cause, insofar as they are bodies or attributes of body.

If this object exists in a place, if its dimensions and extremities exist in a place, then it is going to be perceptible. It if it perceptible, it is possible for us to name the object, or kind of object also; to refer to it, to indicate it, to communicate about it. Thus it can be spoken of and known about. One cannot say this of Spinoza's idea of extension, for it has been proved that this can neither exist, nor be perceived, nor really be referred to or spoken about.

ON THE GEOMETRIC NATURE OF SPINOZA'S ARGUMENT

Spinoza famously modeled his *Ethics* on geometric forms: there are definitions and axioms, stipulated as given; then there are propositions, and scholiums interpreting the propositions for the reader. Commentators are frequently fascinated by a supposed link between Spinoza, Descartes, and even Hobbes with mathematical logic. The great allure and power of the mathematical logic is that, supposedly, it can't be argued with. Who will contest that two plus two equals to four? The appearance therefore of the geometric method in Spinoza's *Ethics* casts a daunting shadow. It intimidates. The definitions are not made open for inspection or proof. Yet appearance is not necessarily substance. Metaphysics is not geometry. Metaphysics certainly must explain the cause of geometry; and it is on these grounds where Spinoza's employment of the geometric form of argument can be contested.

The authority, or the appearance of authority, that the mathematical sciences can lend to philosophical reasoning must be traced back to its foundations. For what is equally notorious about Descartes, Hobbes, and Spinoza is that all three deny the evidence of sense perception. Is it really possible to have mathematics, authoritatively binding mathematical logic, when the evidence of sense perception is indicted a priori? Stuart Hampshire notes that this is what the Enlightenment rationalists base their sciences in, pure or mathematical reasoning.[53] Why? Because nobody can dispute mathematics (except Descartes' God). That one plus one equals two; that the three angles of a triangle must add up to 180 degrees; these truths cannot be challenged. Hence Hobbes's reference to geometry for his science of definition. Hence the very design of Spinoza's *Ethics,* presented deductively in the 'geometric' manner.[54] Yet is it really plausible to argue that Hobbes and Spinoza even have access to geometric method, based upon their metaphysics? Both Hobbes and Spinoza sweep the evidence of perception out of the domain of valid knowledge. In Plato's view, there is no conceivable path to geometry without the knowledge vouchsafed to us by perception.

Richard Kennington observes that there is no way to reach the 'modes,' the perishable objects, by deduction from Spinoza's eternal substance. Kennington opines that in *Ethics* II, Spinoza begins to indicate that the perishable objects must be known *empirically*—which is to say that the deduction stops short at the perishables or 'finite modes.'[55] This deals a blow to Spinoza's claim to be relying on Euclidean geometric models. Spinoza insists that the 'finite modes' or perishable objects cannot be known 'adequately' *as* perishable objects. The finite modes, in order to be known adequately or truly, must be approached from the vantage point of eternal objects. As indicated above, Spinoza offers two alternative avenues to the thinker to reconceive of perishable bodies in an eternal form: that of the great being of *Ethics* I, or that of the 'simple bodies' of *Ethics* II. In order to know a perishable object adequately, in other words, the object must first of all be stripped of its perishable qualities, and converted into a function of eternal being. In my view Spinoza makes available to the reader both the Parmenidean unity of being, and the Epicurean atoms of indivisible being. The point is that in both cases, the dependence of geometry upon sense perception is blocked, upset, preempted. Even more to the point: conceptually, the indivisible natures of either the great being of *Ethics* I or of the simple bodies of *Ethics* II seem to subvert the reality of the extension which geometry is dependent on to measure and abstract from.

According to Plato, the very origin of number is in knowledge obtained by sense perception alone. In the *Republic*, Plato argues that the origin of number is in the perception of *unlike objects*.[56] Given that Spinoza subsumes everything into his 'indivisible' unity, how could Spinoza's philosophy even underwrite the genesis of a competent arithmetic? If there is no 'other' in nature, then there is no duality. The mind would not be able to discern one thing from another. There would be no number.

Thus one cannot have two units to add together in Spinoza. Mathematics loses its foundation in reality, and becomes a mere arbitrary construct of the philosopher's mind.

The geometric format of Spinoza's *Ethics* has long been recognized as a daunting edifice. Mathematics, it is held, is a language that allows for no dispute. Common proofs are enforced, and differences of opinion are thought not to apply. It will not do, however, to associate Spinoza with Euclid. Euclid did not claim that his definitions represent the most real things in the universe. A triangle is not a circle. A square is not a triangle. Spinoza, through the original unity that his argument attempts to enforce, logically vacates the distinctions that the simplest mathematics must make.

If Spinoza's unity was strictly true, it would indeed mean that there is no number in reality, since all things are one thing. Spinoza's first definition in the *Ethics*, his definition of substance, would make it impossible for us to separate existence from what is real. It would require that what is once real is always real, and always the same thing. It requires us to believe that one thing is the only real thing. Number can emerge, in Spinoza's system, as 'ens rationis,' or things of the mind's invention; but the reality of number as *natures* would be subverted. It can equally be proved that extension is made impossible by Spinoza's definitions of a substance.

The only number of which Spinoza is capable of offering proof would be that very same authority of 'conception' or 'invention' which he adumbrates in his *Emendation of the Intellect.* Number, for Spinoza, can be said to be a mere contrivance of reason, a mere supposition or theory of the philosopher. Yet in that case, the indisputable nature of mathematical reasoning would be *vacated. Every* construction of a philosophers' mind is disputable.

Neither, however, do Hobbes and Spinoza truly have access to geometry, their preferred form of mathematical reasoning. Hobbes and Spinoza, following Descartes, affect a belief in something that they call 'extension.' By 'extension' they wish to indicate the dimensions possessed by an object. By dimensions they wish to indicate length, breadth, and depth, that is, the extremities of an object. Yet the theory of body relied upon by both Hobbes and Spinoza is Eleatic: in the case of Hobbes, and in the case of Spinoza, 'matter' is eternal. This claim can be seen to be dependent on the claim that 'body' has no parts.

Spinoza and Hobbes both claim that 'part' is a mere contrivance of the mind. There are no parts in body itself, according to these two philosophers. The mind can simply imagine divisions in the objects it surveys, according to these two sages; but this is a far cry from our being able to say that the objects possess, within themselves, fundamentally different natures which are parts in *reality.* It is the Eleatic logic which maintains that 'being' is indivisible, whether in Parmenides's great unity, or in the atomists' infinite small unities. In both cases, the existence of real and truly distinct natures in the bodies is *denied*. Yet it has already been proved that this argument is untenable. There can be no unity that exists except it have the distinct part of being; and there cannot exist any being which does not belong to some unity. Extension, or dimensions, are parts that cannot even be *reached* until the coequal natures of being and unity are conceded. Yet since Spinoza and Hobbes deny the coequal natures of unity and being; because these two thinkers insist on the indivisibility of real body, their theories have no legitimate way to arrive at the investigation of *any* extremities, or extension.

According to Spinoza and Hobbes both, 'parts' is a mere fiction of reason. The body that has no parts can have no length, no breadth, no depth; no shape, no place, no parts within. It cannot be a container of parts; it cannot have parts of time. The 'extension' or dimensions in this physics is a construct of the philosophers' imaginations, it may truly be said. As a construct such a geometry can make no binding claims about the truth of any external objects.

Hampshire makes the case that Spinoza regarded mathematics as the 'natural light' of reason. Hampshire lashes out at the 'figurative' and 'metaphorical' language of ordinary speech. In this Hampshire follows Spinoza. Hampshire however claims of Euclidean geometry that it does not depend on perceived images; that the mathematician's definitions are sui generis. This is false. Geometry came into existence as the measurement of shapes, of extremities. That a circle is 360 degrees is not comprehensible absent the image of a circle. The dimensions of a triangle are not coherent absent the prior image of a circle and a semi circle; and the concept or definition of a right angle is inconceivable without all these premises. All of which depend on perception, though they abstract from it.

Hampshire is quite in line with the so-called 'reign' of mathematical thought, which he says Descartes and Spinoza 'generalize' to other subject matters. This claim is foundationless. Nothing of geometry is employed in the analysis of nature in Spinoza. From geometry Spinoza claims to borrow the a priori definitions. Yet it is false, as indicated above, to claim that geometric definitions are truly a priori. Geometry, as Plato argues, originates in abstraction from images.[57] Not by the elimination of images, but by the abstraction from their dimensions. This is why Plato regards geometry as an inadequate definition of objects. It cannot define what an object *is*, its cause. The object's cause depends on its form, that part of ordinary speech which Spinoza seeks to banish from his definitions, but which for Plato is the original evidence as to the object's nature.

THE PROOF OF IMMATERIAL REALITIES (FORMS, UNIVERSALS)

The reality of immaterial things in nature is something that has been taken up in this study in the chapters on Thomas Hobbes. It seems that one cannot prove too many times though that immaterial things exist. In point of fact, it is only the unintelligible nature to which Spinoza seeks to ascribe all true being ('natura naturans') which can even lead to this confusion in the first place. For there is no body in nature which can be characterized simply as 'being.' 'Being' is not a body. 'Being,' if it is not abstracted from some prior object, cannot be so much as coherently spoken about. Which is to say that 'being' by itself, contemplated neither as part nor as part of a part, cannot *have* being.

Indivisible being is not an object: it is not nature as a whole, nor is it any object in nature. The human mind cannot know being in any undivided object: the smallest, or the biggest object in which being can exist truly, is a composite object, and the thing that must have being is some unity.[58] To have established this is to have established the only premise that we need to prove, in order to be able to demonstrate the reality of immaterial natures. For once the fact is established that any body in nature whatsoever is a composite whole, then the real nature of parts is established.

To remain focused on the metaphysics of an object (for this is what all attempts to define what an object can be are: metaphysics, that is, sciences of being), it is now proved that irreducibly, a body indicates coequal natures of being and unity. This gives us a whole of parts. This proves the reality of the nature of a whole; and this proves the reality of the nature of a part. We behold this whole of parts then; a unity that has the part of being. May we not now behold the part of being, separately before the mind as it were? Is it not true that the part itself, in order to exist, must have its *own* unity? This is to say: that while the whole is some unity that has the part of being; is not the part in the whole, likewise characterized by *its own* unity? If it were not *some* unity, the part, then it would not be some *one* part. It would not be any part. Therefore it must be one part.[59]

The whole has the nature of unity in it; the part of the whole has the nature of unity in it. Unity in the whole is distinct from the unity that is in the part, *referentially* speaking; yet unity is one nature in both whole and part. Whole and part each partake of the nature of unity, as unity must always indicate the same nature, no matter how many objects have it. Plato calls these natures 'patterns' or 'ideas,' which exist in the innumerable perishables over and over again. Yet each pattern, each nature, each idea is the same as itself, and different from every other nature, always. The natures are thus immaterial, even as they are mixed into matter, that is, forming it. As formed, nature is intelligent. Ordinary speech, which Spinoza and Hobbes attack with all argumentative and rhetorical guns blazing, is also intelligent: for the dreaded universals in ordinary speech just are these patterns. The reality of the immaterial natures is thus proved.

SPINOZA'S 'INDIVIDUALS'

In the very first definition of the second section of the *Ethics*, Spinoza defines 'body.' Body is here firmly anchored in the great indivisible 'being' of *Ethics* I. One of the attributes of Substance adumbrated in *Ethics* I is that it has 'extension.' Extension, as discussed above, concerns the dimensions and shape of an object. It has been established that Spinoza cannot properly reach the attribute of extension based on his definition of substance in *Ethics* I. Nevertheless, *Ethics* II forges ahead towards the perishables, which it does not yet truly indicate.

Richard Kennington, as noted above, has observed that Spinoza seems to be shifting towards an empirical approach in *Ethics* II. The information about these bodies or 'individuals' does not appear to be deducible from *Ethics* I. Yet the bodies on display in *Ethics* II are certainly not the bodies with which non-philosophers are acquainted. The bodies featured in *Ethics* II, and this is as close to an enumeration of the perishables that Spinoza achieves, exist in configurations of mechanism.[60] This is to say that the bodies of *Ethics* II are still indivisible bodies and that they constitute fictitious parts in fictitious wholes. The 'individuals' represented in *Ethics* II are formed by bodies which are separate from one another, but which are related to one another in certain patterns of motion and rest.

Even though the bodies of *Ethics* II do not resemble the bodies with which ordinary opinion is familiar, there can be no doubt about the fact that Spinoza intends this definition of body to apply to the philosophic account of the bodies as ordinarily experienced. The title of *Ethics* II is "On the Nature and Origin of the Mind." Spinoza proceeds to unfold an argument that is almost identical with that of Epicurus: perception, or the contents of an individual mind, will be reduced in this model to the percipient's bodily experience of collision with certain other bodies that are not perceptible.[61] The body is proposed as the sole theater of perceptual experience; which indicates that perception provides information essentially about ourselves, not about the external objects.

Spinoza's account of body in *Ethics* II is fundamental for establishing his basic portrait of the human personality. The human being must be separated from the information of perception, in such a way that the individual can retain authority as the 'inventor' of what shall be construed as fact. The human mind is to be defined in such a way that there are no possible implications of perception which escape a narrowly individualistic signification:

nobody else perceives what some one human individual perceives in this model; and accordingly, there is to be no common moral judgment as to the rightness or wrongness of an individual's response to any facts. Spinoza's definition of body in *Ethics* II abolishes the category of common facts.

The relationship between bodies that are experienced and the human mind is cast by Spinoza into a deep, paralyzing freeze. Whatever an external body may enact, that external body is never to be conceded as the cause of its own action. That external body must be caused to act by some other external body; and that external body in its turn must be caused to act by still another external body, on into an infinite system of mechanistic causation. For the mind of the percipient, a new liberty is obtained. Spinoza denies that the human mind can be the direct cause of the actions of its own body. Free will is a delusion, in Spinoza's view; and it is erased in proportion as one understands the power of invisible causes, that determine the human body itself to behave in various ways.

Lee Rice talks about the 'individual' in Spinoza. In *Ethics* II, Rice observes, Spinoza undertakes to talk about the nature of bodies so that he can better talk about the nature of the human mind. 'Simple bodies' is a phrase Spinoza uses. While Spinoza does refer to a lowest or irreducible level of simple bodies, Rice continues, Spinoza never really attempts to define these. Some scholars have claimed that this is atomism.[62]

'Simple bodies,' if they are atoms, confront us with the exact same set of problems that Spinoza's original great 'being' does. The atoms are indivisible. In the theories of Democritus, Epicurus and Lucretius, the atoms are indestructible bodies which have no genesis. They simply always were and always will be. The atomists find themselves in the same tangle of difficulty that Spinoza does with his one 'being': which is to say that they find that they need to use many more names than just one to characterize the atom.

The atom is indivisible, like Spinoza's unity. It is also being, as Spinoza's one is. The atoms have shapes (extremities, or if you like, 'extension'). They have size and weight. The arguments that we applied to Spinoza's being fully apply in the case of the atom. Neither 'unity' nor 'being' can exist of itself. The irreducibly smallest body must be a composite: a whole of parts, a unity that has the *part* of being. It is established, however, that Spinoza denies the reality of 'whole' and 'part'; that for Spinoza these can only be fictions created by the mind. As proved above, 'whole' and 'part' are *real* natures. They exist whether Spinoza's inventive intellect wishes to concede this or not. Truth is stronger than will.

Spinoza, following Descartes and Hobbes however, formally forswears atomism, as Rice points out. Parkinson too tells us that these simple bodies of Spinoza's are 'atom-like,' but that we are not entitled to 'press the point.' Parkinson proceeds to characterize Spinoza's simple bodies as 'indivisible,' which is the classical postulate of atomist theory. Moreover, the grounds upon which Spinoza forswears atomism do not have to do with the true postulates of the theory. Just as Descartes and Hobbes can be seen to adhere to atomist platitudes, so too can it be seen in Spinoza.

The atoms are indivisible. Spinoza insists, following Hobbes, that all bodies are formally divisible. This is to say, if we define divisibility as a mere imaginary operation performed by the mind upon its own ideas, then we can have a divisibility that does not truly upset the actual indivisibility of the 'simple bodies.' Divisibility is a feature of the imagination for Spinoza. Hence Spinoza's claim that the bodies are divisible does not indicate that these bodies cannot be atoms.

Two things seem clear about Spinoza's theory of the simple bodies. Matter, in Spinoza, is not destructible. If the simple bodies are the irreducible units of matter, then neither are they destructible. In *Letter 4 to Henry Oldenburg*, Spinoza indicates that the matter out of which human beings are formed always existed. Nor can said matter be destroyed. "As to your objection to my first proposition, I beg you, my friend, to consider that men are not created, but only begotten, and that their bodies always existed, but in a different form. However, the conclusion is this, as I am quite willing to admit, that if one part of matter were to be annihilated, the whole of extension would also vanish at the same time" (767).[63]

Spinoza thus follows Hobbes. Hobbes denies that it is proper to use the language of coming into being and passing out of being. There is one indestructible matter, Hobbes argues, and it is organized at one moment in the form of a horse, in another moment in the form of a tree or a man. This is in keeping with the original atomistic postulates. Parmenides's argument that true being can never cease to be animates this physics, and it is in the atomist variant that it does so.

Spinoza tells us that the simple bodies have a certain kind of relationship to one another, and that they do indeed form 'individuals.' An individual, as we have proved above, must be a whole of parts. Yet it is interesting to note that in Spinoza's characterization of the simple bodies gathered into an 'individual,' the only relationship these simple bodies have to one another is 'patterns of motion and rest.' This indicates that even in the case of what Spinoza is anxious to call an 'individual,' the simple bodies retain their separateness. A more atomistic formulation could not be forthcoming from any theorist.

> When a number of bodies of the same or different magnitude form close contact with one another through the pressure of other bodies upon them, or if they are moving at the same or different rates of speed so as to preserve an unvarying relation of movement among themselves, these bodies are said to be united with one another and all together to form one body or individual thing, which is distinguished from other things through this union of bodies.[64]

A body, as simple as you like, must still be a composite. What is truly a composite must be a whole of parts. To finish with our first point though: simple bodies divisible in the imagination of the mind can still be atoms. The mind can simply keep dividing simple bodies into halves. This, however, is not what is truly meant by the divisible. What is truly divisible and thus a true composite has *unlike* parts and natures within it. Especially it has extremities: coming into being and passing away; and Spinoza's simple bodies, caused as they must be by the one great unity of being, are not eligible to pass away or come to be really.

Extension after all is merely an attribute of 'substance' as Spinoza defines it. Secondly, in *Ethics* II, Spinoza defines simple bodies as dependent for their movement or rest on other bodies, in an infinite mechanistic system. "A body in motion or rest must have been determined to motion or rest by another body; which likewise has been determined to motion or rest by another body, and that body by another, and so ad infinitum."[65] Motion is limited to locomotion, and even in this particular the simple bodies cannot move themselves. This is also characteristic of the atoms. Atoms cannot undergo alteration; they cannot come into being or cease to be and they cannot be the cause of their own movement.

Simple bodies are not parts of a unity. When they form what Spinoza calls an individual, they are not contained in any whole of parts. They do not come into being with the birth of an individual, nor do they cease to exist with the passing away of said individual. 'Conatus,' which Spinoza must ascribe to these simple bodies, is in this case a logical absurdity. For that which cannot be destroyed need not attempt to preserve itself. As Hobbes likes to emphasize, 'change' of configuration is the proper way to talk about ceasing to be, since 'matter' is not destructible to any of these early modern thinkers.

Let it be noted too that this is the way Spinoza argues that nature as a whole is one individual. Spinoza offers a vision of the universe on the atomistic model, in addition to the vision of the universe constituted by the one indivisible being of *Ethics* I.

> Now all the bodies in nature can and should be conceived in the same way as we have here conceived of blood; for all bodies are surrounded by others and are reciprocally determined to exist and to act in a fixed and determinate way, the same ratio of motion to rest being preserved in them all taken together, that is, in the universe as a whole. Hence it follows that every body, insofar as it exists as modified in a definite way, must be considered as a part of the whole universe, and as agreeing with the whole and cohering with its other parts.[66]

There are just different layers of simple bodies organized into 'individuals,' each layer 'interacting' with 'individuals' of like size and complexity. Spinoza's simple bodies are atoms, contra Rice. It is true that the atoms do not exist except as fictions, but they exist as fictions for Spinoza in order to resist the truth about what bodies are.

Spinoza's modes are mechanistic, by virtue of which he can insist that our perceptions of perishable bodies be logically reinterpreted 'sub species aeternitatis.' The 'simple bodies' are not the perishables. They do not perish. The simple bodies rather underlay the perishables, in the same way that Spinoza's great 'being' is said to underlay the perishables. The finite modes that Spinoza talks about can only refer to the changing patterns of organization which, it is suggested, cause the perishable bodies. The simple bodies sustain the alleged irrelevance of ordinary perception. It would appear then that even when he is talking about 'finite modes,' Spinoza is not yet talking about the perishables. The perishables themselves do not have direct representation in Spinoza's scheme of knowledge. This is perfectly in line with both Parmenidean and Epicurean doctrine. The point is for the philosopher to retain the power to 'invent' the first facts, as discussed more at length above.

From its beginning atomism regarded perishable objects as less than fully real. This was indeed Democritus's justification for denouncing hot and cold as mere camouflage of atomic movement. It is alleged that perception can't yield truth because the truth (atoms) is not perceptible. Spinoza belongs distinctly to the Epicurean wing of atomism. This becomes unmistakably clear in Spinoza's dispositions towards the non-philosophers' emotions and passions. Spinoza is virulently opposed to the ordinary emotions and their manifestation. His entire theory of knowledge stands or falls with the philosopher's prerogative to deny probative standing to any alleged facts which could authoritatively evoke such passions.

In this version the theory of atoms is not upheld as a strict truth, but as a means for the perishable philosopher to order his own thoughts in such a way that they allow him to live in maximum liberty, what Spinoza again following Hobbes calls 'right of nature.'[67] Thus the entire eternal being hypothesis in Spinoza is an Epicurean mode: it is an instrument for procuring a certain kind of mortal life, as opposed to an orthodoxy which truly regards human life itself as an accident.

'INADEQUATE,' 'CONFUSED,' OR 'MUTILATED' IDEAS

It is appropriate, at this point, to turn to Spinoza's attempt to fashion new categories for thought. In particular, we are interested in the category that Spinoza wishes to superimpose over *ordinary* thought: that kind of thought which points out the existence of horses, deeds of justice, cows, and Theaetetus. The ordinary consciousness, which one supposes Spinoza must have depended upon merely in order to make way to his place of business each day, is dependent on sense perception. Spinoza proposes to label this entire category of thought as 'inadequate.' Inadequate how, exactly? The ordinary ideas are 'inadequate,' Spinoza argues, because they are 'passive.' Unlike the 'one true idea' that Spinoza has invented with his intellect, the ordinary speakers refer to perishable objects, not to some substance the principal attribute of which is existence.

The ordinary people refer to what *Aristotle* regards as substances, as the most real beings. The primary substances in Aristotle's metaphysics are just those horses, cows, acts of courage, and other perishable objects that the non-philosophers are concerned about in their daily language of reference and indication. It has been proved above that Spinoza's allegedly timeless substance is not a substance at all. In fact, it has been proved above that Spinoza's eternal substance cannot so much as possess the part of *being,* unless Spinoza surrenders his definition as to what a substance can be. Yet it is to this timeless and eternal substance that Spinoza wishes to subordinate the entirety of human thought.

This brings one to an absolutely pivotal insight. For Spinoza, following in the tradition of Bacon, Descartes, and Hobbes, has attempted to defame, to discredit, and to intimidate the confidence of ordinary thought. He has gone farther than to classify it as merely 'inadequate.' He furthermore defines it as 'mutilated,' 'confused,' as being *incapable* of truth. Based upon the above proofs, however, it can be seen that the reverse is true. It is Spinoza's language, Hobbes's species of ideas, Descartes' 'clear and distinct ideas,' which are bereft of reality and truth. The ordinary use of names, so far from being the discourse of ignoramuses, proves to be the evidence upon which sound philosophy must *rely*.

There is no call here to romanticize the power or quality of ordinary thought. Neither Plato nor Aristotle ever believed that ordinary thought and speech are capable of ascending the higher ladders of knowledge. In the *Republic*, Plato makes it plain that there is a huge chasm that separates the individual who knows objects merely by perception, as opposed to the individual who knows things truly by their definition.[68] The difference is, that Plato's theory of definition is anchored upon ordinary perception. Spinoza's theory of definition, like that of Hobbes, proposes to relegate the ordinary assignment of name to object as entirely delusive: to sweep it into the rubbish heap, the better to replace it with definitions derived, not from 'passive' insight, nor from analytical probing of passive insights (Plato's mode of definition); but rather to originate definitions wholly out of the mind's inventiveness, quite disposing of the issue of truth and reality.

Spinoza's three kinds of knowledge are indeed anchored in a repudiation of perceptual thought. In *Ethics* I, nay even in the *Emendation of the Intellect*, Spinoza has proclaimed that the 'one true thought' which must guide the mind to other true thoughts, is that of a substance the essence of which is existence. This proposition has been refuted above. It is not a true idea, this one. Yet it can easily be seen that the three forms of knowledge that Spinoza outlines for the reader all take their bearings from that 'one true idea,' which is the invention of Spinoza's 'conception,' and which has been lodged in the philosophic consciousness since Parmenides.

Ethics II indicts ordinary perception wholesale. When Spinoza introduces the definition of body in *Ethics* II, it is conspicuously aloof from the ordinary perception of body. The teaching of *Ethics* II is clear. The ordinary consciousness does not know what the real objects are. Those perceptions which ordinary speakers take to indicate the reality of external objects, the Spinozist declaims as delusions. It is alleged that perception is confined to the bodily theater of the percipient, and the collisions that take place within it, between the unperceived simple bodies which have been discussed above. In order for an ordinary idea to be brought to the light of truth, the perceived perishable must be reconceived, as it were, sub specie aeternitatis. It must be construed as a non-substance, which can exist only insofar as it inheres in Spinoza's eternal substance. Both of the true forms of consciousness that Spinoza talks about—that of the 'adequate ideas,' and that of 'intuition,' are laid upon this foundation.

This study has attempted to indicate the essentially moralistic nature of Spinoza's teaching. Spinoza traces out with unprecedented exactness, in *Ethics* III and IV, how the ordinary perception of objects is linked to customary and communal regulation of the *passions*. Objects perceived in common attract the onus of ordinary praise and blame and its considerable authority. In the Epicurean tradition, which Spinoza's doctrines are well aligned with, this experience of common facts is to be *denied*.[69] If it can be established that the individual experiences none but private facts, then to this degree his disposition towards the facts in question is surrendered to his merely private judgment: to his appetites of pleasure and pain, or Epicurean nirvana. This does indeed appear to be the framework of Spinoza's *Ethics*, after the first section has been dealt with. It is the customary praise and blame which attends to common objects of perception that Spinoza regards as bondage. In other words, that which Aristotle regards as the very original evidence for ethical conduct is dismissed by Spinoza as the principal obstacle to the same.[70]

Once the validity of the eternal substance in *Ethics* I has been properly dealt with, there are only a few terms that are really important for understanding Spinoza. The distinction between inadequate and adequate ideas; the distinction between passive emotions (pain and pleasures) versus active emotion (Desire); these are really the decisive ones. 'Inadequate' ideas concerns the perceptions that people have of perishable objects, and their images. Passive emotions concern the way we are affected by objects external to us. This is the foundation of the unphilosophic mind. Spinoza, it is clear, seeks to obliterate our awareness of the *objects* of 'inadequate' ideas; which are the source of our 'passive' emotions in regard to them. To view a perishable object under the classification of eternity, such as is made available in *Ethics* I, is as if to perform a miracle: the common facts in perceptible objects, the decisive ones for all ethical inquiry, are suspended as if in amber. By subjecting to an eternalizing bath the content of finite perceptual experience, the Spinozist obtains the emotional freedom for conatus, for self-preservation as an ethic of pure power assertion. It is quite a spectacle.

The first thing that needs to be pointed out is that Spinoza's definition of the inadequate and adequate ideas stands in need of *correction.* An idea can only be adequate if it captures its *object.* Those ideas of the common people that Spinoza has described as inadequate, because treating of perceptible, perishable bodies, are accordingly adequate to their objects; while Spinoza's conception of adequate ideas is *inadequate*. Spinoza's concepts of passive and active emotion will also need to be corrected. For our senses, while active, nevertheless must relay information about the external objects to be the conveyance of adequate images and ideas. The feelings that we have must be adjudged based, not on our 'control' of our emotions in the way Spinoza describes, but in terms of the facts that the senses convey to us. Whatever control of our emotions may be properly active and virtuous must be control of the emotions as evoked by the facts that the senses convey, not otherwise.

The perishable object is erased in the transformation of inadequate or confused into adequate ideas. This denial of the common objects enables Spinoza to substitute his 'will,' his 'desire,' for what is commonly perceived. Spinoza's desire is to silence the rest of the world and their inadequate ideas; those inadequate ideas which exist in laws, customary morality, debate; these inadequate ideas, the only ones available to most people, constitute the obstacle to Spinoza's project of self-empowerment. It is quite obvious that Spinoza's desire to 'preserve' himself comes at the expense of denying voice, acknowledgement, or sincere tolerance to the others.

The Spinozist wants an account of facts such that they are subject to his control. When the perishable objects have been rendered unintelligible, mute, and vague beyond the point of recall, this is accomplished. The relationship between objects and names becomes blurred to say the least. The Spinozist, as the Hobbist, has a new vocabulary all prepared. Section III of the *Ethics* dwells on the passions produced by 'inadequate' ideas. This is the point: the common facts predetermine how it is appropriate for us to feel and behave. We are forced by the 'inadequate ideas' to recognize and in part accommodate ourselves to the others. This, it is clear, impedes 'power,' 'conatus.' Steven Smith is incorrect to think that Spinoza is defining imagination merely as lazy thinking, as lazy generalizing from experience.[71] Imagination indicates all ideas for Spinoza insofar as their origin is in the domain of the perishables.

Perception assuredly leads us beyond our bodies, to the forbidden domain of the common objects that the 'vulgar' speak of. Spinoza denounces this domain. He defines it out of existence. There is no doorknob, no blue sky, no can, just the feelings I have from my own body. For this reason we must demote perception to 'inadequate,' 'confused,' or 'mutilated' ideas; because what is perceived is not eternal. To see a cow adequately, we must see it as an instance of eternal 'extension': which means we are to feign that we do not see it, or know it, as it is. In all of these shibboleths, Spinoza wrestles with the common knowledge of fact.

Consider the schoolboyish criticisms that Spinoza employs in his attempt to indict ordinary perception. The role that memory plays in the ordinary assignment of name to object is celebrated by Plato as 'mother of the muses.'[72] To Spinoza, memory is the mother of delusion only. Spinoza observes that the word 'pomum' indicates 'apple.' Yet the sound of the name apple is nothing like the image of the apple. This is 'vague experience' to Spinoza. 'Apple' is one of the dreaded universals for which Spinoza has no use. The sound of the name seems to bear no relationship to the image of the perishable object; hence why should one be accounted to signify the other? Spinoza wants to snip the cord binding name to universal, fusing name to enmattered *pattern*. The real object of the attack is Plato, for this is what Plato's ideas are.[73]

No doubt the apples are of different sizes, shapes, colors, textures, tastes. Spinoza would like to argue that the universal 'apple' itself is a meaningless term, haphazardly put together in the mind of the various individuals, and answering to no common pattern or idea. Yet put your average school-boy into a room filled with a variety of fruits and vegetables, along with just one apple or 'pomum,' and the schoolboy will not fail to retrieve the proper object. Nay. Put a dozen different school-boys in a dozen different rooms filled with all varieties of objects, along with just one apple; and the dozen different school-boys will not fail to retrieve the 'pomum' if asked to do so. This holds true from language to language, from country to country. People do not accomplish this through 'habit.' It is not deep intellect, but it is judgment nevertheless. It is not the highest grade of knowledge, but it is a grade of knowledge. As Plato's Socrates would argue, experience and perception yield 'recollection,' remembrance. The individual possesses a knowledge, and perception reminds him as to what it is he knows, if it is brought to his attention.

Both Nadler and Smith make a great show of this example, as if Spinoza has really made a devastating critique of the ordinary speaker and his use of universals.[74] Spinoza argues that each individual human being's image of a man, or of a horse, is larded with the diverse particulars of the individuals experienced.[75] One man is tall and broad-shouldered, dimpled in the chin, muscular. Another man is short and squat, shaped like a pear. According to

Spinoza, all of our personal images of 'man' are contaminated with the particulars of our experience. As if one person had the image of the tall, broad-shouldered man, in his soul; such that whenever he came across a man, he would wonder: "why is this man not tall and broad-shouldered?" It would be hard to find a ten-year-old who would take this as a serious problem. Individuals, regardless of their experience, so long as they have been acquainted with a human being or horse and have a decent memory, actually learn the *kind*. Spinoza is desperate to deny that there are kinds.

Spinoza does not only indict the ordinary speakers. He indicts the philosophers whose theory of knowledge is firmly anchored in the evidence made available by human perceptual faculties: namely Plato and Aristotle. Aristotle and Plato define man differently, Spinoza suggests: one calling man a two legged featherless animal, the other calling him the rational animal.[76] Each man is a slave to the limits of the men he has met, Spinoza maintains. The definitions don't equate. Plato would say, bring Aristotle, Socrates, and a boy to a farmhouse where are diverse animals and only one man. All will easily pick out the man. Twelve boys on twelve different farms will all pick out the man, though these men differ radically in age, weight, size, color, or even in the number of their legs. This is not due to 'imagination.' It is not due to 'inadequate' knowledge. For the form by which each object is known is a unity. As Aristotle says, it is a formula. Given that the senses are in adequate range of the object, and that said individual has prior experience of this kind of object, Plato holds, and this knowledge is all but infallible.[77] Nobody calls an ox a horse, or odd numbers even, not even in his dreams.

Spinoza is not done with us yet, in his display of fancy examples. There are hoofprints in the sand, Spinoza starts off. A soldier sees the prints and thinks of a horse of war. A farmer sees the prints and thinks of a plow. Thus, from the hoofprints, we have one man looking about for an army, and another man looking about for a farm wagon.[78] What a crisis!

First of all, it is not true that the soldier and the farmer see the hoofprints, and think diversely of an army or a farm. What both see is the hoofprints, and what both think of is a horse. Whether the horse has been carrying a soldier, or dragging a plow, it is still called a horse. Nor would it be easy to find a man who thought that the horse must be fulfilling one office rather than the other, without additional 'signs' or information.

This all becomes relevant if we analyze why Spinoza turns to the eternal, both in his unity and in the finite 'modes' (i.e., the 'simple bodies,' alleged to constitute 'individuals'). One only needs to begin to recognize Spinoza's attitude towards hope and fear, joy and sadness. For the 'ordinary' people, both hope and fear are parts of life. Spinoza wants to be immune to both of them. He wants to live beyond vulnerability. He insists on commanding 'fortune.' He insists on asserting dominion over that which he experiences by superimposing denial upon it.

Facts have a way of evoking, of drawing us out of ourselves, towards interactive emotions, ordinary praise and blame. Spinoza's definition of 'inadequate' ideas is a preemptive strike against one's exposure to the world, psychologically speaking. Spinoza's doctrine of necessitarianism is laid upon the same root. These are the philosopher's dismissal of his bonds with the others, his rejection of common truth, in order to perfect, as it were, his desire, 'adequate knowledge' which is entirely within his control.

'ADEQUATE' OR 'CLEAR AND DISTINCT' IDEAS

Spinoza is often celebrated for the freedom of speech that he espouses in his commonwealth. What we are investigating here, however, is a metaphysics of a science of public opinion. In this metaphysics, Spinoza's model of the human mind is to be employed to exactly contradict the operations of the ordinary human mind. In proportion as ordinary opinion perceives certain facts and registers emotional reactions to these facts, Spinoza's science of public opinion will advance the reverse proposition. It will generate its own facts, and attempt to supplant the passive emotional reactions with the dictates of emotion construed as 'active desire,' that is, suited for conatus.

Scholars tend to get lost in the abstractions of Spinoza's metaphysics. They do not perceive that Spinoza's *Ethics* is setting forth a prototype for mental regulation. Untrained thinkers are not going to be familiar with the origins of this prototype. They are entirely likely to be baffled by it. Yet Locke and Hume both build upon this model. The deep philosophical foundations of the modern study of public opinion is therefore fraught with controversial premises. It can be said that Spinoza's theory of public opinion is itself a *political* science. Like all of the Early Modern philosophies, it is anchored in a false physics.

It really isn't possible to understand Spinoza's category of 'adequate' or 'clear and distinct' ideas exclusive of his enumeration of the kinds of emotion. The implication of perceived fact for the emotions has been pointed out, which Spinoza calls 'passions.' These are 'passive emotions,' which subject the individual to common praise and blame. When Spinoza beholds the world, it is as if from behind a fortress. Not for him the passive experience of the scenes, events, facts of the common world. No: he will not share this passive posture with the others, nor will he subject his emotions to the discipline enacted upon the individual by the community's praise and blame. Spinoza rejects the passive emotions root and branch. Yet in Spinoza's view, it is not sufficient merely to turn himself off from the passive emotions. He must combat the passive emotions with artificially generated emotions, desires galvanized and formed with a view to asserting his power: over, and over, and over again. Thus the dynamic interplay between inadequate and adequate ideas in Spinoza needs to be understood as a form of *combat*.

Spinoza is not the first philosopher to survey this territory. Machiavelli and Hobbes have preceded him in the formulation of independent, highly politicized vocabularies. Yet Spinoza has carried the process much further into articulate speech. Machiavelli gives us almost nothing in the way of theory to exhibit the process. Hobbes, for his part, makes a great show of forming new vocabularies for the emotions, the virtues, political principles and the like. Yet it is only with Spinoza that we get an explication of the emotional dynamic that is underway.

Whenever Spinoza proclaims that an idea is inadequate, this is a trigger mechanism: the inadequate idea must be transformed into an 'adequate idea.' The process is quite simple. Spinoza likes to characterize it as viewing perishable objects 'sub specie aeternitatis': or approaching perishable objects from the 'eternal' point of view. In other words, all of the recognizable properties of the perishable object must be shaved off in analysis. What will be left is no more than the mysterious 'extension' of Spinoza's great 'being'; or else some configuration of 'simple bodies' in specific 'patterns of motion and rest,' which is hardly more informative. To impose the criteria of the 'adequate' idea upon the 'inadequate idea' (perceived image) is to attack its fact-hood: it is like an acid, this process; it dissolves the recognizable images, and in so doing erases the claim of the one-time fact upon the emotions of the philosopher.

Spinoza formulates this process with care. Spinoza tells us that there is only one way to overcome the passive emotions that are evoked in us by perceived facts. That way is to create, within ourselves, a sort of artificial emotion that moves in a contrary direction. "An emotion cannot be checked or destroyed except by a contrary emotion which is stronger than the emotion to be checked," Spinoza writes.[79] The 'adequate' idea is therefore a psychological *assertion*. It is the attempt to manifest within one's own mind the 'conatus' or will-to-power that, according to Spinoza, is the property of God himself. Every object in nature has a 'conatus,' Spinoza argues; and this can scarcely surprise us, since both the great 'being' of his nature or the 'simple bodies' of his nature convert lifelessness into a norm.

Spinoza knows, perhaps better than any of his predecessors in the Epicurean tradition, that the roots of perception and passion are naturally present in himself. His theory of science, of 'adequate' ideas, of 'clear and distinct' ideas, is very much a ritual. It is like a religious rite. Certainly the use of names is the decisive battleground. No matter how abstract Spinoza's terms of nature might seem, he is ultimately a theorist who is aiming to 'perfect' human nature. His theory is an exertion of his 'inventive' power, just as Hobbes advises the philosopher to regard his work as if he were God, reenacting the original creation.[80] Adequate, or clear and distinct, ideas for Spinoza originate in the philosopher's categories of analysis, which we have sufficiently examined above. Their operative force is the philosopher's *will*. As opposed to the passive emotions evoked by the common facts, the philosopher counterposes manufactured 'facts,' which are the manifestation of his 'desire.' 'Desire' and 'passive emotion' are for Spinoza a true diversity in nature. The former yields freedom, the latter 'bondage.'

It is of considerable use to contemplate the analyses of Spinoza provided by one who is quite in sympathy with the project. Stuart Hampshire is eminently qualified for this task. Hampshire takes us through the 'philosophical conversion' that the would-be Spinozist must undergo in order to reach the promised land of desire—the elixir which will protect the philosopher from all varieties of 'fortune,' since he will in fact be in the constant enterprise of manufacturing his own facts, of creating for himself a psychological household within which he can stave off the common perceptions and emotions.[81]

Hampshire, who played such an important role in the development of twentieth-century linguistic philosophy, takes the would-be Spinozist through the logical steps of the enterprise. If perception is really 'confusion,' 'mutilated,' and 'inadequate' ideas, then the individual must become *detached* from the emotions that perception evokes. Based on the critique of perception, one becomes detached from the full range of names that indicate diverse emotional experiences: honor, obligation, reverence, duty, love, loyalty, faithfulness, virtue; but also from words like treachery, betrayal, deceit, exploitation, murder, theft, and the like. Names that convey an emotional charge to the community are the ones that this species of 'enlightened' philosopher seeks to liberate himself from.

Hampshire focuses on the *prescriptive* power of Spinoza's philosophy. Hampshire celebrates Spinozism for being able to increase a man's power by altering his thought. For to read the thought of Spinoza is to be affected in our own thoughts. Hampshire helps to illuminate the revolutionary transformation that Machiavelli had wrought in the Epicurean project. In Epicurus and even in Lucretius, the philosopher seeks to shield himself from the demos, in the emotional and psychological terms discussed above. Beginning with Machiavelli, this line of philosophy is determined to create a society which normalizes and seeks to externalize (in political *institutions*) the extreme conditions under which the proto-Epicurean philosopher prefers to live.

All passions must be redefined in accordance with the will to defend the self against the alien world. As Hampshire and his commentator point out, Spinoza will not allow us to have a true idea without that idea being 'in' the great substance of *Ethics* I. To view things 'sub specie aeternitatis' is to view things from the perspective of the all-inclusive substance. The 'cause' of substance, in Spinoza, is power: the one universal substance is said to be the cause of itself, by willing itself to exist. The existence of substance is said to be an expression of power; and by implication, all of the objects that exist 'in' this substance, must be likewise animated solely and inevitably, by the self-same 'conatus' or assertion of power that characterizes the great substance.[82]

Up until this point in the discussion of Spinoza, I have chosen to characterize his one being as 'substance,' rather than as 'God.' Only at this point in the discussion does it really become meaningful to introduce this name that Spinoza employs to characterize the one substance, the 'God' name. For the psychological orientation of the Spinozistic philosopher, in the methodical suppression of perceived fact, and in the deliberate genesis of manufactured emotions through contrived pseudo-facts, seeks to obtain a sort of serenity. Strauss calls attention to this aspect of Spinoza's thought. 'Beatitude' is the state of mind towards which the Spinozist philosopher migrates.[83] The bliss he feels at being able to procedurally suppress the facts of common experience from articulation as valid truth begins to take on religious aspects for this sort of philosopher. He bows before his philosophical process. This process delivers him, as it were, from the possibility of suffering. It overpowers every passive emotion with an 'active' one, an active desire designed to negate the passively perceived factum.

God's perspective, which this enlightenment philosopher professes to know, is power. Thus the man who detaches himself from the sacred emotions attendant to certain perceived facts, carries out God's work in so doing. Since God is power, the philosopher's very assertion of his own power against the moral feelings of the demos is an act of submission to 'God.' He proudly bears the 'chains of God's love.'

KNOWLEDGE BY 'INTUITION'

'Intuition' marks a different level of escalation in the rhetoric of clear and distinct ideas in the *Ethics*. It conveys a well-masked militance, in fact the crystallization of a new ideological sect. No less a sage than Leo Strauss observes that there is a movement in the *Ethics*, between *Ethics* I and *Ethics* V; which is to say that there is a 'progress.' Intuitive knowledge alone indicates a full grasp of the perishables. In Strauss's view, this movement of Spinoza's prepared the way for German Idealism: it indicates a forward looking direction in nature, a movement in which the human mind attains to awareness of itself, of its spirit.[84]

Spinoza begins the fifth movement of the *Ethics* with a discussion of Descartes. Descartes famously argued that the mind is separate from body. Only knowledge of ourselves as a 'thinking thing' is obtained by Descartes' argument in the *Meditations*. The existence of body is later deduced from this knowledge, but this is not a satisfactory argument from the point of view of

Spinoza. If the existence of body is not proved, or if the independence of the mind from the body is possible, then Spinoza would have no knock-down argument from which to attack the passive emotions.

Spinoza requires that the emotions of the mind be *ascribed* to the affections of the body. The body must be the cause of the emotions for Spinoza, if indeed the philosopher is to have the right to counteract them. If the cause of the emotions of the mind is the body of the percipient, then these emotions, as it were, *belong* to the philosopher alone. The ordinary people believe that the emotions are evoked by perceived external objects and events. Based on that belief, the passive emotions qualify as reality, and obtain authority to guide the emotional household of the percipient. Spinoza, by arguing that emotion is merely an affection of the percipient's body, converts the emotion into a *private* affair. If emotion is caused by affections within the percipient's body, then it is not caused by the externally perceived objects and events at all. The philosopher, after this argument, is then in a position to attempt to counteract or destroy the influence of these passive emotions in his own mind. This is the direction to what Spinoza calls 'blessedness.'[85]

To refer to this sort of knowledge as 'intuition' is larded with significance. It is in *Ethics* V where Spinoza begins to talk of the perishable objects themselves as objects of the mind. 'Intuition' suggests a mysterious mental process, one that is not subject to verification. 'Intuition' is the language that Spinoza elects to employ when his psychological war against his common humanity with the others is reaching its most aggressive stage. In *Ethics* V, the 'bondage' of the passions is renounced; and what is made available is the avenue of power to self-assertion as untrammeled by these common passive feelings.

> If we remove an agitation of the mind, or emotion, from the thought of its external cause, and join it to other thoughts, then love or hatred toward the external cause, and also vacillations, that arise from these emotions will be destroyed.[86] A passive emotion ceases to be a passive emotion as soon as we form a clear and distinct idea of it.[87] So the more an emotion is known to us, the more it is within our control, and the mind is the less passive in respect of it.[88] The highest conatus of the mind and its highest virtue is to understand things by the third kind of knowledge.[89] The more capable the mind is of understanding things by the third kind of knowledge, the more it desires to understand things by this same kind of knowledge.[90]

It is by 'intuition' that the 'modes' or perishable bodies are to be known. In order to be permanently independent of the dreaded passions, Spinoza's armies of philosophy must wage a permanent counter-offensive. For every emotion that is evoked by perceptible matters of fact, Spinoza's science must undertake to destroy the awareness of causes in the popular mind, and to institute countervailing causes to offset the perceptible ones. "The greater the number of things that the mind understands by the second and third kind of knowledge, the less subject it is to emotions that are bad, and the less it fears death."[91] This is the project of permanently mobilized 'active desire,' of relentless conversion therapy.

In *Ethics* V, then, following hard upon the heals of the denunciation of the passions as a state of 'bondage,' Spinoza speaks selectively. He refers to passive effect, evoked emotion, interpersonal emotions. He neglects to mention that his principle of self preservation is an appetite, one through which Spinoza seeks to create or fabricate a species of countervailing feelings sufficient to dominate the passive affects.

The third kind of knowledge is pugilistic. It depends upon the second kind of knowledge, the knowledge of 'adequate ideas' to impose the physics of *Ethics* I and II on the perishable bodies. Yet it moves beyond the second kind of knowledge, in its conceptualization of a program for the confrontation with the passive emotions, and the attempt to break them of their influence over the personality. Spinoza is giving us a portrait that suggests the birth of natural right from the loins of the second type of knowledge in its state of aggressive application: Spinoza would like to assign the lofty name of 'intellect' to the 'conatus' or will to power which is unleashed by the attack on the passive emotions in the philosopher's personality; but this lust for power is clearly simply a baser passion, one quite innate in all human beings, but in most non-philosophic minds kept in check by moral education and self-respect.

To be the master of what one feels, to develop a philosophical technology whereby to bring about this variety of control over the self, is to sever one's bonds methodically to the others. It is no wonder that Spinoza is hostile to the universal 'man.' For Spinoza does not intend to be one among other men merely. For while the others are reacting to their passive emotions and the lines of virtue and vice implicit in them, Spinoza is busy becoming a free agent as it were: his mind converted to worship of the God of *Ethics* I. Spinoza's God is power incarnate. To 'love' God, then, from Spinoza's point of view, is to love the substance in which one exists. If one exists 'in' God, then one must partake of the will to power that *is* God. "From this we clearly understand in what our salvation or blessedness or freedom consists, namely in the constant and eternal love toward God, that is, in God's love toward men." "There is nothing in nature which is contrary to this intellectual love, or which can destroy it."[92]

Spinoza, in *Ethics* IV propositions 67–72, outlines the relationship that he envisions for the philosophers on the one side, and the community on the other. The many will live by the distinction between right and wrong, good and evil, as its indispensable compass. Yet these philosophers reject such an understanding root and branch. Their philosophy of nature does not include a distinction between good and evil. S. Pines indicates that the constraints of good and evil suggest passivity. To be 'free,' Pines reasons, is to no longer be subject to good or evil.[93] Good and evil can indeed, in the first instance, only be manifested in one's relationship to external facts. Reason, in Spinoza, erases these facts. Reason makes one 'autonomous' of the facts. The facts are dissolved into causes that are not perceived. 'Naturalism' is what they call it.

When facts are dissolved by the 'naturalistic method,' the philosopher is transformed into a unitary enterprise: a struggle for power, which he calls 'preservation.' If the attack on perception is false, then and only then does it become possible for us to evaluate whether good and evil are real forms. "Pleasure, therefore, or delight, is the appearance or sense of good; and molestation or displeasure, the appearance or sense of evil."[94]

SPINOZA'S NECESSITARIANISM

For the moment, it is desirable to make some observations regarding the fatalism or necessitarianism in Spinoza's doctrines. Necessity, as Spinoza deploys the concept, is very much a psychological variable. The perspective of 'necessity' comes to life in Spinoza in his war against the passive emotions. The passive emotions felt by the community far exceed the narrow bounds of pain and pleasure that Spinoza adumbrates. Emotions are evoked in us by horrible scenes, beautiful scenes, urgent scenarios and others. As Aristotle explicates in great detail, our passive emotions help to lay out for us the course of action that we must take in order to see to it that the situation into which we have entered has a proper and correct outcome. The passions, from Aristotle's point of view, are in part nature's way of urging us toward fulfillment of the good in interpersonal relationships.

To view all events, all actions, all scenes as 'necessary' is to deprive those objects of their emotional power. That which is necessary cannot be otherwise. Spinoza approaches the doctrine of necessity with precisely this point of view. To proclaim the necessity of all actions and events in human affairs is to shield the individual from any and all expectations that might be imposed upon him and his actions in the context of particular common facts. "Insofar as the mind understands all things as governed by necessity, to that extent it has greater power over emotions, that is, it is less passive in respect of them."[95] "The more this knowledge (namely, that things are governed by necessity) is applied to particular things which we imagine more distinctly and more vividly, the greater is the power of the mind over the emotions, as is testified by experience. For we see that pain over the loss of some good is assuaged as soon as the man who has lost it realizes that that good could not have been saved in any way."[96]

It is important to note that not all things in the universe are governed by 'necessity' for Spinoza. The passive emotions that the non-philosophers feel so strongly, are submitted by Spinoza to the category of necessity for the above cited reasons. Yet the intellect of the philosopher is not subject to this necessity. Rather, the intellect of the Spinozist philosopher is submitted to the category of *freedom.* Virtue, for Spinoza, consists in overcoming the passive emotions, that is, in breaking the bonds of faith with the others. Machiavelli's *Prince* XV, which had such a profound effect on Spinoza, is a palpable influence here.

> Now the power of the mind is defined solely by knowledge; its weakness or passivity solely by the privation of knowledge; that is, it is measured by the extent to which its ideas are said to be inadequate. Hence it follows that that mind is most passive whose greatest part is constituted by inadequate ideas, so that it is characterized more by passivity than by activity. On the other hand, that mind is most active whose greatest part is constituted by adequate ideas, so that even if the latter mind contains as many inadequate ideas as the former, it is characterized by those ideas which are attributed to human virtue rather than by those which point to human weakness.[97]

'Necessity' disconnects the mind from cares. That which could not be otherwise need not be worried about, nor is it required to feel any guilt over such events. Yet there is a deeper cause of the doctrine of necessity that is operative in Spinoza. It is constitutive of the very marrow of Spinoza's argumentative structure. If we go back to the *Emendation of the Intellect*, and to Spinoza's search for the 'one true guiding idea' which could serve as the basis of his deductive system: the 'one true idea' for Spinoza is an exercise of the intellect's *power.* For Spinoza, the intellect *creates* the one true idea. It does not discover it or find it.

Thus *will* displaces the intellect that reflects upon subject matter that is given to it. Spinoza, who refuses to share with us any common facts; who refuses to share with us the original premise of his deductive model—cannot avoid putting will first. 'Necessity' characterizes his reasoning because he cannot demonstrate his arguments to us in a way which would enable him to obtain our rational assent. I think it can be usefully pointed out that Spinoza very early on starts to characterize these movements in his brain as 'necessary' truths. Necessity, compulsion, constraint: they show up as constitutive of Spinoza's philosophy, precisely insofar as they do not, indeed cannot, persuade. This is to say that the driving force in Spinoza's argument is not evidence, not reason, not truth, but will.

NATURAL RIGHT

We approach natural right. As there is but one substance, that is, Spinoza's 'being,' it is defined by its power to cause itself. What does this mean? To say that the one infinite being exists by its own power, indicates that absent its power, it would not exist. Preservation is a meaningless term for that which must always exist. That which never comes to be, has no need of power to sustain its being. It is beyond the reach of inception, decay, past, or future.

If nature is a unity, as Spinoza holds that it is, then it is ignorant to think that man can live in a 'kingdom within a kingdom.' The ordinary language of right and wrong, in other words, remains to be reconciled with the laws observable in nature as a whole. The social contract state envisioned by Hobbes and Spinoza is one that takes natural right as the original power of human beings. Natural right is committed to the alleged laws of nature as a whole. Moral law can only be one, for the little fishes and for men.[98] Thus the social contract state to be built, is to be built out of the bricks and mortar furnished by the transfer of said 'right of nature,' which all are alleged to possess. The social contract state therefore undertakes to enshrine a new morality: and it is towards the establishment of the new morality that Spinoza's knowledge as 'intuition' is dedicated.

The knowledge of right and wrong is first of all the fruit of experience: praise and blame, as both Aristotle and Plato argues, are among the natural barometers of moral truth. Spinoza's attack on these passions is more than an effort to curb 'prejudice.' This is rather the spearhead of an attack to *instill* a prejudice, by force of philosophy: that every man has his own right and wrong; which postulate, once stipulated, is employed to serve as the major premise from which political science operates.

Hobbes never went so far as to call self-preservation a virtue. It is well remarked upon in the secondary literature that Hobbes recoiled upon reading Spinoza's work, regarding it as somewhat too bold. Hobbes masks self-preservation as if it were a moderate impulse, as the opponent of fiery and seditious pride. Truth be told, Hobbes regards the impulse of self-preservation in exactly the same terms that Spinoza does. Spinoza trumpets forth self-preservation as a *virtue*, not as a consolation prize for the abdication of virtues. That which ordinary opinion 'marks' and labels as the basest of the base (boundless selfishness), slouches towards the throne of moral virtue in the artificial culture of the social contract state prepared by Hobbes and Spinoza.

A couple of additional observations are called for at this point. In the *TPT*, Spinoza discusses how dominion can exist in mental power. The human nature of equality, which Aristotle regards as absolute in some dimensions even between parts of society that espouse fundamentally different principles of justice, will be reduced in Spinoza's system to equal right to seek power. The right of nature includes the right to deceive, in Spinoza's system.[99] Indeed, his philosophy itself is not conceivable without this proviso, for the reasons indicated.

The making of contracts, whether by an individual or by a state, is not binding in Spinoza's eyes. One may keep a contract for so long as it is in one's interest to do so; but when this utility can be seen to be vacated, so too is the obligation to fulfill the contract vacated. "If a man has given his pledge to someone, promising only verbally to do this or that, which it was within his own right to do, or not to do, the pledge remains valid for as long as he who made it has not changed his mind. For he who has the power to break faith has in reality not given up his right; he has given no more than words."[100] Strauss contrasts the right of nature as it is manifest in Spinoza and Hobbes. Hobbes, according to Strauss, holds that contracts are binding: and that the peace of the state depends on the enforcement of contracts.[101] Strauss does not point out that in Hobbes, contracts themselves are to be forged out of contesting wills.

The social contract, like society itself, is not voluntary. Human beings cannot live without society; and therefore the social contract model developed by the early moderns is designed to configure the terms upon which social relationships shall revolve. Since society is a necessity for human beings, the social contract's feature of a potentially dissolved society constitutes a trigger for renewing the terms of the social contract. If the terms of the social contract are violated by the government, then natural right is violated. If natural right is violated, then order must be cast to the hazard. The people must be made to adapt to an alternative: either a social contract more firmly anchored in the right of nature, or paralyzing disorder.

Spinoza is heralded by a number of thinkers as the true founder of liberal democracy.[102] This again suggests a strong contrast to Hobbes. In Hobbes's state, individuals do not enjoy freedom of speech, freedom of worship, much less freedom of thought. In Spinoza, individuals will be able to enjoy freedom of speech, but the other two freedoms seem quite dubious in Spinoza's system as well.

It is Spinoza's science which presumes, in the *TPT*, to determine the very nature of what God is. Science is the authority when it comes to matters of truth; and Spinoza's account of the divinity is merely another name for his 'substance.' God does not have passive emotions, in Spinoza's Epicurean point of view. God is only characterized by power, as this is the only thing that befits a superior being. In Spinoza, religion itself is being reconfigured into a merely disciplinary institution. Obedience will be the watchword of religion under the auspices of Spinoza's society; and yet this obedience will be subject to the authority of the philosophy which exercises dominance over religion insofar as issues of *truth* are concerned.

The few, in Spinoza's system, regard themselves not as individuals in the universal class of human beings; but rather as separate species, each individual unto himself. When the layers of ordinary praise and blame are neutralized and suspended by Spinoza's philosophical technology for the negation of the impact of the passive emotions upon the soul, the few become free to assert their individual natures of 'self-preservation' in the aggressive, contract-breaking fashion. The Spinozist lives beyond good and evil, as A.E. Taylor observes.[103] They do not entertain that there is any such thing as a distinction between good and evil, beyond the war that they conduct against the ordinary perception and its concomitant habits of praise and blame.

All beings in Spinoza's cosmos are animated by the nature of the one substance in which all are said to inhere. All beings therefore must be said to observe one moral truth. There is to be no 'kingdom within a kingdom' here. If it is by sovereign natural right that the big fishes eat the little fishes, then human beings are not at liberty to proclaim a different moral code for themselves. If the stones and trees do not have the capacity to live in accordance with ordinary praise and blame, then it is ordinary praise and blame which are excrescences upon the morality of nature: a morality, it can truly be said, in which everything is permitted.

The equality that Spinoza espouses, in rebuttal to the 'anger and ignorance' of the many, is clear enough. Spinoza's moral vocabulary singles out the original human opposition to injustice, that is, indignation, as *the* evil of society. Spinoza only follows Hobbes in this regard. Spinoza intends very much for his philosophy to be tutelary: to disseminate new authorities of praise and blame, to facilitate the dismantling of traditional good manners. Nor ought we to allow the fact that traditional good manners sometimes harbor oppression and brutality to dissuade us from affirming that all true progressive change only ever comes about through appeal to our 'better natures' as opposed to a worse. Spinoza's right of nature, in the individual or as alienated into the collective, is no one's better nature, in a proper manner of speaking.

NOTES

1. Some examples are Alan Gabbey, "Spinoza's Natural Science and Methodology," 146; and Don Garrett, "Spinoza's Ethical Theory," 270. In Don Garrett, editor. *The Cambridge Companion to Spinoza*. Cambridge University Press, 1996. G.H.R. Parkinson, *Spinoza's Theory of Knowledge*. Oxford University Press, 1964. H.A. Wolfson, "Behind the Geometric Method," 15. In Marjorie Grene, editor. *Spinoza: A Collection of Critical Essays.* Notre Dame University Press, 1979. Piet Steenbakkers, "The Textual History of Spinoza's Ethics," 28–29, 33–34; and Valtteri Viljanen, "Spinoza's Ontology," 57. In Olli Koistinen, editor. *The Cambridge Companion to Spinoza's Ethics*. Cambridge: Cambridge University Press, 2009. E.M. Curley, *Spinoza's Metaphysics: An Essay in Interpretation.* Cambridge: Harvard University Press, 1969, 6–11. Aaron V. Garrett. *Meaning in Spinoza's Method.* Cambridge: Cambridge University Press, 2003, 11–12.

2. Cf. Stuart Hampshire. *Spinoza*. London: Penguin Books, 1955, 14–15. E.M. Curley, "Experience in Spinoza's Theory of Knowledge," 58. In Marjorie Grene, editor. *Spinoza: A Collection of Critical Essays*. Notre Dame: Notre Dame University Press, 1979.

3. Spinoza advances dual Eleatic theories: Parmenidean and Epicurean-atomist. *Ethics* I, Propositions 7 and 8, produce a version of Parmenides's 'being.' Parmenides's 'being,' to

elaborate, is stated in the following manner: "It never was nor will be, since it is now, all together, one, continuous" (Frag. 8:5–6). "Thus it must either be completely or not at all. Nor will the force of conviction allow anything besides it to come to be ever from not-being." The translations used are from G.S. Kirk, J.E. Raven, and M. Schofield, *The PreSocratic Philosophers*. 2nd edition. Cambridge University Press: 1983. Spinoza's proposition 7 *in Ethics* I asserts that "existence belongs to the nature of substance." In *Ethics* I proposition 8, Spinoza makes this substance equivalent to all that exists. "Every substance is necessarily infinite." In the proof for Proposition 8, Spinoza states that "there cannot be more than one substance having the same attribute and existence that belongs to the nature of substance." In *Ethics* I proposition 12, Spinoza maintains that "no attribute of substance can be truly conceived from which it would follow that substance can be divided." This was the first thing to be illustrated. In *Ethics* II, Spinoza discusses body in the atomistic language of simple bodies and composite bodies. "So far we have been discussing the simplest bodies, those which are distinguished from one another solely by motion and rest, quickness and slowness." Proposition 13, Lemma 3, axiom 2. Spinoza's definition of the 'simplest bodies' in composite mode does not involve a *fusion* of said simplest bodies. They retain bodily independence even when they are said to be parts of an 'individual.' "When a number of bodies of the same or different magnitude form close contact with one another through the pressure of other bodies upon them, or if they are moving at the same or different rates of speed so as to preserve an unvarying relation of movement amongst themselves, these bodies are said to be united with one another and all together to form one body or individual thing, which is distinguished from other things through this union of bodies" (*Ethics* II, proposition 13, Lemma 3, axiom 2, p. 253). Spinoza goes on to argue that we may view the entire universe as "one individual whose parts—that is—all the constituent bodies—vary in infinite ways without any change in the individual as a whole" (*Ethics* II, proposition 13, Lemma 7, Scholium, p. 255). These translations from Spinoza are all from Samuel Shirley, *Spinoza: The Complete Works*. Edited by Michael Morgan. Indianapolis: Hackett Publishing, 2002. Spinoza never indicates that these 'simplest bodies' can come into being or cease to be. In *Letter 4 to Henry Oldenburg*, 27 September 1661, Spinoza observes that "if one particle of matter were to be annihilated, the whole of extension would vanish at the same time" (767). I take it then that Spinoza's simplest bodies are Eleatic-atomist as well. The thesis that Spinoza employs an atomistic model for the simple bodies is indicated in other passages. For example, Spinoza appears to believe that there is one underlying 'body,' the 'simplest bodies,' which is never destroyed, and out of which the composites emerge. Thus Spinoza speaks of a horse 'changing' into an insect or into a man. *Ethics* IV Preface: "For it is important to note that when I say that somebody passes from a state of less perfection to a state of greater perfection, and vice versa, I do not mean that he changes from one essence to another (for example, a horse is as completely destroyed if it changes into a man as it would be if it were to change into an insect), but that we conceive his power of activity, insofar as this is understood through his nature, to be increased or diminished" (322).

4. Both arguments rest on the premise that the being or body is indivisible. In *Ethics* I proposition 12, Spinoza states that "no attribute of substance can be truly conceived from which it would follow that substance could be divided" (223). In *Ethics* II Spinoza links his definition of 'body' to the solitary substance of *Ethics* I. "By 'body' I understand a mode that expresses in a definite and determinate way God's essence insofar as he is considered an extended thing" (Definition 1). In *Ethics* I, proposition 21, Spinoza states that "all things that follow from the absolute nature of any attribute of God must have existed always, and as infinite; that is, through the said attribute they are infinite and eternal" (230). In *Ethics* II proposition 2, it is stated that "Extension is an attribute of God, that is, God is an extended thing" (245). I take these propositions to establish that the 'simplest bodies' are particles of matter that cannot be divided in Spinoza's theory, which was the thing to be proved.

5. Piet Steenbakkers, "The Geometrical Order in the Ethics," 52–54. In Olli Koistinen, editor. *The Cambridge Companion to Spinoza's Ethics*. Cambridge: Cambridge University Press, 2009. G.H.R. Parkinson. "On the Power and Freedom of Man," 9. In Maurice Mandelbaum and Eugene Freeman, editors. *Spinoza: Essays in Interpretation*. LaSalle: Open Court, 1975. Steve Nadler. *Spinoza's Ethics: An Introduction*. Cambridge: Cambridge University Press, 2006, 42, 47. Wim Klever. "Anti-Falsificationism: Spinoza's Theory of Experience and

Experiments," 124, 126. In Edwin Curley and Pierre-Francois Moreau, editors. *Spinoza: Issues and Directions*. The Proceedings of the Chicago Conference. Leiden: Brill, 1990.

6. *Ethics* I, proposition 12. "No attribute of substance can be truly conceived from which it would follow that substance can be divided." Cf. *Ethics* I, proposition 11: "God or substance consisting of infinite attributes, each of which expresses eternal and infinite essence, necessarily exists."

7. *Ethics* II, proposition 13, Lemma 3, axiom 2: "So far we have been discussing the simplest bodies, those which are distinguished from one another solely by motion and rest, quickness and slowness" (253).

8. *Ethics* V, proposition 36, scholium: "I have thought this worth noting here in order to show by example, the superiority of that knowledge of particular things which I have called 'intuitive' or 'of the third kind,' and its preferability to that abstract knowledge which I have called 'knowledge of the second kind'" (379).

9. *Ethics* IV, Preface, 322.

10. *Ethics* II, proposition 49, corollary 2, 270. In *Ethics* I, proposition 21, proof, Spinoza states: "Furthermore, that which thus follows from the necessity of the nature of some attribute cannot have a determinate existence, or duration." In *Ethics* I, proposition 15, scholium, 226, Spinoza states that all body is part of the 'attribute' of 'corporeal substance,' "which can be conceived of only as infinite, as one, and indivisible." This is very much relevant to the above discussion of Spinoza's Eleatic forbears. A perishable object is indicated by Spinoza, not to *have*, but to *partake* of the essence of substance. *Ethics* I, proposition 24: "The essence of things produced by God does not involve existence" (232). Corollary: "Hence it follows that God is the cause not only of the coming-into existence of things but also of their continuing in existence, or, to use a scholastic term, God is the cause of the being of things (*essendi rerum*). For whether things exist or do not exist, in reflecting on their essence we realize that this essence involves neither existence nor duration, but only God, to whose nature alone existence pertains" (232).

11. Spinoza defines perceptual knowledge as 'imagination.' *Ethics* II, proposition 26: "The human mind does not perceive any external body as actually existing except through the ideas of affections of its own body" (261). Corollary: "Insofar as the human mind imagines an external body, to that extent it does not have an adequate knowledge of it." "When the human mind regards external bodies through the ideas of affections of its own body, we say that it imagines, and in no other way can the mind imagine external bodies as actually existing. Therefore insofar as the mind imagines external bodies, it does not have adequate knowledge of them" (261). Cf. *Ethics* II, proposition 25; *Ethics* II, proposition 17, scholium.

12. "Notes on the Second Part of *Ethics* (I)," 110. In Paul S. Kashap, editor. *Studies in Spinoza: Critical and Interpretive Essays*. Berkeley: University of California Press, 1972.

13. *Ethics* I, definition 3. Cf. Definition 6: "By God I mean an absolutely infinite being, that is, substance consisting of infinite attributes, each of which expresses eternal and infinite essence" (217). In *Ethics* I proposition 7, Spinoza writes: "Existence belongs to the nature of substance. Proof: substance cannot be produced by anything else, and is therefore self-caused; that is, its essence necessarily involves existence; that is, existence belongs to its nature" (219). Cf. *Ethics* I, proposition 12: "No attribute of substance can be truly conceived from which it would follow that substance can be divided" (223). *Ethics* I, proposition 14: "There can be, or be conceived, no other substance than God." Corollary 1: "Hence it follows quite clearly that God is one; that is, in the universe, there is only one substance, and this is absolutely infinite, as I have already indicated in scholium proposition 10" (224).

14. *Metaphysics* V, 1017b10. "All these are called substances because they are not predicated of a subject, but everything else is predicated of them." In book VI, at 1025b17, Aristotle writes: "And since natural science, like other sciences, confines itself to one class of beings, that is, to that sort of substance which has the principle of its movement and rest present in itself, evidently it is neither practical nor productive." At 1039b20: "Since substance is of two kinds, the concrete thing and the formula, (I mean that one kind of substance is the formula taken with the matter, while another kind is formula in its generality), substances in the former sense are capable of destruction (for they are capable also of generation), but there is no destruction of the formula, in the sense that it is ever in the course of being destroyed. . . . For

this reason, also, there is neither definition nor demonstration of sensible individual substances, because they have matter whose nature is such that they are capable both of being and of not being; for which reason all the individual instances of them are destructible." 1039b20–30: "But let us now resume the discussion of generally recognized substances. These are sensible substances, and sensible substances all have matter. The substratum is substance, and this is in one sense the matter (and by matter I mean that which, not being a 'this' actually, is potentially a 'this'), and in another sense the formula or form (which being a 'this' can be separately formulated), and thirdly, the complex of matter and form, which alone is generated and destroyed, and is, without qualification, capable of separate existence." (1043a24–31). In *The Complete Works of Aristotle*. Jonathan Barnes, editor. Volume 2. Princeton University Press, 1984.

15. Steven Nadler. *Spinoza's Ethics: An Introduction*. Cambridge: Cambridge University Press, 2006, 69.

16. "Since God's existence is not self-evident, it must necessarily be inferred from axiomatic truths which are so firm and incontrovertible that there can neither be, nor be conceived, any power that could call them into question. At any rate, once we have inferred from them God's existence, we are bound to regard them as such if we seek to establish beyond all shadow of a doubt our inference from them to God's existence" (447). From The *Theological-Political Treatise*, chapter 6.

17. *Ethics* I, definition 6: "By God I mean an absolutely infinite being, that is, substance consisting of infinite attributes, each of which expresses eternal and infinite existence" (216). Cf. *Ethics* I, proposition 14, corollary; *Ethics* I, Proposition 15, scholium; *Ethics* I, proposition 29, scholium.

18. Some scholars (to be discussed below) insist that Spinoza is not an atomist. The logic they use, is that Spinoza insists that body is divisible. This argument, however, is confused. For the divisibility that Spinoza *allows* to exist, is said to be purely a part of man's 'imagination' or *inadequate* ideas. In *adequate* ideas, Spinoza insists that body is *indivisible*. Spinoza is critical of those who believe that body is *really* divisible. "For corporeal substance, which can be conceived only as infinite, one, and indivisible, they conceive as made up of finite parts, multiplex and divisible, so as to deduce that it is finite. . . . Since, therefore, there is no vacuum in nature (of which more elsewhere) and all its parts must so harmonize that there is no vacuum, it also follows that the parts cannot be distinct in reality; that is, corporeal substance, insofar as it is substance, cannot be divided. If I am asked now why we have this natural inclination to divide quantity, I reply that we conceive quantity in two ways, to wit, abstractly, or superficially—in other words as represented in the imagination—or as substance, which we do only through intellect. If therefore we consider quality insofar as we represent it in the imagination—and this is what we more frequently and more readily do—we find it to be finite, divisible, and made up of parts. But if we consider it intellectually, and conceive of it insofar as it is substance—and this is very difficult—then it will be found to be infinite, one, and indivisible, as we have already sufficiently proved. This will be quite clear to those who can distinguish between the imagination and the intellect" (*Ethics* I, proposition 15, scholium, 226).

19. *Letter 4 from Henry Oldenburg*. Cf. *Ethics* I, proposition 15, scholium: "For example, we conceive water to be divisible and to have separate parts insofar as it is water, but not insofar as it is material substance. In this latter respect it is not capable of separation or division. Furthermore, water qua water comes into existence and goes out of existence; but qua substance it does not come into existence or go out of existence" (227). 'Modal' distinctions in matter are not *real* distinctions, per Spinoza. "Matter is everywhere the same" (226).

20. *Ethics* II proposition 26: "The human mind does not perceive any external body as actually existing except through the ideas of affections of its own body. Corollary: Insofar as the human mind imagines an external body, to that extent it does not have an adequate knowledge of it" (261). *Ethics* I, proposition 28: "the ideas of the affections of the human body, insofar as they are related to the human mind, are not clear and distinct, but confused" (261). Cf. *Ethics*, I proposition 29.

21. *Ethics* I, proposition 41, proof: "In the preceding scholium we asserted that all those ideas which are inadequate and confused belong to the first kind of knowledge; and thus this knowledge is the only cause of falsity. Further, we asserted that to the knowledge of the second

and third kind there belong those ideas which are adequate. Therefore, this knowledge is necessarily true" (268). Cf. *Ethics* II, proposition 44, corollary 2: "It is in the nature of reason to perceive things in the light of eternity (*sub quadam specie aeternitatis*)" (270).

22. Perishable bodies, or 'modes,' exist only in the imagination. Modes are only *adequately* known as *affections* of the indivisible substance, hence as eternal. *Ethics* I, proposition 15.

23. *Ethics* II, proposition 1; proposition 2; proposition 7 scholium. Cf. *Short Treatise on God, Man, and His Well-Being*, part 1, chapter 2, p. 45.

24. At *Metaphysics* 1028a18, Aristotle writes: "While 'being' has all these senses, that which is primarily is the 'what,' which indicates the substance of a thing. For when we say of what quality a thing is, we say that it is good or beautiful." At 1029b12: "The essence of each thing is what it is said to be in virtue of itself. For being you is not being musical; for you are not musical in virtue of yourself. What, then, you are in virtue of yourself is your essence." "Each thing and its essence are one and the same in no merely accidental way, as is evident both from the preceding arguments, and because to know each thing is to know its essence." (1031b18). "Clearly, then, each primary and self-subsistent thing is the same as its essence" (1032a4).

25. *Treatise on the Emendation of the Intellect*, p. 11: "Third, we have learned which is the path our mind should first take in order to make a good beginning, and that is, to proceed to its enquiry by fixed rules, taking as its standard some given true ideal." "Our aim, then, is to have clear and distinct ideas, that is, such as originate from pure mind and not from fortuitous motions of the body" (25).

26. *Emendation of the Intellect*, 18: "As to the second case of fiction, which we said to consist in attending simultaneously, without assenting, to various confused ideas of various things and actions existing in nature, we again saw that a completely simple thing cannot be the object of fiction, but only of intellect. And the same is true of a composite thing, provided we attend to its simplest component parts."

27. *Emendation of the Intellect*: "Our aim, then, is to have clear and distinct ideas, that is, such as originate from a pure mind and not from fortuitous motions of the body" (25). "The clear and distinct ideas that we form seem to depend absolutely on our power alone. But with confused ideas the contrary is the case" (25).

28. *Emendation of the Intellect*: "Next, such that all our ideas may be subsumed under one, we shall endeavor to connect and arrange them in such a manner that our mind, as far as possible, may reproduce in thought the reality of nature, both as to the whole and to its parts" (25).

29. *Ethics* II, definition 3, explication, 244.

30. To Hugo Boxel, 905.

31. *Ethics* II, proposition 44, corollary 2, 270.

32. *Ethics* I, proposition 32, proof: "Will, like intellect, is only a definite mode of thinking, and so no single volition can exist or be determined to act unless it is determined by another cause, and this cause again by another, and so on ad infinitum" (235). *Ethics* II, proposition 35, scholium: "Men are deceived in thinking themselves free, a belief that consists only in this, that they are conscious of their actions and ignorant of the causes by which they are determined. Therefore the idea of their freedom is simply the ignorance of the cause of their actions. As to their saying that human actions depend on the will, these are mere words without any corresponding idea. For none of them knows what the will is, and how it moves the body . . ." (264). Cf. *Ethics* II, proposition 48; *Ethics* III, proposition 2.

33. *Emendation of the Intellect*: "From this we may also see how easily those who have not made a careful distinction between imagination and intellection may fall into grave errors, such as, for instance, that extension may be localized, that it must be finite, that its parts are really distinct from one another, that it is the first and only foundation of all things, that it occupies more space at one time than another . . . all of which are completely opposed to truth, as we shall demonstrate in its proper place" (24).

34. *Ethics* I, definition 6; Ethics I, propositions 8, 9, 16, 21.

35. *Ethics* I, proposition 12, p. 223. Cf. *Ethics* I, proposition 13: "Absolutely infinite substance is indivisible" (224). *Ethics* I, proposition 14, corollary: "that is, in the universe there is only one substance, and this is absolutely infinite" (224).

36. *Ethics* I, proposition 7; proposition 8, scholium: "Since in fact to be finite is in part a negation and to be infinite is the unqualified affirmation of the existence of some nature, it follows from proposition 7 alone that every substance must be infinite." Scholium 2: "I do not doubt that for those who judge things confusedly and are not accustomed to know things through their primary causes it is difficult to grasp the proof of proposition 7 . . . for those who do not know the true causes of things confuse everything" (219–220).

37. *Physics* 206b34–207a: "Thus something is infinite, if, taking quantity by quantity, we can always take something outside. On the other hand, what has nothing outside is complete or whole." In *Aristotle, Collected Works*, volume 1. Jonathan Barnes, editor. Princeton University Press, 1984. "Nothing is complete that has no end and the end is a limit."

38. Martial Gueroult ("Spinoza's Letter on the Infinite (Letter XII to Louis Mayer)." In Marjorie Grene, editor. *Spinoza: A Collection of Critical Essays*. Notre Dame: University of Notre Dame Press, 1979) gets the equation between the infinite and the indivisible backwards. "Infinity and indivisibility are two unique properties of substance which derive immediately from its fundamental property, causa sui. Indeed, whatever necessarily exists of itself cannot, without contradiction, be deprived of any part whatever of its existence; consequently, it is necessarily infinite and excludes any partitioning. Infinity and indivisibility being two sides of the same property, there results a radical antinomy between the infinite and the divisible. If we affirm one we must deny the other" (182). This is actually precisely incorrect. In the first place, Gueroult has not established that Spinoza's 'causa sui' exists. Gueroult conveys Spinoza's position accurately, but the position is untenable. The infinite, as Aristotle argues, is that where there is always some part still 'outside.' As regards what is infinite, all the parts can *never* be together. There are *always* more parts to come. Thus, the infinite and the *divisible* go together. As A. Wolf has noted ("Spinoza's Conception of the Attributes of a Substance." In Paul Kashap, editor. *Studies in Spinoza: Critical and Interpretive Essays*. Berkeley: University of California, 1972, 26), Spinoza cannot afford to *lose* the nature of unity for his substance. Thus Wolf, who correctly appraises the relationship between the infinite and divisibility, insists that Spinoza uses the name 'infinite' to indicate 'complete, or all.'

39. *Short Treatise on God, Man and His Well-Being*: "To this we reply, (1) that 'part' and 'whole' are not true or real entities, but only 'things of reason,' and consequently there are in nature neither whole nor part . . . but extension, being a substance, one cannot say of it that it has parts, since it can neither diminish nor increase, and no parts thereof can be understood apart, because by its nature it must be infinite." "Division, then, or passivity, always takes place in the mode; thus when we say that man passes away or is annihilated, then this is understood to apply to man only insofar as he is a composite being, and a mode of substance, and not the substance on which he depends" (44–45). Cf. *Ethics* I, proposition 15, scholium: "These are the arguments I find put forward by writers who thereby seek to prove that corporeal substance is unworthy of the divine essence and cannot pertain to it. However, the student who looks carefully into these arguments will find that I have already replied to them, since they are all founded on the same supposition that material substance is composed of parts, and this I have already shown to be absurd" (225).

40. Steven Nadler. *Spinoza's Ethics: An Introduction*. Cambridge: Cambridge University Press, 2006, 56–58. Nadler seeks to compare Aristotle's idea of a substance with Spinoza's idea of a substance, but only after attempting to organize the two on some sort of developmental tree, mediated by the Cartesian theory. This construction of the discussion is quite inadequate, to use one of Spinoza's favorite words. Aristotle's substances can only be known through sense perception. Both Descartes and Spinoza banish such knowledge a priori. In order to investigate the differences between the Aristotelian and the Spinozan theories of substances, one must evaluate Spinoza's claim that matter is eternal—a claim that Aristotle does not accept. Matter, in its eternal aspect, for Aristotle, is only potential being. It is *less* than being, not-being in fact. For Spinoza, corporeal substance is indivisible and eternal; and by corporeal substance Spinoza indicates matter, 'the simplest bodies,' the bodies discussed in *Ethics* II. A horse, for Aristotle, is very much a substance. It is a thing of which predications are made. The horse is brown, doe-eyed. 'Brown' and 'doe-eyed' are qualities which cannot exist by themselves. They need a substance to exist *in*. Nadler suggests that Spinoza simply alters a bit of the terminology of substance, so that it now means that which needs nothing else to exist. This is in

fact Spinoza's definition of a substance in the Eleatic mode: whereby substance is inseparable from existence. The horse is unacceptable as a substance for Spinoza, Nadler indicates, because it needs hay to eat and water to drink. Since there are in reality no imperishable primary substances, Spinoza will have to settle for an idea of a horse which can neither drink nor eat hay, and which he cannot ride. Let the reader judge which is the more real thing.

41. *Ethics* III, proposition 6: "Each thing, insofar as it is in itself, endeavors to persist in its own being" (283). *Ethics* III, proposition 7: "The conatus with which each thing endeavors to persist in its own being is nothing but the actual essence of the thing itself" (283).

42. *Ethics* I, proposition 11: "God, or substance consisting of infinite attributes, each of which expresses eternal and infinite essence, necessarily exists" (222).

43. Fragment 8:5–6. In *The Presocratic Philosophers*, op. cit. "Therefore it is right that what is should not be imperfect; for it is not deficient—if it were it would be deficient in everything" (8:32–33). "Thus it must either be completely, or not at all" (8:10–11). "Nor is it divided, since it all exists alike; nor is it more here and less there, which would prevent it from holding together, but it is all full of being. So it is all continuous, for what is draws near to what is" (8:22–25).

44. *Parmenides* 128ab.

45. Each form, for Plato, is a separate nature. There is no 'whole' of nature for Plato. There is radical diversity. The form of unity can only be derived from the absolute nature of unity. In the *Phaedo*, Plato illustrates his theory of causation, while discussing the form of beauty. "It seems to me that whatever else is beautiful apart from absolute beauty is beautiful because it partakes of that absolute beauty, and for no other reason. Do you accept this kind of causality? . . . Well now, that is as far as my mind goes" (*Phaedo* 100c). In Edith Hamilton and Huntington Cairnes, editors. *Plato: the Collected Dialogues*. Princeton University Press, 1999. All of my references to Plato are from the translations in this volume.

46. *Parmenides* 137c.

47. *Parmenides* 141e-142a.

48. *Ethics* II, proposition 13, scholium, axiom 1 (p.252).

49. Steven Nadler. *Spinoza's Ethics: An Introduction*. Cambridge: Cambridge University Press, 2006, 57.

50. *Ethics* I, proposition 6.

51. *Parmenides* 129c.

52. *Parmenides* 143a.

53. Stuart Hampshire. *Spinoza*. London: Penguin, 1955, 17: "Words or symbols within mathematics do not derive their meaning from the figures or images which may be used to illustrate them, but stand for clearly defined conceptions. If a similar clarity, and rigor and certainty are ever to be achieved in philosophy and natural knowledge, their terms must first be purged of all the figurative and subjective associations of ordinary discourse; their vocabularies must be formed from words which are logical counters having a purely intellectual significance, words which stand in this sense for clear and distinct ideas."

54. Leo Strauss notes how the geometric logic of Hobbes and Spinoza delivers a devastating blow to the customary moral opinions which they measure and appraise. *The Political Philosophy of Hobbes: Its Basis and Genesis*. Chicago: University of Chicago Press, 1984, p.137.

55. Richard Kennington, "Spinoza's Methods in the Ethics." In Kennington, editor. *The Philosophy of Baruch Spinoza*. Washington, D.C.: Catholic University of America Press, 1980, 301: "Only on this experienced basis of particulars, at the inception of *Ethics* II, do we acquire knowledge that there are existent modes, and what their determinate properties are." "The realm of finite modes is independent of the divine attributes and infinite modes, as regards the particular determinacy, or 'production' of its members. Precisely because the geometric method begins with definitions of infinite being it reveals itself as not competent to account for the determinacy of finite being" (302). Kennington is obviously correct to make a distinction between the being of *Ethics* I and the bodies of *Ethics* II, since the latter are plural in Spinoza's presentation. However, it is impossible that the bodies of *Ethics* II should be the 'finite modes' if by finite modes we mean to indicate the familiar perishable objects. The objects or bodies of *Ethics* II are *mechanistic*. The individuals discussed in *Ethics* II are not made up of bodies that are fused together, but only which are related to one another in patterns of rest and motion.

Kennington is uncertain as to whether the 'simplest bodies' of *Ethics* II are atomic, but his uncertainty is unwarranted. Spinoza's matter or corporeal substance is indestructible. It is also indivisible, as is the extension in *Ethics* I. As I have indicated, the bodies of *Ethics* II unveil the Epicurean-Lucretian version of Eleatic body. The bodies of *Ethics* II, as Spinoza makes clear, can be said to constitute one individual for the whole of nature at the end of the day. Either or both of these models prepare Spinoza's indictment of the perceptual knowledge as 'inadequate ideas,' which is *the point* of his physics.

56. *Republic* 524b.

57. *Republic* 510d.

58. I wish to make a distinction here, insofar as Plato's forms are concerned. Plato's forms are reached by hypothesis. It is a theory of causation. The mind must first take a form for its object. Forms can only be learned about through perception initially; but pace Plato, it then becomes possible to stipulate that a form exists independently in nature, in order to define it or be capable of defining it. It is Plato's argument that the forms do thus exist independently, and I am in agreement with this view. Therefore it can be said here that all body, all objects that involve body, must be composite in nature. Objects that exist independently of bodies, such as the absolute forms themselves, fall outside of this definition of an object. It is bodies, not objects generally construed, which must be composites.

59. *Parmenides* 144bc.

60. *Ethics* I, proposition 28; *Ethics* II, proposition 13, lemma 3 (252).

61. *Ethics* II, propositions 26, 27, 28, and 29. Cf. my chapter on Epicurus in *Machiavelli and Epicureanism: An Investigation into the Origins of Early Modern Political Thought*. Lanham: Lexington Books, 2012.

62. Lee Rice. "Spinoza on Individuation." In Maurice Mandelbaum and Eugene Freeman, editors. *Spinoza: Essays in Interpretation*. LaSalle: Open Court, 1975. "At the termination of the second axiom following lemma 3 (*Ethics* II proposition 13), Spinoza explains that he has been dealing up to this point only with the most simple bodies, and that attention must be given to composite bodies. The problem of the nature of the corpora simplissima is perennial among commentators. Hicks takes an atomistic interpretation, likening them to Leibnizian monads; and such an interpretation is not uncommon. The central difficulty with this reading is that Spinoza explicitly denies the existence of atoms. Since the nature of corporeal substance for Spinoza consists wholly of extension, and since extension is divisible (though never divided in any ontological sense) there can be no units which are indivisible" (200). Rice has himself indicated the answer to the riddle that he proffers. There is no 'ontological' separation or divisibility, he indicates, for Spinoza's simplest bodies. What does this mean? It means that in terms of the knowledge of *truth*, the simplest bodies are indivisible. The same bodies are 'divisible' through the imagination of human beings, that is, through their 'inadequate' ideas. The same simple bodies understood as 'substance,' as adequate knowledge for Spinoza, are indivisible. Whole and parts are figments of the imagination for Spinoza. Spinoza's simplest bodies are indeed atoms. Rice's confusion is widely shared. Cf. Ruth Saw, "Personal Identity in Spinoza," 91. "As for the simplest bodies, they do not endure, so that they could neither persist nor not persist in their own being." In Paul Kashap, editor. *Studies in Spinoza: Critical Essays*. Berkeley: University of California, 1972. Cf. G.H.R. Parkinson. *Spinoza's Theory of Knowledge*. Oxford University Press, 1964, 75, note 1: "This correspondence with atoms is only rough, and must not be pressed. For Spinoza, although he preferred the Greek atomists to any other of the ancient philosophers, disagreed with the theory that nature was compounded of atoms and void." Spinoza does deny a vacuum, but his simplest bodies are nonetheless atoms that move and rest alternately. The simplest bodies of Spinoza are indivisible nuggets of Eleatic being.

63. *Letter 4 to Henry Oldenburg*.

64. *Ethics* II, proposition 13, definition.

65. *Ethics* II, proposition 13, lemma 3.

66. *Letter 32 to Henry Oldenburg*, 849.

67. *Ethics* II, proposition 20: "The more every man endeavors and is able to seek his own advantage, that is, to preserve his own being, the more he is endowed with virtue. On the other hand, insofar as he neglects to preserve what is to his advantage, that is, his own being, he is

weak" (331). *Ethics* IV, proposition 37, scholium: "Every man exists by the sovereign natural right, and consequently by the sovereign natural right every man does what follows from the necessity of his nature. So it is by the sovereign natural right that every man judges what is good and what is bad, and has regard for his own advantage according to his own way of thinking and seeks revenge and endeavors to preserve what he loves and to destroy what he hates" (340).

68. *Republic* 476c.

69. Leo Strauss. *Spinoza's Critique of Religion.* Translated by Elsa Sinclair. University of Chicago Press, 1997. Strauss is truly a rare philosopher. His unsurpassed learning enables him to move fluidly between the great philosophical epochs of the last 2,500 years. To Strauss the entire Enlightenment obtains decisive aspects of its attitude, to say nothing of its theory of body, from Epicurean and Lucretian atomism. The masterful presentation by Strauss however has not made much impact on Spinoza studies, which perhaps would not disappoint the later esoteric Strauss. The Strauss of *SCR* is however young and almost feverish in pursuit of the truth. This book, written before Strauss was thirty, is raw and honest, much like his early works on Hobbes are. "We may take Democritus as the origin of a theory on which, as on a foundation, the theories of Hobbes, Spinoza, Hume, Holbach, Feurbach, Bruno Bauer, Marx take their stand to characterize and define a great tradition which extends into the present. The school of Epicurus is the first torch bearer of this tradition" (45).

70. *Ethics* IV, proposition 37, scholium 2: "Now if men lived by the guidance of reason, every man would possess this right of his without any harm to another. But since men are subject to emotions, which far surpass the power or virtue in men, they are therefore often pulled in contrary directions and are contrary to one another while needing each other's help." For Aristotle, the definition of virtue does not contemplate *eliminating* passion. *Without* passion, one would not know what virtue *is*. This goes to the root of Aristotle's very different theory of nature. For Aristotle, nature is intelligent; and so therefore are human feelings. Virtue for Aristotle is a mean in between two extremes of passion. Now, the very *basis* for calculating the mean of feeling, depends upon ordering the feeling so that it is commensurate with the *facts*. The facts can only be known through perception, which entire category is the object of Spinoza's unrelenting attack. It is the category of perception that Spinoza derides as 'imagination,' as 'passive,' as the source of 'inadequate' or 'confused' ideas, ideas which Spinoza holds *cannot* be true. "There are also means in the passions and concerned with the passions," Aristotle writes; "since shame is not an excellence, and yet praise is extended to the modest man. For even in these matters one man is said to be intermediate, and another to exceed, as for instance the bashful man who is ashamed of everything, while he who falls short or is not ashamed of anything is called shameless, and the intermediate person is called modest. Righteous indignation is a mean between envy and spite, and these states are concerned with the pain and pleasure that are felt at the fortunes of our neighbors; the man who is characterized by righteous indignation is pained at undeserved good fortune; the envious man, going beyond him, is pained at all good fortune, and the spiteful man falls so far short of being pained that he even rejoices. . . . There are three kinds of disposition, then, two of them vices, involving excess and deficiency and one an excellence, viz. the mean." *Nichomachean Ethics* 1108a31. In *Aristotle, the Collected Works*, volume 1. Edited by Jonathan Barnes. Princeton University Press, 1984. "Now it is a mean between two vices, that which depends on excess and that which depends on defect; and again it is a mean because the two vices respectively fall short or exceed what is right in both passions and actions . . ." (1107a3). For Spinoza, that which Aristotle regards as an excellence, righteous indignation, is dismissed as *necessarily evil.* Aristotle regards ordinary praise and blame as the original evidence for the distinction between right and wrong, while Spinoza persecutes the ordinary opinions, and denies the distinction itself. The ordinary distinction between right and wrong, good and bad are waved away by Spinoza as 'prejudices' (*Ethics* I, Appendix, second paragraph). "Pleasure, pain and consequently the emotions that are compounded of these or derived from them are passive emotions. Now we are necessarily passive insofar as we have inadequate ideas, and only insofar as we have inadequate ideas are we passive" (*Ethics* IV, proposition 56, proof, 307). In *Ethics* IV proposition 20, Spinoza states: "The more every man endeavors, and is able to seek his own advantage, that is, to preserve his being, the more he is endowed with virtue" (331). The Machiavellian cadences

of Spinoza's theory of virtue are unmistakable. Cf. *Prince* XV. *Ethics* IV, proposition 51, scholium states: "Indignation, as we have defined it, is necessarily evil" (347). Cf. *Short Treatise on God, Man and his Well-being*, chapter 4: "We have already said before that all things necessarily are what they are, and that in nature there is no good and no evil" (67).

71. Steven B. Smith. *Spinoza's Book of Life*. New Haven: Yale University Press, 2003. 'Knowledge of the first kind is grounded in the imagination. It consists of random experience (experiencia vaga) acquired through the senses but in ways that are described as 'mutilated, confused, without order' (2Pr40s2). 'This type of knowledge derives from signs derived either from hearsay or reading poor ideas based on memory and recollection' (78). 'Thus the imagination is the source of our mental passivity, of our enslavement to the affective-passionate life' (87). Spinoza does not attack perception because it treats of 'random' nature. He attacks perception because perception discloses the order *in* nature. Spinoza's attack on universals is an indispensable companion piece to Smith's discussion. Spinoza *rebels* against the order in nature as he rebels against the order in human society. Nature, for Spinoza, known *truly*, is purposeless.

72. *Theaetetus* 191d. Spinoza, not surprisingly, ridicules memory. *Ethics* II, proposition 18, scholium, 258.

73. *Ethics* II, proposition 18, scholium.

74. Spinoza discusses memory in *Ethics* II, proposition 18, scholium. "Furthermore, from this we clearly understand why the mind, from thinking of one thing, should straight away pass on to thinking of another thing which has no likeness to the first. For example, from thinking of the name 'pomum' (apple) a Roman will straightaway fall to thinking of the fruit, which has no likeness to that articulated sound nor anything in common with it other than that the man's body has often been affected by them both; that is, the man has often heard the name 'pomum' while seeing the fruit. So everyone will pass on from one thought to another according as habit in each case has arranged the images in his body" (258). Steven Nadler does not contemplate 'pomum' as a valid universal, as the name of a kind of object in nature. He simply climbs aboard Spinoza's general attack on universals as suffering from 'contingency and relativity' (176). *Spinoza's Ethics: An Introduction*. Cambridge: Cambridge University Press, 2006. Nadler follows, or wishes to follow, Spinoza to a new category of ideas in the 'adequate' denomination. The fact that these 'adequate' ideas of Spinoza are the merest delusions, as this chapter undertakes to prove, never crosses Nadler's mind. Smith, responding to a different example of association psychology ridiculed by Spinoza, is similarly docile, unquestioning, obedient: "The imagination is the source of human diversity, and its expressions are as various as are the life experiences of individuals and society. This explains why the soldier who sees a picture of a horse thinks of battle and war, while a farmer thinks of fields and plows. There is no logical connection between these mental acts; it is a loose, psychological association based on memory and experience" (88). Steven B. Smith. *Spinoza's Book of Life.* New Haven: Yale University Press, 2003. Smith doesn't quite get Spinoza's example correct. There are hoofprints on the ground. The soldier is said to think of a horse carrying soldiers; the farmer to think of a horse drawing a plow. The correct answer to this is that both the soldier and the farmer think of the *same* thing: a horse. Any further associations that take place in their minds, they are well aware of, and far from regarding as any part of the fact of the hoofprints, unless additional information is evident.

75. *Ethics* II, proposition 40, scholium 1: "Again, from similar causes have arisen these notions called 'universal,' such as 'man,' 'horse,' 'dog,' etc.; that is to say that so many images are formed in the human body simultaneously (e.g., of men) that our capacity to imagine them is surpassed, not indeed completely, but to the extent that the mind is unable to imagine the unimportant differences of individuals (such as the complexion and stature of each, and their exact number) and imagines distinctly only their common characteristics insofar as the body is affected by them. For it was by this that the body was affected most repeatedly, by each single individual. The mind expresses this by the word 'man,' and predicates this word of an infinite number of individuals."

76. *Ethics* II, proposition 40, scholium 1, continued: "But it should be noted that not all men form these notions in the same way; in the case of each person the notions vary according as the thing varies whereby the body has been more frequently affected, and which the mind more

readily imagines or calls to mind. For example, those who have more often regarded with admiration the stature of men will understand by the word 'man' an animal of upright stature, while those who are wont to regard a different aspect will form a different common image of man, such as that man is the laughing animal, a featherless biped, or a rational animal."

77. *Theaetetus* 190c.

78. *Ethics* II, proposition 18, scholium.

79. *Ethics* IV, proposition 7. Cf. *Ethics* IV, proposition 9: "An emotion whose cause we think to be with us in the present is much stronger than it would be if we did not think the said cause to be with us" (326). *Ethics* IV proposition 14: "No emotion can be checked by the true knowledge of good and evil insofar as it is true, but only insofar as it is considered an emotion" (328). *Ethics* V, proposition 3: "A passive emotion ceases to be a passive emotion as soon as we form a clear and distinct idea of it" (366).

80. Thomas Hobbes. "Author's Epistle to the Reader," first paragraph. In *Works*, volume 1. London: John Bohn, 1839.

81. Stuart Hampshire ("Spinoza's Theory of Freedom." In Maurice Mandelbaum and Eugene Freeman, editors. *Spinoza: Essays in Interpretation*. LaSalle: Open Court, 1975) speaks of a transformation of life that Spinoza's philosophy can bring to an otherwise ordinary person. It is to embrace a new power that the non-philosophers *lack*. "Their powers are limited by purely intellectual errors and by false philosophies of the mind. Their actual freedom and happiness are, in this respect, less than they could be—that is, less than they would be, if they were converted by argument to a less partial understanding of their own powers and interests. For example, they have the power to control many of their passive emotions, and to substitute active emotions for them, provided that they once realize that the objects of their passions are not the true causes of them . . ." (40). Hampshire in other words, is a convert to the 'adequate ideas.' Yet the adequate ideas are mere flotsam and jetsam of philosophical wishes. They have no *truth*. To sunder one's attachments to the 'objects of the passive emotions' in exchange for—in exchange for what exactly?—must prove to be a most serious undertaking. Before he sets foot on that ship, our prospective convert better check his boarding pass, and discuss what class of passenger he really is.

82. *Ethics* III, proposition 7 (p. 283).

83. Leo Strauss. *Spinoza's Critique of Religion*. Translated by Elsa Sinclair. Chicago: University of Chicago Press, 1997, 210.

84. Ibid., 16.

85. *Ethics* V, proposition 42 (382).

86. *Ethics* V, proposition 2.

87. *Ethics* V, proposition 3.

88. *Ethics* V, proposition 3, corollary.

89. *Ethics* V, proposition 25.

90. *Ethics* V, proposition 26.

91. *Ethics* V, proposition 38.

92. *Ethics* V, proposition 37.

93. S. Pines. "On Spinoza's Conception of Human Freedom and of Good and Evil." In Nathan Rotenstreich and Norma Schneider, editors. *Spinoza: His Thought and Work*. Jerusalem: the Israel Academy of Sciences and Humanities, 1983. "Spinoza's position is quite different on this point. According to him, as we have seen, having the notion of good and evil is incompatible with having adequate ideas; that is, true knowledge. The fact that man, according to Spinoza's *Ethics*, is capable of having adequate ideas, thus entails the consequence that he can dispense with the notion of good and evil" (153).

94. Thomas Hobbes. *Leviathan*. Chapter 6, #11. Oxford: Oxford University Press, 1999.

95. *Ethics* V, proposition 6.

96. *Ethics* V, proposition 6, scholium.

97. *Ethics* V, proposition 20, scholium.

98. *Theological-Political Treatise*, chapter 16: "Thus it is by sovereign natural right that fish inhabit water, and the big ones eat the smaller ones. For it is certain that nature, taken in the absolute sense, has the sovereign right to do all that she can do; that is, nature's right is coextensive with her power. For nature's power is the very power of God, who has sovereign

right over all things. But since the universal power of nature as a whole is nothing but the power of the individual things taken together, it follows that each individual thing has the sovereign right to do all that it can do" (527).

99. *Theological-Political Treatise*, chapter 16, fifth paragraph.

100. *Political Treatise*, chapter 2, #12.

101. Leo Strauss. *Spinoza's Critique of Religion*. Translated by Elsa Sinclair. Chicago: University of Chicago Press, 1997. In Strauss's view, Hobbes and Spinoza have distinct theories of natural right. Hobbes, regarding himself as an equal among the ordinary people, any one of whom has the power to kill any other, insists that contracts are binding even in the state of nature. To recommend otherwise would be hubris, since it would announce one's own alleged superiority to the others, the weakest of whom could kill one. Spinoza, who is more distant from the multitude, in Strauss's view, holds that all contracts are based on power and utility. To have the strength needed to break a contract, is to be free to do so (229–236). Let these observations be granted. Hobbes and Spinoza are identical in natural right to the degree that the contracts they *do* make are based on the special philosophical power that enables them to gain the upper hand. For Hobbes himself, contract is an instrument of *dominion*, not fair dealing. In this respect Hobbes and Spinoza are as one.

102. Steven B. Smith. *Spinoza, Liberalism, and the Question of Jewish Identity*. New Haven: Yale University Press, 1997. Smith takes Spinoza's *Theological-Political Treatise* as "one of the great works of autoemancipation. The aim of the work as a whole is the liberation of the individual from bondage to superstition and ecclesiastical authority. Spinoza's idea is the free or autonomous individual who uses reason to achieve mastery over the passions" (17). Smith is right to view the *TPT* and the *Ethics* as linked works. In my view, the *Ethics*' constructions of adequate ideas are themselves delusions, built on delusions. The 'autonomy' that Spinoza celebrates is not anything like that cultural quantum that is embraced in today's liberal-democratic cultures—not if one pays attention to what Spinozistic autonomy actually consists in. Steven Nadler (*A Book Forged in Hell*. Princeton: Princeton University Press, 2011) offers a similarly boilerplate account of Spinoza's project. The Enlightenment is advertised as a revolt against bondage; so therefore Spinoza, as one of the forerunners of the Enlightenment, must be opposed to human bondage too. "The *Treatise* and the *Ethics* are part of the same overall philosophical and political project: to liberate the minds of individuals from superstition and the lives of citizens from ecclesiastical authority. His goal is a tolerant democratic society of individuals whose deeds are guided by the true moral religion" (327). Actually, not so much.

103. A. E. Taylor. "Some Incoherencies of Spinozism (II)." In Paul S. Kashap, editor. *Studies in Spinoza: Critical and Interpretive Essays*. Berkeley: University of California, 1972. "Hitherto I have in effect been arguing that Spinozism can produce no intelligible theory of natural knowledge, because by ignoring the concipient, it has made its 'mind' into one which may be called a theater of 'psychic occurrences,' but really knows nothing. I want now to contend that the system, if consistent, is precluded from having any genuinely ethical doctrine at all but its proposed exclusion of the notion of moral value" (289).

Early Modern Philosophy

Just the Facts

If there is a single mood which characterizes the Early Modern movement in philosophy, it might be conveyed by a flickering candle flame in a dark cellar. This is the new point of origin for thought, courtesy of Descartes. The classical dependence of philosophy on the light of ordinary perceptual experience is hereby swept out of the way: in order to even be eligible to participate in the newly conceived quest for truth and reality, all that is familiar and remembered must be surrendered at the door.

Francis Bacon lashes out at the ordinary usages of speech, at the assignment of names to objects by means of which ordinary life is conducted. These ordinary ways of speech, in Bacon's eyes, are mere primitive 'idols,' superstitions that stand in the way of a competent exploration of the hidden recesses of reality.[1] So Descartes, in order to found his certain knowledge of the *Cogito*, professes that all the information vouchsafed to him by his senses must be rejected, a priori, as unreliable, misleading and without probative value.[2] Hobbes certainly picks up this thread of Descartes' reasoning. To inquire properly after truth, Hobbes insists, one must first of all pretend to oneself that the entire world has been annihilated: leaving only himself and his 'ideas.'[3]

These new philosophers are all politically minded. They seek to alter the ways and means of getting along in the world. A considerable aspect of the new style of authority will consist in the heightened skepticism that will be applied to public opinion, and to the objects of public opinion. It is not necessary for the new scientific authority to directly contradict any ordinary experience of the common objects. It is quite sufficient to call them into question; to treat every statement of fact as a mere alleged account; to buttress all observations with new doses of uncertainty and alternative theories. Early Modern Philosophy seeks indeed to prevent ordinary opinion from functioning in some respects. The mistrust of the senses that the Early Modern philosophies harbor reflect rather severely upon ordinary accounts of reality. This is quite sufficient to raise the role of the passions in culture more generally. When what is obvious to ordinary opinions is presented as merely probable or possible by Early Modern Philosophy, anxiety will be generated. On contested issues, ordinary opinion has no advocate. At least, ordinary opinion has no advocate absent a stronger representation of Platonic and Aristotelian philosophy in the culture at large.

Early Modern Philosophy refuses to demonstrate a pathway from shared facts to general truths. This distance between the emerging discipline of epistemology and ordinary sense perception reveals the roles that rhetoric and passion will play in the emerging political sciences. Plato and Aristotle regarded the passions warily. For Aristotle, virtuous action must escape two forms of viciousness, both of them rooted in an incorrect matching of feeling and deed to fact. By contrast, for Hobbes and Spinoza especially, passion is the desired means for the governing of the populace.

For Thomas Hobbes, the towering figure in modern political thought, fear is the proper orientation, now and always, for the man who would be good.[4] Indeed, the distinction between fear and goodness in Hobbes all but gives way. Goodness is dread, and dread is goodness. He who stands his ground against fear, in the Hobbesian analysis, is both stupid and immoral. By Hobbesian definition, such a man is diseased with conceit or immaturity. He does not possess the maturity of terror; he has not yet figured out that fear is not something that he can guard himself from. For Hobbes, such a man needs to be broken. He needs to be humbled into submission or else, like an angular stone, flung away by the builders of society.[5]

In submission to fear, Hobbesian man is indeed born again. He is not born anew into any ordinary religious catechism. Quite the contrary. By nature, man holds fast to that which he loves. What he loves, and those he loves, are the very antithesis of fear. They savor of the familiar, and of the fellowship in which he is known for himself. For his feelings and character; for his principles and commitments; his past, and his hopes for the future. To Hobbes, this is all a dreamworld which must be brought to an end in order for true righteousness to reign.

If Hobbes's very philosophy begins with the thought experiment that the whole external world has been annihilated, the non-philosophic individual is beckoned into darkness by Hobbes's political science. The pristine geometry of anxiety cannot occur until man has forsaken what he loves. Indeed, he must not merely be driven into desperation, for this does not settle the soul. Rather, the newly disenfranchised individual must be brought to renounce what and whom he loves: he must turn against the private world that resists fear in order to truly become 'righteous.' Only then does man lend himself to the certitude of the possibilities for peace that Hobbes wishes to enact. Only then does man become predictable in the way that Hobbes's science requires. Yet, if he must lose all that he loves in order to become this variable, to what end the 'peace'?

The disconnected cables of scientific fact gathering and ordinary perception lay at the very heart of this political morality and its operations. The postulates of physics upon which the early moderns have attempted to justify their scathing indictments of the senses are, I have attempted to demonstrate, refutable. By this I do not mean that these theories may merely be denied, but that they can be disproven. For this learning I have turned to Plato's dialogue the *Parmenides*, which both contemplates the essence of the atomistic teachings and reveals their impossibility. The Hobbesian political project cannot be identified for what it is until his philosophic indictment of the reputation of sense perception is investigated, penetrated, and refuted. Only then will it become possible to return to a world of shared facts; to a world where authority is subject to the possibility of discourse, speech that is more than isolated individuals mobilizing and unleashing private wills.

As Leo Strauss points out, Hobbes's attack on the authority of sense perception is intimately interwoven with his meta-politics.[6] Metaphysics indicates the study of what is real, with a view to finding out what is most real. 'Being,' for the non-philosophic individual, concerns what Aristotle refers to as *substances*: the individual, perishable objects. A tree, a man, a horse, fire. In Hobbes's evaluation of sense perception, we can have no true knowledge of any such things by our sense faculties. The ordinary person refers to substance so understood every time she speaks; but the new science, as it patiently listens to her, does not regard the uttered names as referring to real objects. Man, in the new physico-psychological conception, can do little more than manifest himself. The domain of common objects is purged from the grid of knowledge. Speech about the common objects does not enact a language that the new sciences regard as competent or enlightened.

What a person sees, smells, or hears, in the new scientific point of view, depends on the percipient merely to be conceded any existence at all. Whatever a person sees, in the new vernacular, this is expressly *her* perception; she is speaking her mind, her history, her body, her attitudes. She may as well not talk, for all the good it will do her. Yet there is one sense that Hobbes concedes a fully objective existence, the sense of touch. The wall will stop us if we attempt to walk through it. The boulder will crush us if we stand in its path. Touch in Hobbes's formulation indicates *resistance*; and it is resistance that Hobbesian political science is determined, at all costs, to make human beings feel. Men will not understand the resistance, which is the source of its devastating power. "And from hence it comes to pass, that men have no other means to acknowledge their own darkness, but only by reasoning from the unforeseen mischances, that befall them in their ways."[7]

Hobbes's metaphysics has its own distinctive conception of 'being.' Only that will be recognized as real in the Hobbesian metaphysics which possesses the power to resist man. Body is real, therefore, and nothing else. This is not to say that the only human experience can be submission for Hobbes. Domination is also a possibility. One's attitude must depend upon a calculation of forces. He who would depart from this monolithic psychology is to that extent unfit for Hobbes's society; and Hobbes's society will undertake to drive such a person toward the margins.

'Substance' is a pivotal term that proves to be a battleground for early modern philosophy. Originated by Aristotle, this term indicates the individual perishable objects, the compounds of matter and form.[8] The modern philosophers prefer not to engage Aristotle's arguments directly: The early moderns' attempt to make Aristotle's theory of substance seem pretentious, ridiculous, a derangement of the imagination. Yet, 'substance' for Aristotle indicates the perishable objects with which everyone is most familiar.

One cannot stress too often how important the language of substance, matter and form is in the Aristotelian legacy. 'Form' is the great discovery of the Socratic line of philosophers: form is that which is *intelligible*.[9] It is the philosophy of form which reveals the intelligence that is operative in the ordinary use of names by the meanest individual. To suppress this language of matter and form, of Aristotelian substance, is to suppress the common origin of Aristotelian philosophy and ordinary uses of speech. That is a rich vein to be mined; but the dismissive attitude in Early Modern Philosophy towards Aristotle seems to have left a residue of reluctance to seriously engage with his works.

The fruits of Early Modern Philosophy can be usefully contemplated in the work of Stuart Hampshire.[10] Hampshire is a relatively recent devotee of Spinozistic ethics. Hampshire provides us with a distinctive language that can be used to characterize the transformation of personality which is advocated by the radical Enlightenment. 'Conversion' away from 'prejudice' is a major theme in Hampshire's prose.

Hobbes and Spinoza prepare a political philosophy for the majority of the human race, which in their judgment will never be enlightened to a satisfactory degree. Hobbes and Spinoza thus attempt to engineer the passions of the majority, to bring their behavior into rough symmetry with enlightened doctrine (since their opinions can not be persuaded). Hampshire provides us with a presentation of the case of the few: of the few who *can* undergo the philosophic enlightenment prescribed by the early moderns, and it is towards a complete liberation from passive emotion that they migrate.

Prejudice is such an important philosophical term. When ordinary people use the name, they wish to indicate conscious bigotry or intolerance of other human beings. Prejudice in ordinary speech indicates the wrong kind of feeling, that is, destructive feelings. For the philosophically enlightened, the entire domain of passive emotion represents prejudice. They seek to evacuate it. The Spinozistic philosopher is serene because he abstracts himself from every emotion: he does not, in abstracting, seek to gain a perspective on it, so much as he seeks to obliterate it with theory and the tracing out of infinite other causes.[11] All of the passive emotions are prejudices to a Spinozist. The ordinary conceptions of right and wrong are swept away in what Hampshire regards as liberation.

For Hampshire, Spinoza's path to enlightenment begins with perception. The ordinary person, in Hampshire's sketch, trusts perception as information that is accurate and true. Hampshire dismisses such thinking as primitive and unenlightened. Perception, in Hampshire's analysis, is packed and marked by individual beliefs, assumptions, interests, values, relations, and other buried causes.

> Any situation which confronts me, and which is not a situation in a game,
> Has an inexhaustible set of discriminable features over and above those which
> I explicitly notice at the time because they are of immediate interest to me.
> Secondly, the situation has features over and above those which are mentionable
> In the vocabulary that I possess and use. I 'take in' the situation, noticing the
> Features that are particularly relevant to me at the time, and I respond to it in
> Accordance with my prevailing desires and purposes and my prevailing
> Beliefs and knowledge about the means to satisfy them. My desires and
> Interests form a vast system and only a few of them are called into play
> At any one time . . . As for the system of my beliefs, it is evident that there
> Is in my mind a vast store of unsurveyable background knowledge and
> Belief; and against this background my specific beliefs about the present
> Situation form themselves.[12]

Thus, while the ordinary individual takes the facts of experience as the original and common moral datum, Hampshire's enlightened individual applies the opposite reasoning. For Hampshire, the mind must deny the evidence of what is ordinarily perceived. He must apply theory to the perception to emancipate himself from its emotional power.

Hampshire views perception as a subconscious value system. There are unconscious *choices* operative in ordinary perception, in Hampshire's view. The enlightened individual thus excuses himself from the 'prejudice' of what is ordinarily perceived and felt. In Hampshire's Spinozistic view, every single cause must be traced back to a prior infinite chain of causes. With each step back into the chain that the philosopher retreats, the emotional power of the perceived fact is divided and weakened. When the philosopher has divided the fact and emotion into a sufficient number of causes to obliterate its presence from his mind, he enters into a state of serenity, of sublime independence.[13] Now he is free to truly decide what he perceives, making sure that his account accommodates choices that he has duly made for his own interest.

Hampshire contrasts the allegedly liberating Spinozistic philosophy with Aristotelian moral philosophy. Hampshire, however, is not quite able to present Aristotle's theory of moral reasoning accurately. No doubt, Hampshire would insist that there is no such objective reading to be had. If perception itself consists in choice (as he believes), then how could one hold a scholar to anything like evidence when it comes to textual interpretation? In any event, Hampshire's theory of perception is dependent on the physics to which he ascribes veracity. The disproof of that physics has been the project of the five chapters of this book, and thus we are resolved to refer the reader to those analyses insofar as the issues of physics and perception are concerned.

For Aristotle, it is true, there are some models of life that are available in the pursuit of fulfillment. Two major choices here are the political and the contemplative. Yet Aristotle does not thereby wish to argue that moral life itself is something that depends on a chosen way of life by the individual. Aristotle, who falls into the category of the unenlightened and the primitive by Hobbes's and Spinoza's lights, believes that ordinary praise and blame is only intermittently operative. Ordinary praise and blame are catapulted into activity by events, occurrences. Perception of the 'primitive' kind identifies them, for Aristotle; and the community, in its reactive praise and blame, is in Aristotle's view the proper guide of virtue and vice.

Aristotle's individual does not, as Hampshire suggests, begin his moral deliberations based on some theory he harbors as to his own good. The fact in question usually concerns someone else, such as an individual subjected to an injustice in our presence. The fact has nothing to do with the Aristotelian individual's choice of a life-plan, or overarching moral philosophy of any kind. The common fact rather indicates man as the *political animal*, that is, as the animal who can discriminate right from wrong, and for whom excellence in relationships is the measure of true happiness or the absence thereof. The theoretical predilections which Hampshire attempts to saddle Aristotle with savor of Spinozistic detachment. This is a grave error. Spinozistic detachment from commonly perceived facts is indicative of radical vice from the Aristotelian point of view. It consists, not so much in greater self-consciousness, as in rejection of one's common bond with the others.

It cannot be denied that, in contrast to the suffering and struggle that is visible in ordinary politics, the Hampshirean convert seems or appears superior. He betrays no angst. He does not lose his composure, or ever seem to struggle. Yet it is the cause of his composure that makes the reader doubt him. Spinozist converts are serene because they are detached. They have not resolved themselves into the virtuous mean of disposition and judgment; it is rather that they employ philosophy to excuse themselves from the entire scene of ordinary travail. In Spinoza's view, this detachment is the mark of civility and strength. From an Aristotelian or Platonic perspective, Spinozistic detachment is indicative of something else.

The early modern movement of philosophy is often on its soapbox, accusing the ordinary run of human beings of engaging in lazy thought, selfish ethics, and inhumane emotions. Yet the reader needs to bring into focus the emotional disposition of the philosophers themselves. For these philosophers refuse to share emotions with the others. They reject the facts that the others undergo; they deny the perceptual evidence upon which discourse can proceed. Early Modern Philosophy exhausts its lash upon those who feel too much; but there is room for viciousness in the opposite direction, in the refusal to feel anything at all.

As Hampshire discusses in his theory of Spinozistic rationalism, the 'passive' emotions simply don't have any place in the enlightened personality. Not for nothing does Spinoza define the entire category of passive perception and related passions as 'inadequate.'[14] For Spinoza, reason does not guide and temper passions: by repudiating the authority of common perceptible fact, Spinoza takes up arms against common feelings. From Hampshire's perspective, this independence reflects the naturalistic point of view. From the vantage point of Spinoza's philosophy, non-philosophers do not know what true causes are.

Bacon, Descartes, Hobbes, Spinoza, and even Locke take infinite pains, and infinite cares, to argue that the philosopher is not bound by the ordinary use of speech. For Hobbes, for Spinoza, for Hampshire, the ordinary run of human beings and philosophers are radically unequal in their capacity to know what is real, especially in the domains of fact and feeling. Both Hobbes and Spinoza are social contract theorists, where equality seems to be an implicit part of the institution. Yet, both Hobbes and Spinoza bitterly deride the concept of free will, which casts the 'contract' in a new light.

Liberated from any obligation to common facts, philosophy can aspire to hitherto unknown dimensions of 'moral justice.' It is possible for philosophy, as in the case of Hampshire, to appear super-virtuous; when in fact, said philosophy denies the most *elemental* bond between human beings: that of truth. What truth? Whose truth? These are the questions which we are asked to answer. Yet these questions do not arise out of the domain of common fact. They arise from the philosophic presuppositions that perception is neither able nor competent to know mutually binding facts. It is not that Early Modern Philosophy wants to improve human morality. It is that it regards human morality as pure artifice, as a mere superstition, unworthy of eliciting more than a distracted yawn from those who have seized the name of enlightened for themselves.

In Hampshire's view, Spinoza's moral philosophy is superior to that of Aristotle because it is more complex. Spinoza recognizes a conflict that exists between the domain of the passions (the familiar), and the broad proclivities of reason and theory on the other. Hampshire holds, as Isaiah Berlin holds, that there are competing virtues and goods; that to choose one is to repress others.[15] Yet this brings us back to Spinoza's definition of choice. Spinoza refuses to admit that he even *has* the power of choice. Whatever he does, whatever he feels, it is to be attributed to *necessity*. "In the mind there is no absolute, free will. The mind is determined to this or that volition by a cause, which is likewise determined by another cause, and this again by another, and so ad infinitum."[16] To admit to making choices entails responsibility for them; it is responsibility for choices which the doctrine of necessity suppresses.

In *De Cive*, Hobbes offers us his own version of the distinction between good and bad men; or between evil and less evil men. There are two parties that Hobbes takes notice of in his quasi-biblical narrative. It is an occasion of mutual antagonism. However, the parties are differently disposed towards one another. One is accordingly less guilty than the other. This less guilty party, in Hobbes's eyes, actually reveres a genuine equality. Yet Hobbes conveys this imagery with surpassing vagueness. Only a little bit of information is given to us.

> All men in the state of nature have a desire and will to hurt, but not
> Proceeding from the same cause, neither equally to be condemned. For
> One man, according to that natural equality which is among us, permits
> As much to another as he assumes to himself; which is an argument of a
> Temperate man, and one that rightly values his power. Another, supposing
> Himself above others, will have a license to do as he lists, and challenges
> Honor and respect, as due to him before others; which is an argument of a
> Fiery spirit. This man's will to hurt ariseth from vainglory, and the false
> Esteem he hath of his own strength.[17]

One party is angry. This party challenges honor and respect. The other party is cool and calculating. The angry party is to blame, for Hobbes. Of course: the angry party feels wronged. He is the victim of vulgar emotion, based upon his perceptions. Perhaps the other party has attempted to take the angry man's property. In the state of nature? There is no property in the state of nature, Hobbes insists. Where property is not, injustice cannot be. The only law in the state of nature is power. To act angry, is to suggest that the other has some obligation to how one feels, to a right which concerns someone other than himself. For Hobbes, this is vainglory, the very cause of war. The cool and calculative party doubtless observes 'the right of nature': everyone has a right to everything, in this version of 'equality.' Hobbes sings this man's tender praises.

Outside Hobbes's state as well as in it, man is subjected to a punishing regimen. It is man's own happiness which must be held at bay. Men feel themselves to achieve their humanity when they are able to rise above calculated self-interest, to act for truly common concerns. Yet in Hobbes's moral epistemology, such a disposition is adolescent at best, or pernicious at worst. The logic of the moral ethos that Hobbes attempts to enshrine in his Leviathan state is one which seeks to bring resistance to man's fundamental experience of himself. He must meet with a world that is not his own, in which he has no natural place. From Hobbes's point of view this is the philosophy of salvation. Yet human misery is not so easily garlanded with even so skillful a rhetoric as that of Hobbes.

NOTES

1. *The New Organon*. Book I, XLI, XLII, XLIII, XLIV, 42–43. Edited by Lisa Jardine and Michael Silverthorne. Cambridge University Press, 2000.
2. *The First Meditation*. In the *Philosophical Writings of Descartes*, volume II. Translated by John Cottingham, Robert Stoothoff, and Dugald Murdoch. Cambridge University Press, 2008, 12–13.
3. *The Elements of Law*, part I. Human nature, cap. 1, article 8, 22. In *the Elements of Law Natural and Politic*. Edited by J.C.A. Gaskin. Oxford University Press, 1994.
4. Ibid., chapter 14, article 6, 78–79: "And forasmuch as necessity of nature maketh men to will and desire bonum sibi, that which is good for themselves, and to avoid that which is hurtful; but most of all that terrible enemy of nature, death, from whom we expect both the loss of all power, and also the greatest of bodily pains in the losing; it is not against reason that a man doth all he can to preserve his own body, and limbs, both from death and pain. And that which is not against reason, men call right, or jus, or blameless liberty of using our natural power and ability."
5. *Man and Citizen*. Chapter 3, article 9. Edited by Bernard Gert. Indianapolis: Hackett Publishing, 1998, 141.
6. *Hobbes's Critique of Religion and Other Writings*. Translated by Gabriel Bartlett and Svetozar Minkov. Chicago: University of Chicago Press, 2011, 101–106.
7. *Leviathan*, chapter 14, article 4. Oxford University Press, 1996. Cf. Strauss, *Hobbes's Critique of Religion*, 103–104. The 'unforeseen mischances' that Hobbes intends is the disappointment of the human expectation of *justice*. In Hobbes's state, each person has her own justice, which all conflict. Thus no ordinary expectation of justice will be possible in the new order. "And because the condition of man . . . is a condition of war of every one against every one; in which case everyone is governed by his own reason; and there is nothing that he can make use of, that may not be a help to him in preserving his life against his enemies, it followeth, in that condition, that every man has a right to everything."
8. *Metaphysics* 1017b10.
9. Thomas Aquinas's discussion of Aristotelian philosophy is however quite reliable. On Aristotle's theory of form, see book 7, article 1419, 474–475. *Commentary on Aristotle's Metaphysics*. Translated by John P. Rowan. Notre Dame: Dumb Ox Books, 1995.
10. *Morality and Conflict*. Cambridge: Harvard University Press, 1983.
11. Ibid., 10, 48, 50, 51, 58, 61, 62.
12. Ibid., 106.
13. Ibid., 152, 162.
14. *Ethics*, book II, proposition 28: "The human mind does not perceive any external body as actually existing through the ideas of affections of its own body." Corollary: "Insofar as the human mind imagines an external body, to that extent it does not have an adequate knowledge of it." In *Spinoza, the Complete Works*. Translated by Samuel Shirley. Indianapolis: Hackett, 2002.

15. *Morality and Conflict*, 159.
16. *Ethics* II, proposition 48, 272.
17. Ibid., 114.

Bibliography

Abbott, Edwin. "The Latest Theory About Bacon." *Contemporary Review* 28(1876): 141–67.
Anderson, F.H. *The Philosophy of Francis Bacon*. Chicago: Phoenix Books, 1966.
Anglo, Sydney. "The Reception of Machiavelli in Tudor England: A Reassessment." *Il Politico* 31(1986): 127–38.
Aquinas, Saint Thomas. *Commentary on Aristotle's Metaphysics*. Notre Dame: Dumb Ox Books, 1995.
Aristotle. *The Collected Works*. Edited by Jonathan Barnes. Princeton: Princeton University Press, 1984.
———. *Nichomachean Ethics*. Oxford: Oxford University Press, 1998.
Bacon, Francis. *Works*. Edited by James Spedding, Robert Leslie Ellis, and Douglass Heath. Boston: Taggard and Thompson, 1843.
———. *The Oxford Francis Bacon VI. Philosophical Studies C.1611–C1619*. Edited by Graham Rees. Oxford: Clarendon Press, 1996.
———. *The New Organon*. Edited by Lisa Jardine and Michael Silverthorne. Cambridge: Cambridge University Press, 2000.
———. *The Major Works*. Edited by Brian Vickers. Oxford: Oxford University Press, 2002.
Baldes, R.W. "Divisibility and Division in Democritus." *Apeiron* 12(1978): 4.
Baumgold, Deborah. "The Difficulties of Hobbes Interpretation." *Political Theory* 36(2008): 827–55.
Berns, Laurence. "Francis Bacon and the Conquest of Nature." *Interpretation* 7(1978): 1–26.
Bodnar, Istvan. "Atomic Independence and Indivisibility." *Oxford Studies in Ancient Philosophy* 16(1998): 43.
Box, Ian. "Politics and Philosophy: Bacon on the Values of War and Peace." *Seventeenth Century* 7(1992): 115.
Brandt, Frithiof. *Thomas Hobbes's Mechanical Conception of Nature*. Copenhagen: Levin and Munksgaard, 1927.
Brown, Alison. *The Return of Lucretius to Renaissance Florence*. Cambridge: Harvard University Press, 2010.
Brown, Keith, editor. *Hobbes Studies*. Cambridge: Harvard University Press, 1965.
Burtt, E.A. *The Metaphysical Foundations of Modern Science*. New York: Doubleday, 1932.
Butterfield, Herbert. *The Origins of Modern Science*. Revised edition. New York: Free Press, 1965.
Clucos, Stephen. "The Atomism of the Cavendish Family Circle: A Reappraisal." *Seventeenth Century* 9(1994): 247–73.
Cogon, Marc. "Rhetoric and Action in Francis Bacon." *Philosophy and Rhetoric* 14(1981): 212–33.

Collingwood, R.G. *An Essay On Metaphysics*. Oxford: Clarendon Press, 1998.
Condren, Conal, Stephen Gaukroger, and Ian Hunter, editors. *The Philosopher in Early Modern Europe: the Nature of a Contested Identity*. Cambridge: Cambridge University Press, 2006.
Cook, Monte. "Descartes' Alleged Representationalism." *History of Philosophy Quarterly* 4(1987): 179–95.
Cottingham, J.C. *The Cambridge Companion to Descartes*. Cambridge: Cambridge University Press, 1992.
———, editor. *Reason, Will and Sensation: Studies in Descartes' Metaphysics*. Oxford: Clarendon Press, 1994.
———. *Cartesian Reflections*. Oxford: Oxford University Press, 2008.
Curd, Patricia, and Daniel Graham, editors. *Presocratic Philosophy: The Oxford Handbook*. Oxford University Press, 2008.
Curley, Edward M. *Spinoza's Metaphysics: An Essay In Interpretation*. Cambridge: Harvard University Press, 1969.
———. *Descartes Against the Skeptics*. Cambridge: Harvard University Press, 1978.
——— and Pierre-Francois Moreau, editors. *Spinoza: Issues and Directions. The Proceedings of the Chicago Conference*. Leiden: Brill, 1990.
Dahlstrom, Daniel, editor. *Nature and Scientific Method*. Washington, D.C.: Catholic University Press, 1991.
Daniel, Stephen H. "Myth and the Grammar of Discovery in Francis Bacon." *Philosophy and Rhetoric* 15(1982): 228.
De Rosa, Raffaella. "Descartes on Sensory Misrepresentation: The Case of Materially False Ideas." *History of Philosophy Quarterly* 21(2004): 261–80.
Dear, Peter. *Discipline and Experience: The Mathematical Way in the Seventeenth Century*. Chicago: University of Chicago Press, 1995.
Descartes, Rene. *The Philosophical Writings of Descartes*. In three volumes. Translated By John Cottingham, Robert Stoothoff, and Dugald Murdoch. Cambridge: Cambridge University Press, 1985.
Dickie, William M. "'Form' and 'Simple nature' in Bacon's Philosophy." *Monist* 33(1923): 428–37.
Dietz, Mary, editor. *Thomas Hobbes and Political Theory*. Lawrence: University of Kansas Press, 1990.
Dungey, Nicholas. "Thomas Hobbes's Materialism, Language and the Possibility of Politics." *Review of Politics* 70(2008): 190–220.
Emerton, Norma. *The Scientific Reinterpretation of Form*. Ithaca: Cornell University Press, 1984.
Epicurus. *The Essential Epicurus*. Translated by Eugene O'Connor. New York: Prometheus Books, 1993.
Farrington, Benjamin. *The Philosophy of Francis Bacon: An Essay on the Development from 1603–1609*. Chicago: University of Chicago Press, 1964.
Faulkner, Robert. *Francis Bacon and the Project of Progress*. Lanham: Rowman and Littlefield, 1993.
Feuer, Lewis Samuel. *Spinoza and the Rise of Liberalism*. Boston: Beacon Press, 1958.
Feyerabend, Paul. *Against Method*. New York: Verso, 2010.
Field, Richard. "Descartes on the Material Falsity of Ideas." *Philosophical Review* 102(1993): 309–33.
Finklestein, Claire, editor. *Hobbes on Law*. Burlington: Ashgate, 2004.
Frankfurt, Harry. *Demons, Dreamers and Madmen: The Defense of Reason in Descartes' Meditations*. Princeton: Princeton University Press, 2008.
Freeman, Kathleen. *Ancilla to the Pre-Socratic Philosophers*. Cambridge: Harvard University Press, 1996.
Garber, Daniel. *Descartes' Metaphysical Physics*. Chicago: University of Chicago Press, 1992.
——— and Steven Nadler, editors. *Oxford Studies in Early Modern Philosophy*. Volume III. Oxford: Clarendon Press, 2006.
Garrett, Aaron V. *Meaning in Spinoza's Method*. Cambridge: Cambridge University Press, 2003.

Garrett, Don, editor. *The Cambridge Companion to Spinoza*. Cambridge: Cambridge University Press, 1996.

Gaukroger, Stephen, editor. *Descartes: Philosophy, Mathematics, and Physics*. Totowa: Barnes and Noble, 1980.

———. *Descartes: An Intellectual Biography*. Oxford: Clarendon Press, 1995.

———. *Francis Bacon and the Transformation of Early Modern Philosophy*. Cambridge: Cambridge University Press, 2001.

———. *The Emergence of a Scientific Culture: Science and the Shaping of Modernity 1210–1685*. Oxford: Oxford University Press, 2006.

Gauthier, D. *The Logic of Leviathan*. Oxford: Oxford University Press, 1969.

Gilespie, Stuart, and Philip Hardie, editors. *The Cambridge Companion to Lucretius*. Cambridge University Press, 2007.

Goldstein, Rebecca Newberger. *Betraying Spinoza: The Renegade Jew Who Gave Us Modernity*. New York: Schocken Books, 2006.

Gorham, Geoffrey. "Descartes on the Innateness of all Ideas." *Canadian Journal of Philosophy* 32(2002): 355–88.

Grene, Marjorie, editor. *Spinoza: A Collection of Critical Essays*. Notre Dame: Notre Dame University Press, 1979.

Guthrie, W.K.C. *The Presocratic Tradition from Parmenides to Democritus*. Cambridge University Press, 1965.

Hall, A.R. *The Scientific Revolution 1500–1800: The Formation of the Modern Scientific Attitude*. 2nd edition. Boston: Beacon Press, 1966.

Hampshire, Stuart. *Spinoza*. London: Penguin, 1955.

———. *Morality and Conflict*. Cambridge: Harvard University Press, 1983.

———. *Justice is Conflict*. Princeton: Princeton University Press, 2000.

Harrison, Charles T. "The Ancient Atomists and English Literature of the Seventeenth Century." *Harvard Studies in Classical Philology* 45(1934): 4.

Hatfield, Gary. "Metaphysics and the New Science." In David C. Lindberg and Robert S. Westerman, editors. *Reappraisals of the Scientific Revolution*. Cambridge: Cambridge University Press, 1990.

Hattab, Helen. *Descartes on Forms and Mechanisms*. Cambridge: Cambridge University Press, 2009.

Herbert, Gary. *Thomas Hobbes: The Unity of Scientific and Moral Wisdom*. Vancouver: University Press of British Columbia, 1989.

Hobbes, Thomas. *Works*. Edited by William Molesworth. London: John Bohm, 1839.

———. *The Elements of Law, Natural and Politic*. Edited by Ferdinand Tonnies. London: Simpkin, Marshall and Co., 1889.

———. *Thomas White's De Mundo Examined*. Translated by Harold Whitmore Jones. London: Bradford University Press, 1976.

———. *The Elements of Law Natural and Political*. Edited by J.C.A. Gaskin. Oxford: Oxford University Press, 1994.

———. *Leviathan*. Edited by J.C.A. Gaskin. Oxford: Oxford University Press, 1996.

———. *Man and Citizen*. Edited by Bernard Gert. Indianapolis: Hackett, 1998.

———. *On the Citizen*. Edited by Richard Tuck and Michael Silverthorne. Cambridge: Cambridge University Press, 1998.

Holden, Thomas. *The Architecture of Matter: Galileo to Kant*. Oxford: Clarendon Press, 2006.

Hooker, Michael, editor. *Descartes: Critical and Interpretive Essays*. Baltimore: Johns Hopkins University Press, 1978.

Horton, Mary. "In Defense of Francis Bacon: A Criticism of the Critics of Inductive Method." *Studies in the History and Philosophy of Science* 4(1973): 241–78.

Huenemann, Charlie, editor. *Interpreting Spinoza: Critical Essays*. Cambridge: Cambridge University Press, 2008.

Hunter, Graeme, editor. *Spinoza: The Enduring Questions*. Toronto: University of Toronto Press, 1994.

Israel, Jonathan. *Radical Enlightenment: Philosophy and the Making of Modernity, 1650–1750*. Oxford: Oxford University Press, 2002.

Jardine, Lisa, and Alan Stewart. *Hostage to Fortune: the Troubled Life of Francis Bacon.* New York: Hill and Wang, 1999.

Johnston, D. *The Rhetoric of Leviathan.* Princeton: Princeton University Press, 1986.

Jones, Richard F. *Ancients and Moderns: A Study of the Rise of the Scientific Movement in Seventeenth Century England.* St. Louis: Washington University Press, 1961.

Kargon, Robert Hugh. *Atomism in England From Hariot to Newton.* Oxford: Clarendon Press, 1966.

Kashap, Paul S. *Studies in Spinoza: Critical and Interpretive Essays.* Berkeley: University of California Press, 1972.

Kavka, G. *Hobbesian Moral and Political Theory.* Princeton: Princeton University Press, 1986.

Kendrick, Nancy. "Why Cartesian Ideas of Sense are Innate." *Southern Journal of Philosophy* 38(2000): 413–28.

Kennington, Richard, editor. *The Philosophy of Baruch Spinoza.* Washington, D.C.: Catholic University of America Press, 1980.

———. *On Modern Origins: Essays in Early Modern Philosophy.* Lanham: Lexington Books, 2004.

Kirk, G.S., J.E. Raven, and M. Schofield, Editors. *The PreSocratic Philosophers.* 2nd edition. Cambridge: Cambridge University Press, 1983.

Koistinen, Olli, editor. *The Cambridge Companion to Spinoza's Ethics.* Cambridge: Cambridge University Press, 2009.

Koyre, Alexandre. *Metaphysics and Measurement.* Cambridge: Harvard University Press, 1968.

Kristeller, Paul Oskar. *Renaissance Thought and the Arts.* Princeton University Press, 1990.

Kuhn, Thomas. *The Essential Tension.* Chicago: University of Chicago Press, 1977.

———. "Reflections on My Critics." In Imre Lakatos and Alan Musgrave, editors. *Criticism and the Growth of Knowledge.* Cambridge: Cambridge University Press, 1999.

Laird, John. *Hobbes.* New York: Russell and Russell, 1968.

Laird, W.R. "The Scope of Renaissance Mechanics." *Osiris* 2nd series 2(1986): 43–68.

Leijenhorst, Cees. *The Mechanisation of Aristotelianism.* Leiden: Brill, 2001.

Linden, Stanton. "Francis Bacon and Alchemy: The Reformation of Vulcan." *Journal of the History of Ideas* 35(1974): 547–60.

Luciani, Vincent. "Bacon and Machiavelli." *Italica* 24(1947): 26–40.

Lucretius. *On the Nature of the Universe.* Translated by Ronald Melville. Oxford: Oxford University Press, 2008.

Machiavelli. *The Prince.* Translated by Harvey Mansfield. Chicago: University of Chicago Press, 1985.

———. *The Chief Works and Others.* Translated by Alan Gilbert. In three volumes. Durham: Duke University Press, 1989.

———. *Discourses on Livy.* Translated by Harvey C. Mansfield and Nathan Tarcov. Chicago: University of Chicago Press, 1996.

———. *The Sweetness of Power: Machiavelli's Discourses and Guicciardini's Considerations.* Translated by James B. Atkinson and David Sices. Dekalb: University of Northern Illinois, 2002.

Makin, Stephen. *Indifference Arguments.* Oxford: Blackwell, 1993.

Mandelbaum, Maurice, and Eugene Freeman, editors. *Spinoza: Essays in Interpretation.* LasSalle: Open Court, 1975.

McCreary, Eugene. "Bacon's theory of Imagination Reconsidered." *Huntington Library Quarterly* 36(1973): 317–26.

Meinel, Christopher. "Early Seventeenth Century Atomism: Theory, Epistemology, and the Insufficiency of Experiment." *Isis* 79(1988): 68–103.

Mintz, S.I. *The Hunting of Leviathan.* Cambridge: Cambridge University Press, 1962.

Nadler, Steven. *Spinoza's Ethics: An Introduction.* Cambridge: Cambridge University Press, 2006.

———. *A Book Forged In Hell.* Princeton: Princeton University Press, 2011.

Najemy, John, editor. *The Cambridge Companion to Machiavelli.* Cambridge University Press, 2010.

Nolan, Lawrence. "The Ontological Sketch of Cartesian Natures." *Pacific Philosophical Quarterly* 78(1997): 169–94.

———. "Descartes Theory of Universals." *Philosophical Studies* 89(1998): 161–80.

Oakeshott, Michael. *Hobbes on Civil Association*. Indianapolis: Liberty Press, 1975.

Parkinson, G. H. R. *Spinoza's Theory of Knowledge*. Oxford: Oxford University Press, 1964.

Perez-Ramos, Antonio. *Francis Bacon's Idea of Science and the Maker's Knowledge Tradition*. Oxford: Clarendon Press, 1988.

Pesic, Peter. "Wrestling with Proteus: Francis Bacon and the 'Torture' of Nature." *Isis* 90(1991): 81–94.

Plato. *The Collected Dialogues*. Edited by Edith Hamilton and Huntington Cairns. Princeton: Princeton University Press, 1999.

Pelonen, Markuu, editor. *The Cambridge Companion to Francis Bacon*. Cambridge: Cambridge University Press, 1996.

Peters, Richard. *Hobbes*. New York: Penguin, 1956.

Rahe, Paul. "In the Shadow of Lucretius: The Epicurean Foundations of Machiavelli's Political Thought." *The History of Political Thought* 1(2007): 30–66.

Rees, Graham. "Matter Theory: A Unifying Factor in Bacon's Natural Philosophy." *Ambix* 24(1977): 110–25.

———. "'Atomism' and 'Subtlety' in Francis Bacon's Philosophy." *Annals of Science* 27(1980): 549–71.

———. "Mathematics and Francis Bacon's Natural Philosophy." *Review Internationale de Philosophie* 40(1986): 399–426.

Reiss, Timothy J. *Knowledge, Discourse and Imagination in Early Modern Europe*. Cambridge: Cambridge University Press, 1997.

Renn, Jurgen, editor. *Galileo in Context*. Cambridge: Cambridge University Press, 2001.

Richardson, R. C. "The 'Scandal' of Cartesian Interactionism." *Mind* 91(1982): 20–37.

Robertson, Croom. *Hobbes*. London: William Blackwood, 1910.

Rogers, G.A.J., and Alan Ryan, editors. *Perspectives on Thomas Hobbes*. Oxford: Clarendon Press, 1990.

Rorty, Amelie Oksenberg, editor. *Essays on Descartes' Meditations*. Berkeley: University Of California Press, 1986.

Rossi, Paolo. *Francis Bacon: From Magic to Science*. Translated by Sacha Rabinovich. London: Routledge and Kegan Paul, 1968.

Rossky, William. "Imagination in the English Renaissance: Psychology and Poetic." *Studies in the Renaissance* 5(1988): 49–73.

Rotenstreich, Nathan, and Norma Schneider, editors. *Spinoza: His Thought and Work*. Jerusalem: The Israel Academy of Sciences and Humanities, 1983.

Rozemund, Marleen. "Descartes on Mind-Body Interaction: What's the Problem?" *Journal of the History of Philosophy* 37(1999): 435–67.

Sacksteder, William. "Hobbes: Teaching Philosophy to Speak English." *Journal of the History of Philosophy 16*(1978): 33–45.

———. "Hobbes: Geometrical Objects." *Philosophy of Science* 48(1981): 573–90.

———. "Three Diverse Sciences in Hobbes: First Philosophy, Geometry, and Physics." *Review of Metaphysics* 45(1992): 739–72.

Salles, Richard, editor. *Metaphysics: Soul and Ethics in Ancient Thought*. Oxford: Clarendon Press, 2005.

Sarasohn, Lisa. "Pierre Gassendi, Thomas Hobbes and the Mechanical Worldview." *Journal of the History of Ideas* 46(1985): 363–79.

Shapere, Dudley. "Meaning and Scientific Change." In Ian Hacking, editor. *Scientific Revolutions*. Oxford: Oxford University Press, 1981.

Shapin, Steven. *The Scientific Revolution*. Chicago: University of Chicago Press, 1998.

Simmons, Alison. "Descartes on the Cognitive Structure of Sensory Experience." *Philosophy and Phenomenological Research* 67(2003): 549–79.

Skinner, Quentin. *Visions of Politics*, volume 3: *Hobbes and Civil Science*. Cambridge: Cambridge University Press, 2002.

———. *Liberty Before Liberalism*. Cambridge: Cambridge University Press, 2006.

Smith, Kurt. "Descartes' Ontology of Sensation." *Canadian Journal of Philosophy* 35(2005): 563–84.

———. *Matter Matters: Metaphysics and Methodology in the Early Modern Period.* Oxford: Oxford University Press, 2010.

Smith, Steven. *Spinoza, Liberalism, and the Question of Jewish Identity.* New Haven: Yale University Press, 1997.

———. *Spinoza's Book of Life*. New Haven: Yale University Press, 2003.

Solmsen, Friedrich. "Abdera's Argument for the Atomic Theory." *Greek, Roman and Byzantine Studies* 29 (1988): 50.

Solomon, Julie Robinson. *Objectivity in the Making: Francis Bacon and the Politics of Inquiry.* Baltimore: Johns Hopkins University Press, 1998.

——— and Catherine Gimelli Martin, editors. *Francis Bacon and the Reconfiguring of Modern Thought.* Adlershot: Ashgate, 2005.

Sorabji, Richard. *Time, Creation and the Continuum*. Ithaca: Cornell University Press, 1983.

Sorell, Tom. *Hobbes*. London: Routledge and Kegan Paul, 1988.

———, editor. *The Cambridge Companion to Hobbes*. Cambridge: Cambridge University Press, 1996.

——— and Luc Foisneau, editors. *Leviathan After 350 Years*. Oxford: Oxford University Press, 2004.

Spinoza, Baruch. *The Complete Works*. Translated by Samuel Shirley. Edited by Michael Morgan. Indianapolis: Hackett, 2002.

Spragens, Thomas. *The Politics of Motion: the World of Thomas Hobbes*. Lexington: University of Kentucky Press, 1973.

Springborg, Patricia, editor. *The Cambridge Companion to Hobbes's Leviathan*. Cambridge: Cambridge University Press, 2007.

Leo Strauss. *The Political Philosophy of Hobbes: Its Basis and Genesis*. Translated by Elsa Sinclair. Chicago: University of Chicago Press, 1984.

———. *Persecution and the Art of Writing*. Chicago: University of Chicago Press, 1988.

———. "Notes on the Concept of the Political." In Carl Schmitt, *The Concept of the Political.* Translated by George Schwab. Chicago: University of Chicago Press, 1996.

———. *Natural Right and History*. Chicago: University of Chicago Press, 1997.

———. *Spinoza's Critique of Religion*. Translated by Elsa Sinclair. Chicago: University of Chicago Press, 1997.

———. *Hobbes's Critique of Religion and Other Writings*. Translated by Gabriel Bartlett and Svetozar Minkov. Chicago: University of Chicago Press, 2011.

Talaska, Richard. "Analytic and Synthetic Method According to Hobbes." *Journal of the History of Philosophy* 26(1988): 207–37.

Taylor, C.C.W. *The Atomists Leucippus and Democritus*. Toronto University Press, 1999.

Verdun, Michael. "On the Laws of Physical and Human Nature: Hobbes's Physical and Social Cosmologies." *Journal of the History of Ideas* 43(1982): 653–663.

Vickers, Brian. *Francis Bacon and Renaissance Prose*. Cambridge University Press, 1968.

Warhaft, Sidney. "Science against Man in Bacon." *Bucknell Review* 7(1958): 158.

Warren, James. *The Presocratics*. Berkeley: University of California Press, 2007.

Watkins, J.W.N. *Hobbes's System of Ideas*. London: Hutchinson University Library, 1965.

Weinberger, Jerry. "Hobbes's Doctrine of Method." *American Political Science Review* 69(1975): 1336–1353.

———. "Science and Rule in Bacon's Utopia: An Introduction to the Reading of *the New Atlantis. American Political Science Review* 70(1976): 865–885.

———. *Science, Faith and Politics: Francis Bacon and the Utopian Roots of the Modern Age.* Ithaca: Cornell University Press, 1985.

Weissman, David, editor. *Rene Descartes: Discourse on Method and Meditations on First Philosophy.* New Haven: Yale University Press, 1996.

Whitehead, Alfred North. *Science and the Modern World.* New York: Free Press, 1967.

Whitney, Charles. *Francis Bacon and Modernity*. New Haven: Yale University Press, 1986.

Wilson, Catherine. *Epicureanism at the Origins of Modernity*. Oxford: Clarendon Press, 2010.

Yovel, Yirmayahu, editor. *God and Nature: Spinoza's Metaphysics*: Papers Presented at the First Jerusalem Conference. Leiden: Brill, 1991.

Zagorin, Perez. *Francis Bacon*. Princeton: Princeton University Press, 1999.

———. *Thomas Hobbes and the Law of Nature*. Princeton: Princeton University Press, 2009.

Index